TP
374
2

FOOD PACKAGING

A Guide for the Supplier, Processor, and Distributor

other AVI books

FOOD PACKAGING

A Guide for the Supplier, Processor, and Distributor

by

Stanley Sacharow, B.A., M.A.

Project Director, Laminated Products
Packaging Research Division
Reynolds Metals Company
Richmond, Virginia
Professional Member,
The Packaging Institute

and

Roger C. Griffin, Jr., B.S., M.S.

Manager, Paper and Plastics Laboratories
Packaging Research Division
Reynolds Metals Company
Richmond, Virginia
Professional Member,
Institute of Food Technologists

Westport, Connecticut
The AVI Publishing Company, Inc.
1970

© *Copyright 1970 by*

AVI PUBLISHING COMPANY, INC.

Westport, Connecticut

REGISTERED AT STATIONERS' HALL
LONDON, ENGLAND 1970

Printed in the United States of America

DEDICATION

This book is dedicated to my wife, Beverly Lynn. Without her encouragement, inspiration and belief, it would never have been brought to fruition. To my parents, Max and Fannie Sacharow, I hope that I have accomplished their fondest dream. And to my sons, Scott Hunter and Brian Evan, my apologies for the many hours not spent playing with them. Someday they may realize that this was written to create a better world for their generation.

S. Sacharow.

DEDICATION

My father was born in 1883 and died in 1956. During his lifetime the electric light, telephone, radio, automobile, airplane, television, moving picture, rocket, atom bomb and many other marvelous creations revolutionized mankind's way of life. A few years before his death on a whim of the moment I asked him what one thing he had seen in his opinion had the most beneficial influence on our society. To my surprise he instantly named food packaging, for he could remember the cracker barrel and flour barrel of the country store. I dedicate this work to his generation who saw so much change, to my children, Roger and Alysia, who will see even more amazing changes, and to my wife Phyllis, whose steadfast belief that I shall someday accomplish something of importance has encouraged me to keep trying.

R. C. Griffin, Jr.

Preface

Ever since man appeared on this Earth, he has fought for survival. One of his greatest battles has been against famine. Over the centuries food packaging has played an ever increasing role in the war against starvation. The final crucial battle was won with the 19th century discoveries in the fields of food processing and microbiology, and the 20th century discoveries in the science of agriculture.

Twentieth century food packaging is less concerned with man's survival and more dedicated, rather, to his comfort and convenience. Today's affluent society and modern technology have combined to bring a myriad of new packages into the marketplace.

During the last 50 years packaging has evolved from an art to a science with all the agonies that ensue in such a transition. During the past several decades the food industry has recognized the packaging function as an important facet of their operations and has created packaging departments staffed with professional packaging scientists and engineers. Colleges and universities have added packaging courses to their curricula and at least two now grant graduate degrees in packaging science.

The pioneer workers in this field were so busy creating that few took time to record their accomplishments. There is no recorded comprehensive history of the development of packaging machinery. The reports on food packaging developments are found scattered in many diverse journals. No text exists which covers the whole field of commercial food packaging in such a way as to serve as a guide for new or inexperienced workers, whether in industry, government, or the academic world, whose duties involve them in the creation, marketing, or use of food packages. It is hoped that this book will serve this purpose. Like all first efforts it is bound to have imperfections. The authors hope each reader will submit criticisms or additional material that can be used to improve subsequent editions and make them more useful to all.

Appreciation is here expressed to all who have helped us complete this book. We expressly acknowledge our gratitude to Reynolds Metals Company whose pioneering efforts in the creation of food packages have resulted in an organization that is both knowledgeable and dedicated; to A. I. Totten, Jr., General Director of Packaging Research, and his staff who have encouraged our efforts; and to our Librarian, Mrs. Marie Davis Walker, who has unflaggingly tracked down elusive references; and to many who have contributed information, photographs, reference material, or just plain enthusiasm.

STANLEY SACHAROW
ROGER C. GRIFFIN, JR.

Richmond, Virginia
September 1, 1969

CONTENTS

CHAPTER 1

The Evolution of Food Packaging

PRIMITIVE PACKAGING

The foods we eat have come to us from many lands and many cultures. A great proportion of them predate written history. Food packaging is similarly entwined with earlier civilizations and its earliest beginnings are rooted in man's advancements in the production and processing of foods.

In Paleolithic times, food was consumed where found, and, when needed, man used natural containers such as hollow tree trunks, gourds, hollow rocks, shells, leaves, and pieces of bark. In later times he learned to fashion containers from natural materials. He deliberately hollowed-out logs or stones, and utilized animal parts, such as bladders, skins, horns, bones, sinews and hair. Earliest fabrics evolved from furs. First, fibers were matted into felts, later they were plaited and woven. Weaving of thin vines or wooden withes into shelters led to the making of baskets out of rushes and grasses. Mesolithic man stored food surpluses in this type of basket. Neolithic man fashioned metal containers and discovered pottery. Some of the earliest metal "cups" were drinking horns shaped like natural horns. Pottery provided an infinite number of clay vessels of many shapes and sizes from tiny amphorae to huge urns. Herodotus recorded that in 530 BC the Persians supplied conquered Egyptian cities with water and wine in earthenware jars. These were collected later and returned for reuse. Glass was known in Sumerian times and the small glass jar or bottle was known in 1500 BC. Hollowed alabaster was used for small quantities of precious liquids or ointments. In 79 AD the Romans were using glass bottles as well as ceramic pots and jars, but due to fragility they preferred leather bags for large quantities of liquids and solids. The leather bag wineskin was referred to by Jesus of Nazareth (Matthew 9:17) "Neither do men put new wine into old wineskins; else the skins burst, —"

The barrel was invented by Alpine tribes, according to Pliny. Its carefully fitted staves and heads were fashioned from wood, interlocked by mortises and the whole bound together with iron hoops. Wooden chests were made in similar fashion from fitted and pegged boards because nails were hand made and too expensive for such use.

From prehistoric times until about 1200 AD the status of packaging could be summed up as follows:

1

Material	Package Form and Use
Leather	Wrappings, bags, bottles
Cloth	Wrappings, sacks
Wood	Barrels, boxes, kegs, chests
Grass, or split wood	Baskets, matting
Stone	Small pots or jars
Earthenware	Pots, jars, urns, ewers, bowls, etc.
Metal	Pots, bowls, cups, etc.
Glass	Jars, bottles, cups, bowls, etc.

ORIGINS OF PACKAGING MATERIALS

Paper

Paper and papyrus were developed originally as writing materials to replace parchment (animal skin) and vellum (skins of newborn calves, kids, or lambs). Papyrus was formed from flattened strips of the center pith of the papyrus reed. It was used by the ancient Egyptians and

Courtesy of Reynolds Metals Co.

FIG. 1.   REPRESENTATION OF ANCIENT EGYPTIANS WRAPPING
POULTRY IN PALM LEAVES FOR PROTECTION
AGAINST CONTAMINATION

Greeks at least as early as 500 BC. The earliest paper was made in China from mulberry bark about 200 BC. The Arabs learned the art when a Chinese army attacked Samarkand in 751 AD and some paper makers were captured. Papers were first made from flax fiber and later from old linen rags. Although some say the Crusaders brought paper home with them from the Holy Land (not unlikely), it is certain that the Arabs introduced it to Sicily and thence to Italy and South Germany and the Moors introduced it to Spain in the 12th century, from whence it spread through France, West Germany, the Netherlands, Belgium and England. Earliest records of English paper making date to about 1310 AD. The first paper making in America was in 1690 in Germantown, Pennsylvania. Cardboard was also a Chinese invention of about the 16th century. The first paper-making machinery, in which fibers were laid down on a moving wire cloth, began production in 1799 in England. The use of wood pulp in paper making was introduced in 1867. Corrugated paper was invented in the mid 1800's and shipping cartons made from faced corrugated paperboard began to replace wooden crates and boxes about the turn of the century.

Glass

Glass is believed to be an offshoot of pottery. First usages were as glass glazes on beads or amulets. Green glazed stone beads dating to 12,000 BC and a blue pure glass amulet dating to 7000 BC were found in Egypt. By 1550 BC glassmaking was an important industry in Egypt and various small bottles and ornaments were produced in many colors—but none was clear. The art of pressing glass in molds to produce bowls and cups dates to 1200 BC. The Phoenecians were believed to have invented the blowpipe about 300 BC which vastly speeded up production, nevertheless even in later Roman times (400 AD) fine glassware was more valuable than silver or gold. Clear transparent glass was discovered about the beginning of the Christian Era and in the 3rd century Roman glass makers were casting window glass on flat stones. During the following thousand years glassmaking spread over Europe and the art of making stained glass windows for cathedrals was developed. Up to the 18th century, Venetian and other Italian glass makers were considered the world leaders, but then the Bohemian and English glass began to challenge. Cut glass was invented in 1600 by a court jeweler in Prague (Bohemia). Still, glass remained expensive until improved techniques of the 18th and 19th century brought the price of bottles to a more reasonable level.

Tinplate

Sometime around the year 1200 AD the artisans of Bohemia discovered a hot dip process for plating tin onto thin sheets of iron which

Courtesy of Walter Landor and Associates

FIG. 2.　CERAMIC　MARMALADE　JAR　ILLUSTRATING　EARLY
METHOD OF PRINTING

This package is still used for this type of product.

had been hammered out by hand from rods. This secret was guarded zealously for 400 years, but in 1620 the Duke of Saxony managed to steal it. The Iron Age brought new metals and machinery for continuous rolling and plating. Steel has replaced iron and stronger alloys have made it possible to use thinner gauges of metal. Changes in processing techniques have led to thinner and thinner coatings of tin. Developments of the past decade have produced tinless cans, steel foil, and aluminum-clad steel alloys.

Tin and Lead and Other Metals

In ancient times boxes and cups were made from silver and gold but these were too valuable for common usage. We know the Romans used lead in many ways, including water pipes. The only known evidence in packaging is their application of lead seals to ointment jars. It is likely they found ways to hammer the soft metal into thin foils. Not knowing its poisonous nature they could have wrapped foods in it. Leadfoil was

Courtesy of Walter Landor and Associates

FIG. 3. LITHOGRAPHED TIN PLATE CANISTERS

These were very popular for dry products.

used to line tea boxes in 1826, and in cigarette packaging as late as the 1930's.

Tinfoil is an all embracing term. Most tin alloys were made from tin and either lead, antimony, zinc, or copper. Pure tin is difficult to roll. While "tin" foil was used for wrapping chocolates in the 1840's, cheese wrap foils used as late as the 1930's were tin plus antimony. Tinlead and tin clad lead foils were used in packaging also.

Aluminum

Aluminum had its start in 1825 when Oersted produced the first particles. The metal was so difficult to extract from its ore that as late as 1852 it cost $545 a pound. French Emperor Napoleon III used aluminum forks and spoons for special guests, others had to dine off of silver or gold. In 1854, Deville and Bunsen developed more efficient processes, which by 1885 had brought the price down to $11.33 per pound, but this was still too high. In 1886, Heroult and Hall discovered the modern electrolytic process for extracting aluminum from alumina, and, in 1888, Bayer found a less expensive way to extract alumina from bauxite ores.

By 1892, the price had dropped to $0.57 a pound, and it kept dropping to a low of $0.14 in 1942.

As soon as the price became low enough, aluminum found its way into many uses. Its properties lent well to easy fabrication. The first commercial aluminum foil was rolled sometime after 1910, and during the early 1920's techniques for rolling and printing were perfected.

Plastics

A search for a substitute for ivory billiard balls led to an investigation of cellulose nitrate, which itself was first described in 1845. In 1862, molded articles made from "parkesine" were displayed at the Great Exposition in London, and in 1870 Hyatt patented "celluloid," a mixture of cellulose nitrate and camphor. This was the first commercially successful man-made plastic, and also one of the first examples of a solid plasticizer.

Celluloid and other cellulose nitrate products were so flammable that investigators sought other less flammable materials. Cellulose acetate dissolved in acetone was widely used in cements and lacquers. The latter were used to "dope" the fabrics on World War I airplane wings.

The *Viscose* or cellulose xanthate process for making regenerated cellulose fibers and films was developed in 1892. The first commercial applications of viscose films were in Europe between 1912 and 1914. "Cellophane" film was introduced in the United States in 1924. The addition of coatings of cellulose nitrate added heat-sealability and better resistance to moisture.

Casein plastics were introduced in 1899, and casein-formaldehyde buttons and buckles were used extensively during the 1920's. Casein plastics are still used today in the familiar casein-latex emulsion adhesives. Other plastics which were made from polymerization with formaldehyde in the latter part of the 19th century were phenol-formaldehyde, urea-formaldehyde, and melamine-formaldehyde. During the 1930's and 1940's, melamine-formaldehyde resins were introduced as wet-strength additives for papers.

Styrene was first distilled from a tree balsam in 1831 and its spontaneous polymerization was noted. In 1866, it was produced from coaltar benzene. The first commercially successful polystyrene was introduced in 1930. Molded polystyrene articles were widely used but found to be brittle. The hazard from shattered toys or food containers led in the 1950's to the blending of polystyrene with synthetic rubbers to gain impact resistance. Polystyrene films and molded polystyrene containers achieved commercial acceptance thereafter. The final evolution occurred in the past decade when styrene, acrylonitrile, and butadiene were copolymerized to produce the ABS plastics. Also in 1950 expanded polystyrene foam was introduced and soon found a market as an insulation and cushioning material. Housewives and store decorators found it was great for Christmas decorations.

Studies on natural rubber led to the discoveries of rubber coatings and rubber cements. In 1934, rubber hydrochloride films were introduced. Even before this scientists isolated the monomer of natural rubber, isoprene, and then polymerized it. In 1915, a synthetic methyl rubber (2, 3 methyl butadiene) was produced in the laboratory and was followed by a series of conjugated diene rubbers. In 1931, neoprene was discovered, and in 1933 butadiene-styrene rubber was introduced—the familiar Buna S of World War II. Its cousin Buna N, butadiene acrylonitrile rubber, was also a war-baby. By the 1950's, rubber chemistry had made great strides, a whole spectrum of elastomers was available. In packaging they have been used widely as adhesive components.

Vinyl chloride (VC) monomer was first prepared in 1835. The first polyvinyl chloride (PVC) was reported in 1872. By 1912, commercial processes were available for making cable insulation. Polyvinyl alcohol was prepared in 1924 from hydrolyzed polyvinyl acetate. The acetates were found to be too soft, and the chlorides too hard, so in 1928 a copolymer of vinyl acetate and vinyl chloride was prepared. The copolymers proved to be very useful. With the addition of plasticizers a series of coated and calendered fabrics were produced. The first shower curtains and raincoats of the early 1940's had strong odors from their migratory plasticizers, however, as technology improved, better polymers and better plasticizers were found. In 1958, heat shrinkable PVC film was introduced commercially—the first strong entry of vinyls into the packaging field. VC-acrylic copolymers have made excellent rigid molded articles. Vinyl chloride and vinyl ether copolymers have found use in lacquers.

Polyvinylidene chloride (PVDC) was discovered in 1838. The pure monomer was synthesized in 1872 and homopolymers were made in 1916. They were found to have very narrow melting ranges and to be very hard and brittle. In 1936, a copolymer of VC and VDC was prepared, the first of the "Sarans." Emulsion polymerization was developed between 1940 and 1947.

The first polyester was prepared in 1847. Acrylate esters were made in 1873 and methyl acrylate was polymerized in 1880. First commercial production of polyacrylates was in 1927 and of methyl methacrylate in 1933. The latter was just in time for service as cockpit covers and gun turrets on World War II aircraft. Alkyd polyesters were discovered in 1901 and eventually found extensive use in paints, lacquers and coatings. Polyethylene-glycol-terephthalate was first prepared in 1941 and marketed in the 1950's as "Mylar" film, other polyester films have since become available.

One of the most important plastics developments as far as packaging is concerned occurred in 1933 when Fawcett and Gibson of ICI ran some experiments on ethylene and found a white waxy solid coated on the wall of the pressurized vessel. This was the first high-pressure reacted polyethylene. The first commercial use of this material was as a film wrap for submarine telephone cable and during World War II as a radar cable wrap. In the early 1940's, Karl Ziegler began research which led to a catalytic method for producing polyethylenes by low pressure reactions. The Ziegler (or Philips) process produced higher density polymers than were achieved by the ICI process. These high density polyethylenes were commercialized in 1956 and very rapidly achieved high volume sales.

In 1954, Natta in Italy used a stereospecific catalyst to produce polypropylene and firmly established the principle that polymer chains can be tailored to fit a preconceived pattern. In recent years, a whole family of copolymers and homopolymers of ethylene, propylene, and butylene have been streaming into the marketplace.

Other plastics of interest in packaging include the polyamides (nylons), which were first developed in 1937 as fibers and graduated to packaging films in the late 1950's and early '60's; and the polycarbonates marketed in 1959.

As time goes on, more and more new plastics are being commercialized. Current trends are not only adding new families such as ionomers and polyimides, and new family members such as nylon 13 and polybutene, but also second generation blends and combinations, such as PVDC coated polyester, polyethylene coated polyamide, rubber modified polystyrene, ethylene-vinyl-acetate-copolymers, and the new coextruded films such as polyethylene-polypropylene-polyethylene. Possibilities appear to be unlimited.

PACKAGE FORMS

The barrel, wooden box, ceramic jar or pot, leather bag and cloth sack are as old as civilization. Some of the other important milestones in the historical development of package forms are the following:

Metal Boxes

With the invention of tinplate in 1200 AD the fabrication of soldered metal boxes was made possible. There are however no records of such containers until about 1764 when tobacconists in London began selling snuff in metal canisters. In the early 1830's, both matches and cookies

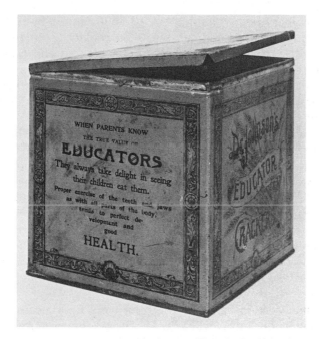

Courtesy of Walter Landor and Associates

FIG. 4.   EARLY CRACKER BOX

This is an excellent example of an early cracker box made from
soldered tin plate.

were sold in tin boxes. Some boxes were designed with embossed lettering or patterns, others carried printed paper labels. Between 1850 and
1900, processes were developed for printing the metal itself (see discus-
sion of Graphic Arts later in this chapter). The first printed metal boxes
in the United States were made in 1866 for cakes of Dr. Lyon's tooth
powder.

Metal Cans

The cylindrical tinplate can was devised by Peter Durand in 1810.
First cans were soldered by hand leaving a 1-1/2 in. hole in the top.
After the food was forced through the hole (not without damage), the
hole was closed by soldering a patch over it. Sometimes a small hole was
punched to let let air escape during cooking and then this was closed
with a drop of solder. Can makers could turn out 60 cans a day when
making them by hand. Various tools and jigs were devised to make the
fabrication job easier. In 1868, enamels were put on can interiors to halt
corrosion or discoloration of the food. The Max Ams double seamed
method of closure using sealing compound was introduced in 1888 and

by 1900 machinery was available to make 2500 cans an hour. With such production rates the soldered can was soon replaced, but the hole and cap can didn't disappear until 1922. Some soldered end cans are still used today (for condensed milk). In 1957, aluminum cans were introduced for lubricating oil and by 1963 were being used for beer. In the past few years cans have been introduced with side seams that are heat-sealed using a hot melt as the sealant.

The first cans were opened using a hammer and chisel. The key-wind metal tear-strip can was invented in 1866 and was carried to the extreme in the sardine can where nearly the whole top panel could be torn out by the key. The can opener based on the lever principle was first introduced in 1875. All manner of variations on the lever principle including key wind, crank wind, and motor driven rotary can openers have been developed over the past 75 years. The pop-top or tear tab can lid of the 1950's depended upon aluminum for its success. Other vacuum packed cans are now being fabricated with tear tapes made from plastic or laminated foil and plastic, which provide both the seal and a means for easy opening.

Aerosol cans were invented in 1889 but were first successfully marketed as "bug bombs" in World War II. Today, they are used for a wide variety of products, including foods of suitable consistency.

Bottles and Jars

Glass bottles and jars date back to antiquity. Glass jars were first fashioned from ropes of molten glass coiled into cylindrical shapes and fused together. The invention of the blowpipe made the manufacture of round containers and bottles much easier. The invention of the split mold permitted irregular shapes and raised decorations to be made. Most bottles of the 17th and 18th centuries utilized these features to identify the bottle maker or the product contained. The first fully automatic rotary bottle-making machine was patented in 1889. A modern Owens machine can produce 20,000 bottles a day. The latest trends in both jars and bottles are the substitution of PVC and acrylic plastics for glass or ceramics.

Closures have come a long way from whittled wooden plugs and hand carved corks which were used as far back as 1000 BC, according to Horace. For best seals the plugs were daubed with wax or pitch. Corks remained the chief closure used for bottles until the invention of the screw cap in 1875 and of the crown seal bottle cap in 1892. At first the crown cap liners were plain cork, then linoleum, and then in 1945 composition cork which lasted until very recent years. In the mid 1930's, experimentation was conducted on use of rubber latex and plastic liners and in 1941 a polyethylene liner was tried. In 1955, vinyl plastisols had

From the private collection of Roger C. Griffin, Jr.

FIG. 5.   HAND BLOWN GLASS BOTTLE

An example of early American blown glass is shown here in a
bottle from Wistar glass works (early 1700's).

some success and in 1957 cellular vinyl compositions were used. Aluminum facings were introduced in 1960. The crown itself has remained essentially unchanged, however in the past few years bottle mouths have been modified to permit the crown to be twisted off. Other noncorrugated capseals have been used as far back as 1906. The latest types are made of aluminum and so designed that they are torn off, or when they are twisted off, a lower ring tears away to indicate the package has been opened. This latter type is common on whiskey bottles.

For nonpressurized products, e.g., ketchup, bottles have been closed with molded plastic or metal screw-on caps.

Wide-mouth jars have been closed with wax and waxed cloth on paper, and with metal foils. Tighter closures have been made with rubber gasketed metal screw caps or pry-off lids. Some jars today have a heat sealed membrane barrier initial closure plus a protective screw-on cover. The first provides product freshness, the second provides protection and reclosure.

Metal Tubes

Collapsible tubes were first made from soft metal for artists paints in 1841 and replaced animal-bladders which heretofore had served the purpose. They found early use in dispensing glues, medicinal salves, and toothpastes, but little use was made of them for food products until the past decade. Plastic collapsible tubes have come on the market containing sandwich pastes, cake icings, pudding toppings, and the like. The first closure was by a (male) metal screw with double head. The (female) molded screw cap was a later innovation probably after the turn of the century.

Trays, Pans and Other Containers

In this category must be listed packages that are formed into dishes, cups, bowls, pans, or trays, such as pie pans, and the now familiar TV dinner tray. Molded paper and pulpboard picnic plates both rectangular and round with and without compartmentation have been made for several decades. Pulpboard trays are also used to package meats and produce in supermarkets. Aluminum foil containers were test marketed in 1948 and soon thereafter were available in many shapes. Convolute and spiral wound canisters made from paper were marketed toward the latter part of the 19th century and were used extensively, with asphalt

Courtesy of Reynolds Metals Co.

FIG. 6.   FORMED ALUMINUM CONTAINERS WITH CRIMPED LIDS
USED FOR FROZEN ENTRÉES

providing moisture barrier, as shippers for war materials and artillery shells in World War I. Few changes were made until the 1940's when aluminum foil labels were used to provide superior moisture barrier. Still later, foil was incorporated as an innerliner, thereby making it possible to package liquids. With the introduction of plastics that could be thermoformed, a wide variety of molded plastic boxes, trays, pans and the like became possible.

Boxes of Wood or Cardboard

In earliest times, if you wanted a box you made it yourself. Later box making became a trade. In the 1630's, boxes made of thin wood were advertised. Similar boxes, some of wood and some of paperboard, were still being manufactured by hand labor in the 19th century. By this time printed labels were being pasted on the outside. The oldest box surviving today is a wooden box made in 1801.

Commercial box making is supposed to have begun in England in 1817. The oldest surviving cardboard box is believed to have been made prior to 1825. In the United States first box makers set up their trade in 1810 in Philadelphia. Because of the difficulty in making corners, wooden boxes and cardboard boxes of the early 19th century were usually round.

Courtesy of Container Corp. of America

FIG. 7. WOODEN SHIPPING CONTAINERS

Wooden boxes were used extensively as shipping cases. Shapes could be round, tapered round, or rectangular.

Courtesy of Walter Landor and Associates

FIG. 8. ONE OF THE FIRST FOLDED CARTONS USED
FOR BREAKFAST CEREALS

By 1855 machinery was being used for cutting and creasing boxboard. The setup box was first used for jewelry, later for pills and drugs, and still later (about 1840) for candies.

When stores began to supply boxes to their customers, folding boxes were introduced to save storage space. The first are believed to have been made in America in 1850 to package tacks. In 1870, Robert Gair who had already made a success from paper bag manufacture invented the first automatic cutter-creaser, and by 1897, there were over 800 patents relating to folding boxes. By the early 1900's cereals, such as Grape-Nuts (1898) and Post Toasties (1904), and the original Uneeda biscuits were being packaged in waxed paper lined unprinted cartons with printed overwraps with the overwrap carrying the product brand name and advertising copy.

Wrappers and Labels

Wrappers have been used since primitive days when leaves were employed. In the days of handmade paper, the sheets were so expensive the papermakers wrapped them with cast-offs to protect against soil and damage. As far back as the 1550's or earlier such wrappers were being printed with the manufacturer's name. Medicinals, and tobacco were sold in "sealed" paper wrappers in the 1660's and by the early 1700's one could buy papers containing pins, tobacco, tea, and wig powder. With the coming of machine-made paper and lithography, printed labels were introduced. These were applied to boxes, bottles, jars, and cans of all manner of products. Soon food products were being marked with brand names so that the consumer familiar with the quality of the contents would be able to identify his future repurchase.

Parisian bonbon vendors wrapped their wares in twists of colored paper in 1847, and even as early as 1840, metal foil was used to wrap chocolate "cakes" (probably wafers). Anatole France described such a package, which included a printed paper overwrap sealed with wax. This could have been lead foil, but tinfoil was known and used at that time for both tea and coffee. In the 1850's, Cadbury's drinking chocolate was wrapped in tinfoil, greaseproof paper, and a printed paper outer wrapper.

During the latter half of the 19th century, paper wrappers were used for more and more products including butter and margarine. In 1877, a candlemaker, who liked to eat fish, founded a company to make waxed

Courtesy of Walter Landor and Associates

FIG. 9.   AN EARLY TEA PACKAGE

This is an early type of flexible package for loose tea. Note the metal foil showing at the top.

paper. He didn't like the mess he encountered in carrying fresh fish home from the market wrapped in a newspaper. Newspaper coated with candlewax however solved his problem. By 1894, paraffin waxed paper was being used as a carton inner liner for crackers; by 1900, for candy wraps; by 1902, as a bread wrapper; and in the same year rolls of waxed paper in a cutter edged carton were sold to housewives.

Waxed papers proved the most commonly used moisture barrier material until the introduction of cellophane in 1912 and aluminum foil about the same time. It wasn't long before wax, paper, and foil were combined to produce superior-barrier wrappers. In the late 1950's came the introduction of heat sealable, heat shrinkable, plastic films such as polyvinylchloride, polyvinylidene chloride, polyethylene, and polypropylene.

Bags

The making of bags out of paper dates back at least to the Thirty Years' War (1618–1648). The first commercial English paper bag manufacturer was established in 1844 in Bristol. Printing was done by way of a steam driven litho press. The first bag making machine was developed in 1852 by Francis Wolle in the United States. By 1873, the gusseted bag design had been patented. Still, as late as 1902 some bags were being made by hand. In the 1870's, the glued paper sack was invented to replace the cotton flour sack, but the multiwalled paper sack for larger quantities did not replace cloth until 1925 when a means of sewing the ends was invented. As previously noted, by 1905 machinery was available for automatically producing in-line printed paper bags. When plastic films came on the scene, machinery was adapted to make bags from them also.

GRAPHIC ARTS

Decoration and ornamentation predate history. Ancient artifacts show that primitive man learned to paint, carve, and sculpt. For many thousands of years all such work was applied directly to the object that was to be decorated. No one knows when the first experimentation with indirect methods occurred. Whoever first applied a color or ink to one surface and then pressed it against another to transfer the color was the first printer. We have evidence that the first printing on paper from carved wooden blocks took place in China about 868 AD. Also in China, in about 1041 AD individual wooden blocks for printing of characters were used. The art of printing on paper from wooden blocks undoubtedly existed in Europe prior to Gutenberg's use of movable type in 1454, and probably was predated by block printing on textiles but the earliest surviving printed package was a wrapper used by a German papermak-

Courtesy of Reynolds Metals Co.

FIG. 10. THE SHOPPING BAG

Before modern packaging began to exert its influence with prepackaging of produce the shopper carried food home in the
utilitarian paper sack.

er, Andreas Bernhart, in the 1550's. During the next hundred years, the
use of printed paper wrappers spread to other products from patent
medicines and dentifrices to tobacco and various foods. In the early
1600's few, if any, products were identified by the manufacturer's name.
As various new items were imported to England through overseas trade,
as for instance tobacco (1610), tea (1614), coffee (1648), and chocolate
(1655), the opportunity for cheating the public by adulteration or by
disguising inferior quality was seized upon by the unscrupulous. Honest
merchants began to mark their wares in a distinctive manner, first by
newspaper advertising, then by printed "trade cards," and finally by
printed wrappers or labels. In the mid 1700's, engraved copper or steel
plates were being used in place of wood blocks for printed labels. A
transfer method was invented for printing on pottery. Paper was first

printed from engraved plates and then the colored side was adhered to the unglazed, unfired clay, which absorbed the colored ink. After the paper was washed away, the subsequent firing burned the color in permanently. In 1798, the principles of lithography were discovered by Senefelder in Bavaria. This, together with the advent of cheaper machine-made paper, gave the graphic arts a strong impetus. Lithography was used for transfer printing on pottery in more than one color, and color printing on paper became increasingly popular. Some printers, as for example Currier and Ives, printed one color from a lithoplate and then added the other colors by hand painting. As the 19th century progressed, printers vied with one another in designing new movable type faces and fancier labels. In the 1830's printed color was introduced on match box labels. In 1835, George Baxter produced multicolor work using a copper plate for the base color and wooden blocks for the registered colors.

In 1859, papermakers were selling colored papers, and by then many printers were using the Baxter process. In 1870, Richard Cadbury painted landscapes in oil and then had them reproduced in lithographed labels for the Cadbury chocolate box tops. Powered presses in 1865 were flat bed and required very large lithograph stones. About 600 impressions an hour were standard because of the reciprocal action. The discovery (in 1820) that zinc could be substituted for stone made the rotary press possible and increased production to 1100 impressions per hour.

Direct printing on primed tinplate was not too successfully done in 1864 as the engraved printing plates or lithograph stones would not make good contact. The transfer method held sway for about ten years. Here the ink was printed on thin paper which was applied to the tinplate and then the paper was washed off. In 1875, Robert Barclay in England invented offset lithography. He applied slow drying inks to a glazed cardboard cylinder from which they were transferred to the tinplate. Later the cardboard was replaced with rubber covered canvas, and still later by a rubber roll. In 1903, this principle was converted from flat bed to rotary presses and up to 4000 impressions an hour became possible.

Flexography was also born in England about 1890 in an attempt to keep up with the speed of paper bag making machines. Here the inks were fast drying "aniline" types and were applied to resilient rubber blocks or plates, from which they were printed. By 1905, a bag machine and press had been combined for an inline operation.

During the 19th and 20th centuries, photographic techniques were invented whereby printing plates could be prepared and chemically etched. Process color separation by photography lowered costs of plate preparation and reduced the number of colored inks required. By the turn of the century (1900) four color lithography was commonplace and six colors were possible.

Modern high speed rotary presses with their automatic controls all spring from these earlier beginnings.

PACKAGING AND CONVERTING MACHINERY

It is difficult to establish dates for the various developments in packaging and converting machinery. The industry was so busy growing it forgot to record its accomplishments.

Earliest machines were confined to making hand operations easier. Foot treadle operated or steam powered, they each contributed to increases in output and to lower costs. As time went on several separate machines might be combined to produce a still more efficient in-line operation.

Laminating equipment was probably an offshoot of wax-paper making equipment where a second web could be applied to the first web by heat and pressure. However, laminators may also have borrowed features from box gluing machinery and label applicating machinery, all of which were developed during the decades just before and after World War I. One of the first automatic gluer of labels for set up boxes was marketed in the United States just after 1918 by the New Jersey Machine Corporation.

Extruders, which are widely used today, originated in the dim past perhaps in the manufacture of extruded foods such as spaghetti or macaroni, or building materials such as brick and tile. We know one early extruder was patented before 1840 to extrude lead pipe. Between 1840 and 1860 extruders were used to apply gutta-percha rubber to electrical cables and wires. First extruders were ram types and therefore discontinuous. This was all right for molded parts, but not for continuous sheeting or tubing. The screw extruder was invented in 1879. In 1912 a longer-screw, heated-barrel extruder was developed for rubber. It had a ten to one ratio of length to barrel diameter (l/d), and incorporated air cooling. Little success was achieved in attempts to extrude plastics until the late 1950's when screw design was improved to allow efficient and uniform melting and l/d ratios were increased. This was the double channeled screw with separate mixing and melting section and subsequent metering section. Other screwless extruders have since been developed but some time will be required to gain experience before they will seriously compete with the screw types. The extruder has provided the packaging material converter with a means for making a wide variety of combinations of laminated and coated materials, tailored to meet the demands of nearly any food product.

By the latter part of the 19th century, all of the ingredients for modern packaging machinery was available. Packaging materials were being made by machine. Packages were being fabricated completely by machine, as with paper bags, or in several separate operations as with cans. One by one various hand operations were studied and replaced by machinery. These included: package setup; product metering; application of package to the product or vice versa; package closing; collecting of packages; preparation (setup) of shipper container; placement of packages in shippers; closing of shippers.

Weighing, metering, and counting equipment date to before 1900, as also do carton setup and gluing equipment and bag-making equipment. In the next decade machines were developed for wrapping chewing gum, and overwrapping cartons. One machine in 1910 would form, cut, and twist-wrap candy kisses. Sometime after 1913 machinery was developed to fold biscuit cartons with paper liners. Between the two World Wars, there was a difficult period during which food manufacturers resisted the change over from bulk packaging and the machinery manufacturers tried to persuade them to the advantages of high speed unit packaging.

Most machines during this period were designed to fit individual requirements and secrecy was the order of the day. It wasn't until after World War II that serious efforts were made to mechanize fully and integrate entire packaging lines. Pouchmaking equipment came on the scene in the late 40's and early 50's. Vacuum packed pouches in the mid 50's.

Today, few manufacturer's of food products are without automated high speed packaging lines. Without them the cost of packaging in a highly competitive marketplace would be prohibitive.

MODERN PACKAGING MATERIALS AND FORMS

Rigid Packaging Materials and Package Forms

Glass Containers.—Glass containers are one of the stalwarts of food packaging. Over 58 billion containers are used annually by food and beverage packagers. Narrow-necked glass containers are called bottles and wide-necked are called jars.

Glass is made from limestone, sand, soda ash and alumina. Colorants may be added to the melt or applied later to the surface. Glass is strong, rigid, and very chemically inert. It has complete FDA clearance as do many of the standard coatings. Glass does not appreciably deteriorate with age and is an excellent barrier to solids, liquids and gases. It is therefore excellent protection against odor and flavor contaminations. It is very low cost (7¢ per lb in finished, delivered container). However strength requirements may demand a heavier weight ratio. Its chief disadvantages as a packaging material are weight and fragility. It is also

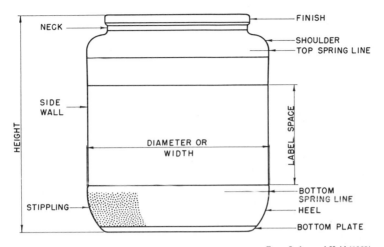

From Joslyn and Heid (1963)

FIG. 11.   NOMENCLATURE FOR GLASS CONTAINERS FOR FOODS

not easy to dispose; consequently glass is frequently designed for multi-trip or multi-use. Significant improvements in strength weight ratios have recently been achieved and new coatings have reduced damage and enabled higher handling speeds.

The transparency of glass gives excellent product visibility, and new methods for surface decoration yield highly attractive finishes. Glass is extremely versatile as to size and shape of the final container. New computerized furnaces have improved production rates and dimensional tolerances.

Glass containers include the following:

Bottles.—Bottles are the most extensively used type of glass container. They may be many different shapes but the neck is always round and much narrower than the body. The neck facilitates pouring and reduces the size of the closure required. Principal uses are for liquids, or small sized solids.

Jars.—Jars are really very wide mouthed bottles and usually have no appreciable neck. The opening permits the insertion of fingers or a utensil to remove portions of their contents. They may be used for liquids, solids, and nonpourable semi-liquids such as thick sauces and pastes.

Tumblers.—These are like jars but they are open ended. They have no neck and no "finish." They are shaped like a drinking glass and are used for products like jams and jellies.

Jugs.—These are large-sized bottles with carrying handles. Necks are usually short and narrow. They are usually used for liquids in 1/2 gallon and larger sizes.

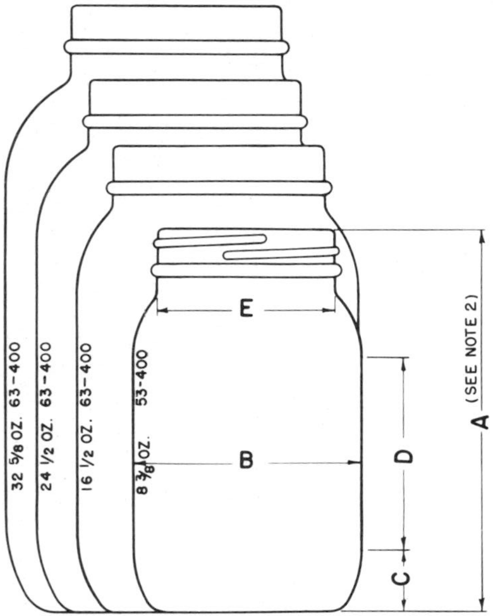

GCMI NUMBER	CAP. OF FLOW OZS.	WT. FL OZS MAX.	A	B MAX	C	D	E	SPECIMEN FINISH
10-14	4 1/4	3 3/4	3 25/64	2 1/16	35/64	1 49/64	1 3/4	48 – 400
10-20	6 1/4	4 1/4	3 55/64	2 9/32	5/8	2 1/16	1 15/16	53 – 400
10-24	7 3/4	4 1/2	4 1/8	2 25/64	21/32	2 5/16	2 3/32	58 – 400
10-26	8 3/8	4 3/4	4 9/32	2 1/2	11/16	2 9/64	1 15/16	53 – 400
10-27	8 3/8	4 3/4	4 9/32	2 29/64	11/16	2 3/8	2 3/32	58 – 400
10-28	8 3/4	5 1/2	4 31/32	2 1/2	5/8	2 31/64	2 3/32	58 – 400
10-30	9 1/2	5 1/2	4 31/64	2 35/64	41/64	2 9/16	2 3/32	58 – 400
10-36	11 1/2	6	4 25/32	2 3/4	49/64	2 7/16	2 3/32	58 – 400
10-40	12 1/2	6 1/4	4 27/32	2 53/64	49/64	2 1/2	2 3/32	58 – 400
10-48	15 1/2	7	5 1/8	3 1/32	13/16	2 11/16	2 5/16	63 – 400
10-50	16 1/2	7 1/2	5 7/32	3 1/8	27/32	2 47/64	2 3/16	60 – 440
10-51	16 1/2	7 1/2	5 7/32	3 1/8	27/32	2 47/64	2 5/16	63 – 400
10-52	17	7 1/2	5 11/32	3 7/64	27/32	2 55/64	2 5/16	63 – 400
10-53	18 5/16	8 1/4	5 3/8	3 17/64	29/32	2 49/64	2 5/16	63 – 400
10-60	22 3/4	9 1/2	5 7/8	3 29/64	15/16	3 11/64	2 5/16	63 – 400
10-62	24	10 1/8	5 7/8	3 35/64	61/64	3 3/16	2 5/16	63 – 400
10-63	24 1/2	10 1/2	5 6/63	3 9/16	31/32	3 7/32	2 5/16	63 – 400
10-67	27 1/2	11	6 1/4	3 43/64	1	3 25/64	2 5/16	63 – 400
10-71	30 1/4	11 1/2	6 9/16	3 47/64	1 1/32	3 39/64	2 5/16	63 – 400
10-72	31	11 1/2	6 9/16	3 25/32	1 1/32	3 39/64	2 5/16	63 – 400
10-75	32 5/8	11 1/2	6 3/4	3 53/64	1 1/16	3 45/64	2 5/16	63 – 400
10-77	34	12 3/4	6 13/16	3 59/64	1 3/32	3 47/64	2 5/16	63 – 400
10-81	48 3/4	18 1/2	7 11/16	4 13/32	1 1/4	4 17/64	2 5/8	70 – 400

NOTES :-

1. WHEN OTHER FINISHES ARE USED, CAPACITY, WEIGHT AND HEIGHT SPECIFICATIONS ARE ADJUSTABLE WITHIN THE REQUIREMENTS OF THE FINISH USED. SEE NOTE 2.

2. HEIGHT (DIMENSION 'A') IS BASED UPON USE OF SPECIMEN FINISH SHOWN.

3. THE SPECIFICATIONS SHOWN MAY VARY MODERATELY ACCORDING TO COMMERCIAL TOLERANCES AND INDIVIDUAL MANUFACTURERS PRACTICE.

4. 'B' DIMENSION IS VARIED TO MAINTAIN CAPACITY.

From Joslyn and Heid (1963)

FIG. 12. STANDARD DIMENSIONS OF PLAIN ROUND JARS

GCMI ITEM NO.	CAP. O'FLOW FL.OZS.	WT. MAX. OZS.	A	B MAX.	C	D	E	SPECIMEN FINISH
29-16	5	$3\frac{9}{16}$	$3\frac{17}{32}$	$2\frac{7}{64}$	$\frac{1}{16}$	$2\frac{7}{32}$	$1\frac{19}{64}$	48-1740 OR 48-870
29-25	8.2	$5\frac{1}{8}$	$3\frac{7}{8}$	$2\frac{17}{32}$	$\frac{41}{64}$	$2\frac{39}{64}$	$2\frac{5}{32}$	58-1740 OR 58-870

NOTES:—

1. WHEN OTHER FINISHES ARE USED, CAPACITY, WEIGHT AND HEIGHT SPECIFICATIONS ARE ADJUSTABLE WITHIN THE REQUIREMENTS OF THE FINISH USED. SEE NOTE 2.

2. HEIGHT (DIMENSION 'A') IS BASED UPON USE OF SPECIMEN FINISH SHOWN.

3. THE SPECIFICATIONS SHOWN MAY VARY MODERATELY ACCORDING TO COMMERCIAL TOLERANCES AND INDIVIDUAL MANUFACTURER'S PRACTICE.

4. 'B' DIMENSION IS VARIED TO MAINTAIN CAPACITY.

From Joslyn and Heid (1963)

FIG. 13.   STANDARD DIMENSIONS FOR BABY FOOD JARS

Carboys.—These are very heavy shipping containers shaped like a short necked bottle and having 3 gallon or more capacity. Typically they have been used with a wooden crate holder. Other outer protective frames are now finding use.

Vials and Ampoules.—These are small glass containers. The latter are principally used for pharmaceuticals. Vials are sometimes used for small quantities of foods such as spices, or food colorants.

When selecting a glass container for a food it is important to be careful in choosing dimensions and "finish" so that the correct volume will be available, the product can be easily filled and dispensed, and a proper closure can be selected. "Finish" refers to the type and dimensions of neck and mouth of the container, i.e., thread, lug, friction, snap-cap, roll-on, etc. There are many standard finishes listed by the Glass Container Manufacturers Institute.

Other important factors in selecting a glass container for food are its color which can influence the type of light reaching the food, and its ability to resist thermal shock. Some glasses cannot withstand sudden changes in temperature, i.e., filling a hot product into a container and then plunging it into cold water. Special glasses are available for this purpose.

Metal Cans.—The average American family uses over 250 cans yearly—or a total of 57 billion units. Of these a large percentage is for food and beverages. There are over 600 different styles, shapes, and sizes of cans to choose from.

Traditionally cans have been made from soldered tinplate steel. More recently aluminum cans have been introduced. Today there are several more choices available: standard tinplate; lightweight, double reduced tinplate; tinfree steel (coated); vacuum-deposited aluminum on steel; and aluminum.

Can bodies can be soldered, welded, or cemented. Steel bodies can be combined with aluminum ends. Many new easy open devices are available for cans ranging from pop-tabs for beverages to complete removal of lids or panels for frozen or meat products.

Can coatings are now regarded as vital components—especially for foods and beverages. Coatings must be nontoxic and free from odors or taste. They must not deteriorate or come loose from the can wall during food processing storage. Interior coatings are made from acrylics, alkyds, butadienes, epoxyamines, epoxy-esters, epoxy-phenolics, oleoresins, phenolics, and vinyls depending upon the type of food and process. Outside coatings include acrylics, alkyds, oleoresins, phenolics, and vinyls, and are usually pigmented. They are less exposed to food contact

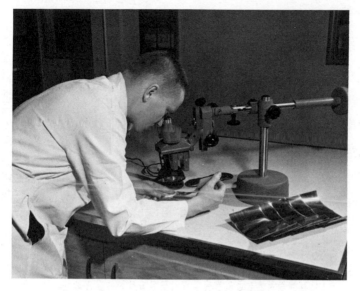

Courtesy of American Can Co.

FIG. 14. ANALYSIS OF CAN COATING

Enameled can bodies from packed cans are examined under magnification for coating and tinplate performance.

but must survive processing and be receptive to further decorative coatings and inks.

In 1959, aluminum was used for the first time commercially for beer in seven-ounce cans. Aluminum cans are supplied as the conventional cemented side seam three-piece can, shallow draw, deep draw and impact extruded varieties.

Conventional cemented side seam three-piece, aluminum cans may be square as well as round and there is no limitation on diameter and height. Side seams are sealed with a special cement. Usage of these cans is limited to nonprocessed products because the cement is not capable of standing processing after the can is filled. An interior coating is usually necessary with aluminum cans. However with some foods, such as orange juice, a coating is not required.

Shallow drawn aluminum cans are limited in size but can be fabricated in irregular shapes as well as round. They can be used for packing processed foods. Some products successfully packed in this type can include cod, roe, tuna and sardines.

Deep drawn aluminum cans have been made only in round shape.

TABLE 1

SOME COMMON CAN SIZES[1]

Can Name	Can Dimensions[2]	Approx. Net Wt.[3]	Net Contents Liquid Product[4]	Some Products for Which Ordinarily Used
2z Mushroom	200 x 204	$3^1/_4$ oz.	$3^1/_4$ fl. oz.	Mushrooms
5z Baby food	202 x 214	$4^3/_4$ oz.	$4^1/_4$ fl. oz.	Baby foods, chocolate syrup
6z	202 x 308	6 oz.	$5^1/_4$ fl. oz.	Tomato paste, tomato sauce, juices
$6^1/_2$z	202 x 314	6 oz.	6 fl. oz.	Frozen concentrates, juices
Evaporated milk	206 x 208	6 oz.	6 fl. oz.	Evaporated milk
Meat spread	208 x 109	3 oz.	$2^3/_4$ fl. oz.	Meat spreads
.....	208 x 208	$^1/_2$ lb.	$4^3/_4$ fl. oz.	Meat products
4z Pimiento	211 x 200	$4^3/_4$ oz.	4 fl. oz.	Pimientos, chopped olives
.....	211 x 208	7 oz.	$5^1/_2$ fl. oz.	Cranberry sauce
4z Mushroom	211 x 212	$6^3/_4$ oz.	$6^1/_2$ fl. oz.	Mushrooms
8z Short	211 x 300	$7^3/_4$ oz.	7 fl. oz.	Baked beans, tomato sauce, shrimp, specialties
8z Tall	211 x 304	$8^1/_2$ oz.	$7^3/_4$ fl. oz.	Vegetables and fruits, juices, specialties
No. 1 Picnic	211 x 400	$10^1/_2$ oz.	$9^1/_2$ fl. oz.	Vegetables, some fruit juices, soups, meat, fish, specialties
Beer	211 x 413	.....	12 fl. oz.	Beer and carbonated beverages
Half-quart beer	211 x 604	.....	16 fl. oz.	Beer
16z Domed	211 x 604	.....	8 fl. oz.	Whipped cream, toppings
No. 211 cylinder	211 x 414	13 oz.	12 fl. oz.	Fruit juices, nectars, tomato juice
Evaporated milk	215 x 315	.....	$14^1/_2$ fl. oz.	Evaporated milk
4z Flat pimiento	300 x 108	4 oz.	$3^3/_4$ fl. oz.	Pimientos
.....	300 x 308	$11^1/_2$ oz.	$11^1/_4$ fl. oz.	Pork and beans
Chocolate syrup	300 x $315^1/_2$	1 lb.	13 fl. oz.	Chocolate syrup
.....	300 x 402	$14^1/_2$ oz.	13 fl. oz.	Infant formulas
No. 300	300 x 407	$14^1/_2$ oz.	$13^1/_2$ fl. oz.	Vegetables, some fruits, juices, soups, meat, fish, pet foods, specialties
.....	300 x 409	1 lb.	14 fl. oz.	Meat products
No. 300 cylinder	300 x 509	1 lb. 3 oz.	1 pt. 1 fl. oz.	Soups, pork and beans, specialties
$^1/_4$ lb. Flat	301 x 106	$3^3/_4$ oz.	$3^1/_2$ fl. oz.	Salmon
No. 1 Tall	301 x 411	1 lb.	15 fl. oz.	Fruits, some vegetables, juices, fish, specialties
No. 303	303 x 406	1 lb.	15 fl. oz.	Most commonly used size for vegetables, fruits, juices, soups, specialties
No. $^1/_2$	307 x 113	7 oz.	$5^3/_4$ fl. oz.	Tuna
$^1/_2$ lb. Flat	307 x $200^1/_4$	$7^3/_4$ oz.	$6^1/_2$ fl. oz.	Salmon

Can	Dimensions[2]	Net weight[3]	Volume[4]	Products
No. 1 Flat	307 x 203	9 oz.	8 fl. oz.	Pineapple
	307 x 208	......	9 fl. oz.	Sausage, fish flakes, coffee
No. 2 Squat[5]	307 x 302	½ lb.	......	Nuts
12z Vacuum	307 x 306	12 oz.	13 fl. oz.	Vacuum packed corn
No. 95	307 x 400	1 lb. 1 oz.	1 pt.	Breads, sea foods
No. 2	307 x 409	1 lb. 4 oz.	1 pt. 2 fl. oz.	Vegetables, fruits, juices, soups, specialties
Jumbo	307 x 510	1 lb. 9 oz.	1 pt. 7 fl. oz.	Pork and beans, mushrooms
32z (Quart)	307 x 710	2 lb. 2 oz.	1 qt.	Fruit juices and drinks
No. 1¼	401 x 207.5	14½ oz.	12½ fl. oz.	Pineapple
Shortening[5]	401 x 307.5	1 lb.	......	Shortening
No. 2½	401 x 411	1 lb. 13 oz.	1 pt. 10 fl. oz.	Fruits, some vegetables and juices, meat products
	401 x 509	2 lb. 3 oz.	1 qt.	Frozen concentrates
	401 x 602	2 lb. 6 oz.	36¾ fl. oz.	Spaghetti, beans in tomato sauce
	404 x 309	1 lb. 8 oz.	22 fl. oz.	Meat products
No. 3 Cylinder	404 x 700	3 lb. 2 oz.	1 qt. 14 fl. oz.	Fruit and vegetable juices
Vacuum Coffee[5]	502 x 308	1 lb.	......	Coffee
Vacuum Coffee[5]	502 x 607	2 lb.	......	Coffee
Shortening[5]	502 x 514	3 lb.	......	Shortening
No. 10	603 x 700	6 lb. 10 oz.	3 qt.	Institutional size for vegetables, fruits, juices, meat, and fish products, soups, specialties
	603 x 812	8 lb. 4 oz.	1 gal.	Soft drink syrups
12z Oblong[5]	314 x 202 x 303	12 oz.	......	Meat products
Pullman Base[5]	402 x 310 x 608	3 lb.	......	Meat products
Pullman Base[5]	402 x 310 x 1208	6 lb.	......	Meat products
Miniature Base	414 x 410 x 1100	8 lb.	......	Meat products
Ham[5]	512 x 400 x 211	1 lb. 8 oz.	......	Ham, pear-shaped
#1 Base Ham[5]	710 x 506 x 300	3 lb.	......	Ham, pear-shaped
#1 Base Ham[5]	710 x 506 x 312	4 lb.	......	Ham, pear-shaped
#2 Base Ham[5]	904 x 606 x 308	5 lb. 3 oz.	......	Ham, pear-shaped
#4 Base Ham[5]	1010 x 709 x 412	10 lb. 3 oz.	......	Ham, pear-shaped
¼ Drawn	405 x 301 x 0145	3¼ oz.	......	Sardines
No. 1 Oval	607 x 406 x 108	15 oz.	......	Sardines, sea foods

[1] Adapted from American Can Co., *The Canned Food Reference Manual*.
[2] In inches and sixteenths of inches. Dimensions vary slightly within manufacturing tolerances. Diameter is listed first, followed by height.
[3] The net weights of various foods in the same size can will vary with the density of the product. The weights cited are for foods of average density, except where the container is largely used for one specific class of product.
[4] The volume figures cited are average commercial fills.
[5] Key-opened cans.

TABLE 2
GENERAL CLASSES OF FOOD PRODUCTS AND TYPES OF STEEL BASE REQUIRED[1]

Class of Foods	Characteristics	Typical Examples	Steel Base Required
Most strongly corrosive	highly or moderately acid products, including dark colored fruits and pickles	apple juice berries prunes cherries pickles	type L
Moderately corrosive	acidified vegetables mildly acid fruit products	sauerkraut apricots figs grapefruit peaches	type MS type MR
Mildly corrosive	low acid products	peas corn meats fish	type MR or MC
Noncorrosive	mostly dry and nonpro- cessed products	dehydrated soups frozen foods shortening nuts	type MR or MC

[1] From Ellis (1963).

They can be used for packing processed foods. Products packed include beer, powdered milk and condensed milk.

Impact extruded aluminum cans have usually been fabricated in round shapes, but irregular shapes may also be made. Since these cans withstand high pressures, they can be used for processed foods and hot packs.

Advantages of metal cans as packages are strength, speed of manufacture, filling and closing. Disadvantages include weight, ease of reclosure and ease of disposal.

Composite Containers.—A composite container is a container made from two or more constituent materials. It usually consists of a paperboard body with metal or plastic ends.

Two basic types are available. Spiral wound containers are made in cylindrical shapes where two or more plies of board are glued together around a mandrel. Convolute-wound composites are produced by straight winding. Body materials used are chipboard and kraft paper. Linings used are vegetable parchment, wax laminates, aluminum foil, glassine and polyethylene coated paper. Other linings can also be used depending on the product to be packaged.

Composite cans are closed by either a snap-on lid, plug-in lid or a lever lid. In the non-detachable type of closure, perforated tops and string-opening devices are used as well as double seamed ends.

Specific advantages in using composite cans are ease of disposal and economics. In recent years, composite cans have been widely used for refrigerated dough and other food products.

Aerosol Containers.—Aerosol containers are used to dispense a product by means of a pressurized gas or liquid that is held in the same container. The basic components of the package are the container, the valve and the protective cap.

The container must be gas-tight and may be constructed from steel cans, aluminum cans, glass, plastics, or a combination. Choice of material must be related to safety (ability to hold pressure and lack of fragility), size, and product compatibility.

There are several types of aerosols which can dispense products as fine mists, sprays, dusts, or foams depending upon the type of valve used and the product/propellant arrangement. A single phase aerosol contains a liquid product layer and a compressed gas propellant layer. A two phase aerosol contains a liquid product with dissolved propellant layer plus a compressed gas propellant layer. A three-phase aerosol contains a layer of liquid propellant, a layer of product and a layer of propellant vapor. Removal of product causes the liquid propellant to boil and replenish the vapor phase.

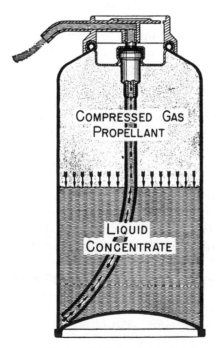

FIG. 15.   AEROSOL COMPONENTS

One type of aerosol container used for food.

Piston aerosols contain a flexible plastic barrier between the product and the propellant gas. Co-dispensing aerosols disperse two products through the same nozzle.

Advantages of aerosols are convenience, premixing, lack of evaporation, and exclusion of air even when partly used. Disadvantages are high cost, risk of explosion, breakage, or leakage. Aerosols have been used for beverage concentrates, cocktail mixes, cake icings, pancake mixes, syrups, salad dressings, and seasonings.

Rigid Plastic Packages.—A wide variety of rigid plastics can be used in the form of thermoformed, injection-molded or blow-molded containers.

Thermoformed containers are formed by exposing the plastic sheet to heat and forming into a female or around a male mold. Vacuum forming, pressure forming and several other techniques are used.

Thermoformed trays are used in many food applications ranging from margarine tubs to formed polystyrene foam egg cartons. The plastics used for thermoformed trays are polyethylene (high density), polyvinyl chloride, polystyrene, polypropylene, ABS (acrylonitrile butadiene styrene) and cellulose acetate. The specific plastic selected is dependent on the food packaged and storage requirements needed. Polystyrene tends to be brittle at low temperatures while linear polyethylene trays are not brittle, but tend to appear cloudy in appearance. All thermoformed trays are covered with heat-sealable, or snap-on lids made of plastic film or aluminum foil laminates.

Injection-molded containers are used in high volume applications for jars, bottles and tubs. Most widely used plastics are polystyrene and polypropylene. A significant advantage in the use of polystyrene is its outstanding clarity. The overall injection-molded container tends to have a thicker wall than one formed by thermoforming techniques.

Blow molding is generally used to produce containers where the neck diameter is small as contrasted to the overall container diameter. Plastics used are polyvinyl chloride, polypropylene, polycarbonate, cellulose acetate, polystyrene, polyethylene and polyacetal. Blow molded bottles have mostly been made from polyethylene but increasingly PVC and polystyrene are being used. Final selection of the specific plastic is determined by product compatibility, economics and ease in processing.

Advantages of plastic containers are low cost and ease of fabrication. Disadvantages may include lack of product compatability, low barrier qualities, plastic deterioration, lack of resistance to high heat, fragility at low temperature.

Solid and Corrugated Fiberboard Containers.—Solid and corrugated fiberboard materials are used to fabricate shipping cartons and cases used extensively in wholesale and industrial shipping. They are not usually used as direct containers for foods, but are extensively employed as outer shippers for food packages, i.e., cans and bottles.

Both are made from heavy fibrous kraft paperboards either cylinder or Fourdrinier. Solid fiberboard is made by gluing several plies of paperboard together. By using asphalt, or special resin adhesives, such as ureaformaldehyde, enhanced moisture resistance may be built in. Selection of weight, fibrous construction, and number of plies is related to the desired burst, tear, puncture and bend resistance. Corrugated fiberboard is made from similar base materials but is generally thinner, as it is then structured by combining facings (flat sheets) and liners (corrugated or fluted sheets) by means of adhesives. There are five major types of corrugated paperboard: unlined, single-faced, double-faced, double-walled, and triple walled. The first two types are used for wrappings of fragile objects or as interior padding of boxes. The latter two types are used where exceptional strength and rigidity are essential. Double-faced corrugated is the most commonly used type for boxes, liners, and partitions.

	A Flute	B Flute	C Flute	E Flute
Flutes/foot	35–37	50–52	41–45	90–96
Flute height (in.)	0.185	0.105	0.145	0.085
Thickness (in.)	$3/16$–$7/32$	$4/32$	$5/32$	$5/64$

Each flute has particular properties with respect to load support in each of the three possible directions. In double wall construction different flutes can be combined.

Advantages for corrugated board packages are versatility, light weight, strength, disposability and low cost.

Disadvantages include low wet strength; however, newer techniques of manufacture and new coatings are eliminating this disadvantage.

Corrugated containers are now available with easy open tear strips, self locking assembly, and smooth white liners permitting flexographic printing on the exterior. Special reinforcements can be employed. Where strength is less important molded pulp has been used for liner material. Other special corrugated boards or fiberboard constructions include foil laminated facings and plastic foam fillers between paper facings.

Wooden Boxes and Crates.—When lumber was plentiful and inexpensive many shipping containers were made in the form of nailed or wirebound wooden boxes and crates. With rising costs of lumber, labor, and freight rates, and the availability of cheaper substitutes—fiberboards, composition boards and the like—use of wood containers has been on the decline.

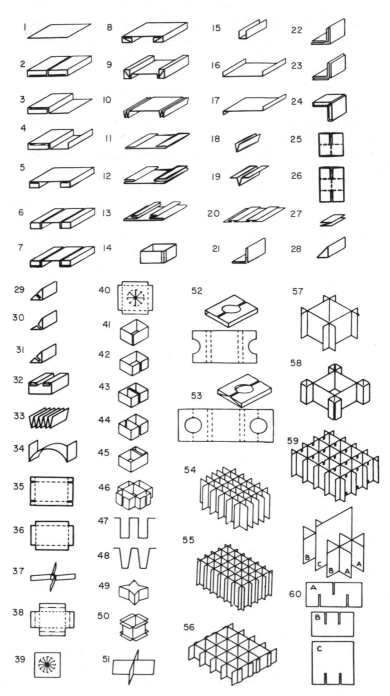

Courtesy of Hinde & Dauche Div., W. Va. Pulp & Paper Co.

The type and cure of wood used will affect the weight, strength, and ease of fabrication of the container. Soft woods are easier to nail but not as strong as hardwoods. Green lumber is excessively heavy, weaker, and will warp and shrink causing loosening of nails or other fasteners.

Boxes are usually solidly walled, rectangular shaped, nailed wooden containers and will vary in construction and in extra cleats and braces as may be required by the load. The top, bottom, and sides of a box provide the main structural strength.

Crates are similar to boxes but may be of lighter weight and more open construction—that is spaces may be left between boards or the crate may be fully enclosed or sheathed. A crate differs from a box in that the frame members carry the load. The sheath merely encloses, hence sheathing may be corrugated fiberboard, or thin plywood or light weight lumber.

Other joining methods may be used for boxes and crates. These include metal fasteners, glues, and wires or wire tapes. When using wires, thinner side, top, and bottom sheathing can be utilized as the wires add strength. Cleated ends and stiffeners provide the structural strength required.

Advantages of wooden boxes and crates depend on the relative cost, strength, and weight ratios involved. Where lumber is cheap or boxes can be reused, they may prove advantageous. However, in most food uses today, wooden containers are being phased out and solid or corrugated fiberboard containers are replacing them. Some wood is still used for reinforcing cleats and bottoms.

Cylindrical Shipping Containers.—Cylindrical containers have high stacking strength and can be rolled in handling. The latter is now less important due to mechanical handling devices. Cylindrical containers may be made from fiberboard, glass, metal, plastic or wood. Glass containers have already been discussed. Plastic containers are principally in the form of liners for other shells made from steel, aluminum, fiberboard, or wood.

A barrel is a cylindrical container of greater length than breadth, having two flat ends of equal diameter and bulging at the waist. A drum has straight sides and flat or bumped ends. A pail is a cylindrical or tapered (truncated cone) shaped container with or without a wire handle or bail. Small pails may be called cans. A keg is a small barrel. A cask is a large tight wooden barrel.

Barrels may be made of wooden staves bound together with hoops and may be tight or slack. Metal barrels are made of steel or aluminum.

FIG. 16.   (OPPOSITE) STYLES OF INNER PACKING

Drums may be made of metal, plywood, or fiberboard. Fiberboard drums may have wood, metal, or fiber ends.

Fiber drums and cans are made from spiral or convolute wound paper or paperboard bodies and may be closed with metal, plastic or fiber ends. With interior facings of special papers, foils, or plastics such as glassine, parchment, aluminum foil, polyethylene, or pliofilm they may be used successfully for bulk shipment of foods as the linings prevent contamination of the food or weakening of the container. Other barrier components can be built into the side wall at time of lamination. These may be asphalt, foil, or polyethylene to achieve a moisture barrier.

Steel drums are used as single trip or returnable containers. Drum heads may be removable or fixed. Fixed heads may be fastened to the body by brazing, welding, or double seaming. Some drums are formed in two halves and joined circumferentially at the waist.

Drums may be fitted with removable covers which are of the friction lid type or which may be held in place by locking rings or lugs. Tight head drums or pails are furnished with small capped or screw plug openings for ease of pouring. Additional vents and drains may be specified.

Aluminum drums are designed to have similar features to steel drums. Stainless steel drums are available where corrosion resistance or high levels of sanitation are required.

Latest types of drum constructions use blow molded polyethylene drums as the primary container and a steel shell for added protection and rigidity.

Containerization of Freight Shipments.—The basic aim of containerization is to transport goods from the source of manufacture to the sales point safely and economically. The concept uses a freight container which is delivered directly to the factory for loading. In other cases, the goods are sent to a container base to be loaded on the container. At the point of use, the container is directly unloaded. Freight containers are made of various materials including aluminum. Specially designed handling equipment is used to accommodate containers.

Semi-rigid Packaging Materials and Package Forms

Aluminum Containers.—Containers fabricated from plain aluminum foil have differing degrees of rigidity depending upon thickness, temper, alloy, and container design. Some could be called rigid as they are not easily deformed, others are more delicate, hence they are included here as semi-rigid—that is a package which is intended to maintain a definite form or shape and is not influenced by the shape or bulk of the contents but which can rather readily be bent or dented.

Aluminum containers provide extreme convenience in preparation and serving of foods, they withstand extremes of temperatures. Foods may be frozen in the package or cooked in it. Aluminum protects the food against moisture, gases, and light. General categories are compartmented, folded end, ovals, pie plates, rectangulars, rounds, squares, and specialty items. Folded end cartons have wrinkle-free sides and bottoms with folded corners. Seamless formed containers made with matching male and female dies have wrinkled side walls and flanges. Newer techniques of forming are now producing smooth flanges and smooth side walls. With smooth flanges flat-web materials can be hermetically sealed on as closures. Colored containers can be decorated attractively to permit the container to be placed on the table as a server. Soft butter and margarine products are now found in such packages.

Setup Paperboard Boxes.—Four basic components are used to make setup paperboard boxes; paperboard, adhesive, corner stays, and covering. Paperboard is selected to give the right weight and smoothness for the size box required. Sheets of the boxboard are cut and scored, the sides are folded up to make a tray, and the corner stays are adhered. This is repeated for the cover. Finally the covering material is glued on. Coverings may be colored papers, foil laminates, or highly coated embossed and printed litho papers.

Boxes can be made in almost any shape and with a wide variety of lid arrangements—separate or hinged. Semi-rigid plastics can be used as lids for better product display, or die cut windows can be similarly employed.

Advantages of setup boxes as packages include convenience, individuality, strength, reusability, and excellent product protection and display. Equipment required is minimal and low cost. Boxes are shipped set up—hence no setup time. Small quantities are no more expensive than large quantities making large inventories unnecessary.

Disadvantages are the generally higher cost in comparison to folding cartons produced in large quantities.

Folding Paperboard Cartons.—A folding carton is a container made from bending-type boxboard by die-cutting and scoring it properly to fold into the desired form. It is supplied by the maker as a flat blank, preglued, or partially glued, and collapsed. It is erected, filled, and closed by the packer. Because they are supplied in knocked-down form (flat), folding cartons are easier to pack than setup boxes and provide economies in transportation.

Paperboards used in folding cartons must be capable of being bent and scored. There are many grades and thicknesses ranging from cheap lined or unlined chipboards to manila, kraft, laminated, and clay coated solid bleached sulfate boards. The latter being best for high quality printing. Boards may also be coated with plastics such as polyethylene, ethylene vinyl acetate, wax, or blends of resins. Foil laminates are used both for aesthetics and for added product protection.

There are two common styles of folding cartons and a large number of special constructions. Tube types are one piece cartons that are bent into a tube (generally square) with a longitudinal glued body seam. End flaps are glued shut, tucked, or selflocking.

Tray types may be one or two piece with or without a lid. They are shipped flat and are set up and glued to form the tray and/or lid in the packager's plant. Some tray types are glued by the box manufacturer and folded flat along diagonal score lines. They can be snapped open to set up. Carriers for cans or bottles are special types of folding cartons.

Cartons may be printed, embossed, or die cut prior to blanking. Printing may be by letterpress, offset lithography, or rotogravure.

Folding cartons are widely used for both solid and liquid foods. Advantages are low cost, ease of automatic high speed setup, filling and closing, good stackability, easy opening and reclosure and excellent graphics.

Molded Pulp Containers.—When a fibrous material is mixed with water and molded, a molded pulp container results. These containers are made of various virgin and chemical wood pulp or wastepaper pulp. The process used closely resembles conventional papermaking techniques.

Molded containers can be formed by pressure, injection, or suction molding methods. They are usually water sensitive but low cost. Uses in the food industry are for egg containers and various produce containers.

Flexible Packaging Materials

Flexible packages are made from combinations of flexible materials. These include the basic substrate, laminating adhesives, protective or decorative coatings, and decorative inks. There are literally millions of possible combinations. The flexible packaging engineer achieves his objective by selecting appropriate components and tailor-making a laminate to fit the packaging needs of a particular product.

Paper.—In flexible packaging, the basic papers used consist of bonds, tissues, litho, krafts, glassines, parchment, and greaseproof.

Bond papers are uncoated sheets made of bleached chemical pulps in weights ranging from 20 to 70 lb. They are supplied with a wide variety of finishes and may be purchased with high wet strength and tear strength and as printing and laminating grades. Machine finished papers are usually not smooth enough for fine printing. Machine glazed bond papers are used for gum wraps and printing applications.

Tissues are lightweight papers made of semi- and fully-bleached chemical pulps in basis weights ranging from 8 through 20 lb with open or close fiber formation. Special treatments may be added such as wet strength and mold resistance. Where a bright, white tissue is required, machine glazed bleached tissues are used. For applications requiring "bleed," porous tissues are required.

Litho papers are sheets coated on one or both sides in basic weights ranging from 29 through 60 lb. These papers are especially prepared to produce a smooth printing surface. They are not as strong as bonds, but have a better surface for reproduction of printing and are preferred for label stocks. Two-side coated litho paper is often used in publication advertising where both sides of the sheet must be smooth for printing. Groundwood-free and groundwood-content sheets are used. Many beer labels use a groundwood-content paper incorporating a special finish.

Kraft papers are extremely strong and may be used in bleached or unbleached forms. Basis weights range between 25–80 lb. A wide variety of finishes and wet strength levels are available. Krafts are classified as "Southern" or "Northern." "Southern" grades are made from softwoods grown in the southern part of the United States. They are usually porous, nonuniform in formation and comparatively rough. They also have high tear strength. "Northern" kraft papers are supplied from the northern part of the United States and are more uniform, less porous and have high tensile and mullen strength values.

Due to the strength of kraft papers (kraft in German means strength), they are most useful in many applications where bonds and lithos are not strong enough. In addition, they are usually much cheaper than other stocks. A high wet strength natural "Northern" kraft is used for printed cannister labels where a smooth and strong sheet is required.

Glassine papers are supercalendered chemical pulp sheets in basic weights ranging from 15 to 45 lb. They have high resistance to air and grease. Genuine glassine pulp is beaten much longer than the imitation grade and therefore the genuine glassines have the highest resistance to grease. Imitation glassines offer greater resistance than bond papers. All glassines are stronger and stiffer than most other papers and have a smooth, glossy surface. Some are fully bleached while others have a greenish, grayish, or bluish cast. A pouch paper is made in a similar manner to glassines but plasticizer is added to give the paper softness and machinability. Glassines are used in candy wraps, and other food applications.

Parchment papers are made of bleached chemical pulp which is immersed in a bath of sulfuric acid. They have good grease resistance and high wet strength. Greaseproof papers are essentially nonsupercalendered glassines. Both parchment and greaseproof papers have a rougher and more opaque appearance than glassines. They are supplied in grades ranging from 15 to 27 lb. Applications include butter and oleo wrappers.

Stiffness affects packaging machine conditions. Papers remain an important factor in flexible packaging because they contribute strength, stiffness, smoothness and low cost.

Films.—A film is a thin flexible plastic sheeting having a thickness of .010 in. or less. Although solid in its finished form, at some stage of its manufacture it is capable of being formed into shape through the application of heat and pressure, or by chemical reaction, or by casting from a solvent solution.

The first commercial flexible film was cellophane. In 1924, the E. I. DuPont de Nemours Company acquired rights from La Cellophane (France) to produce the film in the United States. Cellophane is manufactured from highly purified cellulose derived from bleached sulfite pulp. The cellulose is treated with sodium hydroxide solution and carbon disulfide to produce "xanthate" which, when dispersed in sodium hydroxide, produces viscose. The viscose is extruded through a slot into acid-salt baths to produce a regenerated cellulose film. It is then washed, desulfured, bleached, softened, dried, and wound up as plain, nonmoisture-proof PT film. By incorporating various coatings and modifications, over 100 different grades of cellophane are now available. Nitrocellulose coated cellophane offers non-blocking, moisture vapor resistance (MVT), slip, flexibility, and heat sealability. Saran coated cellophane offers superior protection from oxygen and better MVT. Most cellophanes are used in packaging baked goods, vegetables, confectionery, meats, and overwraps. Polymer coated (Saran) varieties are useful for oily and greasy products. Over 400 million lb of cellophane are produced annually and its present commercial strength is based on its wide range of coated varieties.

Cellulose acetates are derived from cellulose treated with acetic acid anhydride. The cellulose triacetate is partially hydrolyzed. Additives include plasticizer, antiblocking agent and sometimes a uv-absorber. It is used where stiffness, high gloss and dimensional stability are important. Cellulose acetate is also used for window applications in boxes since it has a low static attraction of dusts. Its high rate of gas and water transmission make it useful in produce packaging.

Polyethylene is the largest volume single film used in the flexible packaging industry. It is a polymer of ethylene and is obtained by two different processes. High pressure polyethylene is obtained by exposing ethylene to temperatures between 302° and 392°F (150° and 200°C) at a pressure of about 1200 atm in the presence of traces of oxygen. This film is known as low density polyethylene. Low pressure or high density polyethylene is produced at temperatures between 140° and 320°F (60° and 160°C) and a pressure of 40 atm with alkylmetal catalysts.

Polyethylene (low density) is a low cost film with moderate tensile strength and clarity. It is a good moisture and poor oxygen barrier and

TABLE 3

GENERAL CHARACTERISTICS OF PACKAGING FILMS[1]

Film Material	Thickness, in.	Max. Width, in.	Yield, 0.0001 in. Thickness, sq. in./lb.	Clarity	Specific Gravity, at 73°F.	Normal Performance		Limit Conditions		
						Max. Temp., °F.	Min. Temp., °F.	Sunlight	Aging[2]	Flammability, Temperature of Combustion, °F.
Cellophane-plain	0.0008–0.0016[6]	...	20,000[6]	Transparent[2] or colored	1.40–1.50[3]	300	...	No effect	...	155–216[2]
NC coated[4]	0.0009–0.0017	60	11,500–21,000	Transparent	1.40–1.50	300	About 0	No effect	...	Slow burning
Saran coated[4]	0.002–0.0014	46	19,500	Transparent	1.44	180	About 0	...	...	Same as newsprint
Polyethylene coated[4]	0.002 or more	60	11,800 (2 mil combination)	Transparent to translucent	1.20	180		Excellent	Good	Slow burning
Cellulose acetate	0.0005–0.010	40–60	21,000–22,000	Transparent or colored	1.28–1.31	150–220	Brittle at 0	Good	Excellent	Slow burning
Polyethylene Low density	0.0004–0.010	480	30,000	Transparent to translucent	0.910–0.925	200	−50	Fair to good	Excellent	Slow burning
Medium density	0.0004 or more	480	29,500	...	0.926–0.940	220	−60	Fair to good	...	Slow burning
High density	0.0004–0.010	60	29,000	...	0.941–0.965	250	−70	Fair to good	...	Slow burning
Polyester (Mylar Scotchpak, Videne)	0.00025–0.0075	50–55	20,000	Transparent or opaque	1.38–1.39	300	−80	Moderate	Excellent	Slow burning
Polypropylene	0.0005 or more	60	30,900–31,300	...	0.885–0.90	190–220	−60	Fair to excellent	...	Slow burning
Polystyrene (oriented)	0.00075–0.020	43–60	26,100	...	1.05–1.06	175–220	−60	Good to air	...	Slow burning
Rubber hydrochloride (Pliofilm)	0.0004–0.0025	60	25,000	Transparent to opaque	1.11	180–205	−20	Fair	Good in dark	Self extinguishing
Vinylidene Cryovac	0.0008–0.003	...	16,700	...	1.64	Softens 270	Depends on plasticizer	Excellent to good	Excellent below 76°	Self extinguishing
Saran	0.0005–0.010	40–54	16,000–23,000	Transparent to opaque or colored	1.20–1.68	150–200 (dry) 300 (wet)	−25 to −50	Excellent to good	Excellent below 76°	Self extinguishing
Vinyl chloride	0.001–0.010	54–84	20,000–23,000	Transparent to opaque or colored	1.20–1.45	150–200	Good depends on plasticizer	Good	Excellent	Self extinguishing
Nylon 6[5]	0.0005–0.01	60	24,000	Transparent or colored	1.12	400	−80	...	...	Self extinguishing

[1] All values by permission Modern Plastic; Encyclopedia Issue for 1962, unless otherwise noted.
[2] By permission Modern Packaging Encyclopedia 1958. Copyright 1957 by Packaging Catalog Corp., 770 Lexington Ave., New York 21.
[3] Stone and Reinhart (1954).
[4] Test data assume coated side toward product.
[5] Miscellaneous sources.
[6] Miller (1959).

several grades are unaffected by mineral and essential oils. Its greatest benefit is its ability to be fusion welded to itself to give excellent liquid tight seals. High density polyethylene offers excellent moisture protection and increased stability to heat.

Polypropylene is produced by the polymerization of propylene. It is more rigid, stronger, and lighter than polyethylene. The film has low water vapor permeability, good resistance to greases, high temperature stability and good gloss. The cast unoriented film has fair low temperature stability and is a poor gas barrier. Polypropylene is usually used in the oriented form. By stretching the film, monoaxially or biaxially, tensile strength and abrasion resistance are greatly increased.

Polyamides are obtained by polycondensation of ω-amino acids or by polycondensation of diamines with diacids. Various grades of polyamides (Nylon) are available. Nylon 6 offers ease in handing and abrasion resistance. Nylon-11 and Nylon-12 are superior barriers to oxygen and water and have lower heat seal temperatures. Nylon-66 is very high melting and difficult to heat seal.

Polyesters are condensation products of a polyalcohol with a diacid or its anhydride. They offer excellent tensile strength, tear resistance and good ageing properties. Although expensive, they may be used in thin gauges to be competitive.

Polyvinylchloride films are prepared by polymerizing vinylchloride in the presence of suitable catalysts. By the addition of plasticizers, flexible films are obtained. Vinyl copolymer films are used as shrinkable oriented films for dairy, meat, confectionery, and beverage packaging as well as a laminate component.

Polyvinylidene chloride is usually produced as a copolymer with 13–20% vinylchloride. It contains a low per cent of plasticizers, slip agents, and stabilizers. "Saran" films are clear, have excellent mechanical resistance and extremely low water vapor and gas transmission rates. The film is used for cheese, meat, sausage, and dried fruit wrappers.

Rubber hydrochloride (pliofilm) is produced from natural rubber by the addition of hydrochloric acid. It is stretchable, nontoxic, resistant to oils and greases, unaffected by acids or alkalies and does not support combustion. Extended storage leads to discoloration and brittleness. It is heat sealable and aroma-proof. The film is used for self-service packages of meat and cheese. Bags lined with pliofilm are used for coffee, spices and cookie packaging.

Polyvinylacetate is produced by block polymerization of vinylacetate with benzoyl or acetylperoxide, or by emulsion polymerization with potassium persulfate or hydrogen peroxide. Its main use is as a coating on paper; however, a copolymer is used for packaging cheese and meats in direct contact, such as seamless casing.

Polyvinylalcohols are made by hydrolysis of polyvinylacetate and are soluble in water and insoluble in many organic solvents. They are used as a lining on paper for oil and fat packaging.

New Films.—*Amylose film.*—Now also available for use in a wide variety of food packaging applications. Sold by American Maize Products as "Ediflex" film, it is made from corn starch and has full FDA approval.

Its most unique property is that it is edible. Amylose films are also very soft, slightly hazy, stretchable, and water soluble. The film is a poor moisture barrier, however, it is satisfactory as a gas and flavor barrier. It is outstanding in its resistance to oil and grease.

Possible uses for amylose films include an inner wrapper for frozen foods. The film would dissolve during thawing. Other possibilities are its use in portion packs, dehydrated soups and other foods that are dissolved in water.

Ionomers.—Polymers used for packaging applications are usually constructed of covalent bonds holding together various C, H, or O linkages. If ionic bonds are used, bond strengths increase and properties differ. The introduction of *Surlyn A* by E. I. DuPont de Nemours Company marked the development of a new family of thermoplastics called ionomers. Based on low density polyethylene, the ionic bonds in ionomers serve to increase overall bond strengths and yield superior oil, grease, and solvent resistance.

At normal temperatures, Surlyn A is tougher than polyethylene and is more sensitive to water absorption. Its high melt strength makes Surlyn A extremely suitable for extrusion coating, vacuum forming, and skin packaging. Applications requiring deep "draw" vacuum forming are potential markets. Skin packaging, blister packaging, and blow-molded bottles are likely areas of development.

Due to the complex bonds of Surlyn A, the cost of production is increased. It is competitive to polyethylene only where its special properties can be used to their fullest advantage. Spicy and greasy products can be packaged in Surlyn A laminates since grease hold-out is greater than low density polyethylene. In Mexico, Surlyn A has been used as the inner component of a laminate for Tetra-Pack container holding olive oil. Excellent oil resistance was required to contain the olive oil. Since Surlyn A has superior hot melt strength as compared to polyethylene, a thinner extrusion coating enables a converter to reduce any higher raw material cost.

Future development may possibly include the development of ionic copolymers based on high density polyethylene. As an extension of prior research, these ionomers would offer good rigidity allowing thinner walls in injection and blow molded applications.

TABLE 4

PERMEABILITY AND CHEMICAL PROPERTIES OF PACKAGING FILM

Film Material	Gas Transmission[2] cc./100 sq. in./24 hr. 72°F., 1 mil.			Water Vapor Transfer Gm./100 sq. in./24 hr. 76 cm. Mercury 100°F. 90% R.H., 1 mil.	Water Adsorption in 24 Hr. Immersion Test, %, 1 mil.
	Oxygen	Nitrogen	Carbon Dioxide		
Cellophane-plain	Low (dry); variable (moist)[3]			6.86–12.8[3]	44.7–114.8[3]
NC coated	Very low (dry); variable (moist)			1.4–2.7	45–115
Saran coated[4]	Very low			1.2	...
Polyethylene coated[4]		Low		1.2	...
Cellulose Acetate	100–140	25–40	500–800	160	3.6–6.8
Polyethylene Low density	500	200	1350	1.4	0–0.8
Medium density	240[2]	...	500[2]	0.7	Nil
High Density	100–150	50–60	300–400	0.3	Nil
Polyester (Mylar) Scotchpak, Videne)	5–10	0.50	7.30	1.0–3.0	0.5
Polypropylene	200–300	3–100	700–800	1.2	0.005 or less
Polystyrene (oriented)	200–300	40–100	500–1000	4.0–8.0	0.04–0.06
Rubber hydrochloride (Pliofilm)	38–3,250[2a]	...	288–13,500[2a]	0.5–15.0	5.0
Vinylidene Cryovac	0.1–0.2	0.025	0.30	1.0–1.4	Negligible
Saran	0.1–0.2	0.025	0.30	0.1–0.3	Negligible
Vinyl chloride	100 (est)	25–50 (est)	500 (est)	3.5–13.0[1]	Negligible
Nylon 6[5]	1.0	...	2.0	8.1	...

[1] All values by permission Modern Plastics Encyclopedia Issue for 1962, unless otherwise noted.
[2] Data courtesy Modern Plastics, 1961 all values unless otherwise noted.
[3] Stone and Reinhart (1954).
[4] Test data assume coated side toward product.

Ethylene-vinyl Acetate Copolymers (EVA).—Copolymerization offers polymer chemists a method of obtaining "tailor-made" polymers with specific properties. Ethylene-vinyl acetate copolymer results from the copolymerization of high pressure (low density) polyethylene and vinyl acetate. EVA is more flexible than polyethylene; however, it is also more permeable to water vapor and gases. Impact strength is excellent and the film is clearer than polyethylene. EVA's are thermally instable at high heats but very stable and versatile at lower temperatures and have a wide heat seal range.

A significant disadvantage is EVA's tendency to block due to high surface friction. The polar nature of the side groups present in the copolymer cause blocking characteristics. The addition of slip and antiblock additives partially overcome blocking. In applications requiring high friction, EVA may prove extremely useful. As a coating for bags and industrial sacks demanding stackability, high surface friction is desirable. Future applications include use as a shrink film. Bags can be produced to stretch wrap products in the form of a contour wrap. The

TABLE 4 (Continued)

| | | Resistance To | | |
Acids	Alkalies	Greases and Oils	Organic Solvents	Water
Poor to strong acids[6]	Poor to strong alkalies[6]	Impermeable[6]	Insoluble	Moderate
Poor to strong acids	Poor to strong alkalies	Impermeable	Coating attached	Moderate
Excellent except H$_2$SO$_4$ & HNO$_3$	Good except ammonia	Impermeable		...
Excellent	Excellent	Like polyethylene	Like polyethylene	
Poor to strong acids	Poor to strong alkalies	Good	Soluble except in hydrocarbons	Good
Excellent	Excellent	May swell slightly on long immersion	Good except hydrocarbon and chlorinated solvents	Excellent
Excellent	Excellent	Good	Good	Excellent
Excellent	Excellent	Excellent	Good	Excellent
Good	Good	Excellent	Excellent	Excellent
Excellent	Excellent	Good	Good	Excellent
Good	Excellent	Good	Excellent to poor	Excellent
Good	Good	Excellent	Good except in cyclic hydrocarbons chlorinated solvent	Excellent
Excellent	Good except ammonia	Excellent	Good to excellent	Excellent
Excellent except H$_2$SO$_4$ and HNO$_3$	Good except ammonia	Excellent	Good to excellent	Excellent
Good	Good	Moderate to good	Poor to good	Excellent
Poor	Excellent	Excellent	Excellent	Excellent

[5] Miller (1959).
[6] By permission Modern Packaging Encyclopedia Issue 1960. Copyright 1959. Packaging Catalog Corp., New York.
[a] Relative humidity, 0%.

extremely high flexibility of EVA enables it to stretch tightly without the use of a shrink tunnel. In addition, EVA's high impact strength allows the use of thinner gauge films for similar applications than polyethylene.

No pre-treatment is necessary to achieve excellent ink adhesion in printing; the resin can be blow molded, injection molded and extruded. Stress cracking resistance is good thus allowing EVA to be used for detergent bottle production. For this use, the proper gauge material must be selected since EVA is somewhat flexible. At present, EVA copolymer is expensive due to raw material costs and low production volume. When larger production facilities are constructed, the price gap between EVA and polyethylene may be significantly reduced.

Polypropylene Copolymers.—An area of research receiving considerable attention is the development of polypropylene modifications. Polypropylene offers excellent water, gas, and strength properties; however, its cold temperature impact strength is poor. In an effort to overcome this defect, a second monomer, usually ethylene, is introduced into the propylene structure. As a film, propylene-ethylene co-polymers are hazy due to light scattering. The low temperature impact strength of polypropylene is improved by the addition of ethylene and potential uses include blow molding applications.

Polypropylene copolymers are particularly useful in shrink wrapping. Due to the reduction in crystallinity, a lower melting point is achieved. Orientation thus takes place at fairly low temperatures. Additional copolymers of polypropylene include the introduction of 1-pentene, 1-butene, or 1-hexene. The subsequent films are clear and offer improved impact strength compared with the polypropylene homopolymer. Prices are still fairly high; however, they may be reduced in the near future.

TABLE 5

MECHANICAL PROPERTIES OF PACKAGING FILMS

Film Material	Tensile Strength,[2] 100 p.s.i.	Elongation,[2] %	Tearing Strength,[2] g./mil.	Bursting Strength, 1 mil. thick, p.s.i.	Folding Endurance/ 1 mil. thick, no. of folds × 10[3]	Heat Sealing Range,[2] °F.
Cellophane-plain	104–186[3]	14–36[3]	275–426[3]	...	...	...
NC coated	70–150	10–50	2–10	55–65[1]	Over 15[5]	200–300
Saran coated	70–130	25–50	2–15	...	...	225–350
Polyethylene coated	50 and over[1]	15–25[1]	16–50[1]	40–50[5]	Good[5]	230–300[1]
Cellulose acetate	54–139[1]	25–45[1]	2–25[1]	50–85[5]	0.25–0.4[1]	350–450
Polyethylene						
Low density	13.5–25	200–800[1]	150–350	48[5]	Very high[5]	250–350
Medium density	20–35	150–650[1]	50–300	...	...	240–350
High density	24–61	150–650[1]	15–300	...	...	275–350
Polyester (Mylar Scotch-pak, Videne)	170–237	70–130	10–27[1]	45–50[3]	20[3]	275–400
Polypropylene	45–100	400–600	32–1750[1]	Very high[1]	...	325–400
Polystyrene (oriented)	80–120	3–20	20–30	23–60[1]	...	250–325
Rubber hydrochloride (Pliofilm)	35–50	200–800[1]	60–1600	Stretches[3]	10–1000[5]	250–350
Vinylidene						
Cryovac	60–120[1]	50–100[1]	15–20[1]	...	Over 500[5]	275–300
Saran	80–200	40–80	10–100[1]	20–40[3]	...	280–300
Vinyl chloride	30–110	5–250	30–1400[1]	25–40[5]	250[1]	200–350
Nylon 6[4]	138–170[4]	200[4]	50[4]	No burst[4]	...	400–450[4]

[1] Data by permission Modern Plastics Encyclopedia Issue for 1962.
[2] Data by permission Modern Plastics, 1961 all values unless otherwise noted.
[3] Stone and Reinhart (1954).
[4] Miller, (1959).
[5] By permission Modern Packaging Encyclopedia Issue 1960. Copyright 1959 Packaging Catalog Corp., New York.

TPX.—This new ethylene derived polymer has recently been introduced. Chemically, it is closely related to polyethylene and is based on the polymerization of 4-methylpentene-1. The material is transparent, has the lowest specific density of any plastic material (0.83) and has outstanding electrical properties. Its clarity is excellent and the crystalline melting point of 474°F enables it to be used for hot-filling and cooked food applications. The resin can be extruded, blow-molded and injection molded.

A major disadvantage in TPX is its high permeability to water vapor and gases. Presently, the resin is expensive since it is only being produced in development quantities. The first production facility by ICI, Ltd. (United Kingdom) was established in 1967; however, plans include the construction of a larger plant fairly soon. Prices should decrease with larger production.

Halogenated Polyethylene.—Polyethylene has been produced incorporating chlorine in the monomer. The addition of chlorine enables polyethylene to be printed without any pretreatment and does not effect film flexibility. The film is called halogenated polyethylene. Since the film is nontoxic, it has applications in conventional polyethylene markets. The higher film price must be evaluated relative to potential uses.

Coextruded Structured Films.—Laminations of two films such as cellophane and polypropylene have been available for some time as also have been coated films such as Saran coated polyesters, nylons, and polyolefins. A new type of films called structured films is now being produced by a process of coextrusion (simultaneous extrusion through complex dies). Intermixing of the polymers is avoided through the maintenance of laminar flow throughout the molten phase. Both flat die and blown structured films are being manufactured. One of the first commercial products was a polyethylene-polypropylene-polyethylene film designed for breadwrap. Another structure now available is a copolymer-homopolymer-copolymer polypropylene film. In Denmark and England, a combination of white-black-white pigmented polyethylene is used for packaging milk.

Polycarbonate films are heat sealable and have excellent clarity, high impact strength and good temperature resistance. Although they have been commercially available for several years, their high cost has retarded most flexible packaging applications. The film is resistant to weak acids and alkalies but is attacked by aromatic and chlorinated hydrocarbons and permeability to gases and water vapor is poor.

Possible markets for polycarbonate films include rigid and semi-rigid applications such as blister packs and thin-walled bottles. In Japan, a thin wall polycarbonate bottle is currently used for packaging salad oil. Another Japanese development consists of a soda siphon which contains plain water and a carbon dioxide charge. Polycarbonates withstand the effect of high internal pressures.

The raw material costs inherent in the production of polycarbonate resin are expensive. Future price reductions will probably not bring polycarbonates into the price range of polyethylene.

Phenoxy Films.—Since phenoxy films are chemically related to epoxy resins, they are extremely hard, scuff-resistant, and exhibit superior heat resistance than most other thermoplastics. Bottles, films and extrusion coatings made of phenoxy are very resistant to oils and greases. Permeability to gases, moisture, and odors is low. One major

TABLE 6

AMOUNTS OF DIFFERENT GASES TRANSFERRED IN 24 HRS. BY DIFFERENT FILMS AT ROOM TEMPERATURE

Cubic centimeters per 100 sq. in. of film at 15 lbs. pressure differential

Type of Film	Temp.—R.H. Variations	Nitro-gen	Oxygen	Carbon Dioxide	Air	Ethyl-ene	Ethyl-ene Oxide	Sulfur Dioxide
Cellophane (MSAT)[1,2]	...	32.0	43.0	111.0	35.0	...	...	...
Cellophane (MSAD)[5]	2% R.H.	...	6.30	37.04	...	...	...	...
	100% R.H.	...	205.4	346.8	...	...	...	...
Vinylidene[1] (Cryovac)	...	3.1	11.1	27.1	5.1	...	...	...
Cellulose Acetate[3]	25°C.	...	352.5	1912.0	...	549.2	19,500	18,510
(P-903)	0°C.	...	132.2	809.0	...	...	5,590	7,848
Cellulose Acetate[3]	25°C.	...	250.0	932.0	...	112.7	4,560	5,445
(P-912)	0°C.	...	112.7	421.0	...	...	...	...
Rubber Hydrochloride[4]								
(Pliofilm 100 FMI)	...	...	144.0	924.0	...	...	...	...
0.0005 in. Polyester[5]	2% R.H.	...	18.67	24.65	...	...	...	...
(Mylar)—	100% R.H.	...	49.67	86.50	...	...	...	...
0.0005 in. Polyester[5]								
(Mylar)—	2% R.H.	...	7.70	...	...	...	...	...
0.002 in. Polyethylene	100% R.H.	...	15.52	...	...	...	...	...

[1] Values by Cryovac Co., Food Processing 17, No. 8 (1956).
[2] With moist gas at high relative humidity.
[3] Values by Celanese Corp. on two types of film differing in flexibility and toughness (converted from investigator's values expressed in cc. per sq. meter per 24 hr. per cm. mercury pressures).
[4] Values by Goodyear Tire and Rubber Co.
[5] Values by Nagel and Wilkins (1957) converted from investigator's values expressed in grams per sq meter per hr.

problem is alcohol resistance. Phenoxy films are subject to stress-cracking with products containing more than 40% alcohol.

The outstanding clarity of phenoxy films coupled with good impact resistance and rigidity made blown bottles a feasible outlet. Several cosmetic items such as shampoos have already been packaged in blown phenoxy bottles.

The price of phenoxy resin is fairly high. Most curing type plastics are expensive due to the cost of curing systems. They are unlikely to become significantly cheaper; however, phenoxies are most useful for highly specialized application.

Acrylic Films.—XT, acrylic copolymer exhibits excellent clarity, oil, grease resistance, and gas and odor permeability. Its rigidity is claimed to be four times that of high density polyethylene and the material exhibits good impact strength. It has poor resistance to aromatic and chlorinated hydrocarbons, and alcohol containing products cannot be packed in XT polymer.

Potential markets for this newly formulated polymer include rigid bottles for foods and cosmetics. The excellent clarity of XT makes direct printed labeling possible.

Polyurethane Films.—The most important properties of polyurethane film are toughness and strength. It is a very soft and abrasion resistant film. Due to polyurethane's excellent oil and grease resistance, it is currently being used as a material for motor oil packaging. Other possible applications include boil-in-bags and flexible meat packages. Solvent resistance is good toward aromatic hydrocarbons and gasoline.

Aluminum Foil.—Aluminum foil is defined as a solid sheet rolled to a thickness less than .006 in. Plain aluminum foil refers to foil, in either roll or sheet form, that is not combined with another material such as paper, film, or cloth. In speaking about plain foil, bare foil, unmounted foil, unsupported foil, and free foil are used synonymously. The bulk of converting and packaging applications utilize foil in rolls.

Aluminum foil is available in different tempers and alloys. The high purity grades of alloys (1,000 series) are normally used for foil packaging applications. The single most widely used alloy for packaging foil is 1235. For drawn and/or formed containers, 3003 alloy is used. It has a 1.0–1.5% manganese content.

"O" temper is produced by subjecting the foil to controlled heat followed by controlled cooling. Foil with "O" temper is referred to as dead soft and has the lowest physical properties. It has good folding characteristics and is widely used in converting operations.

All "H" temper foils are produced by strain hardening the metal during normal foil rolling. There may also be present additional thermal treatments which serve to soften the material partially. H-12 or H-14 temper metal are strain hardened by rolling to temper after an intermediate anneal. The physical characteristics are improved as contrasted to soft foil; however, close adherence to certain properties is difficult to control. Full hard foil is H-18 which is fully strain hardened by rolling. There is no intermediate or final anneal and definite physical property limits are attained. Full hard metal is the direct opposite of dead soft metal.

Superhard foil is H-19 which is fully strain hardened by rolling and controlled to yield maximum physical properties. H-24 metal is an intermediate temper strain hardened by rolling and then partially stress relieved by annealing. It is used in various container applications. Most foil sold and used by packaging converters is soft or "O" temper.

Steel Foil.—Steel foil made from tin coated steel .002 inches thick was brought to market in 1964. In recent years the tin has been eliminated, various gauges from .001 to .0045 in. have been made available and different levels of temper (or anneal). Steel foil can be laminated to paper, paperboard and other packaging materials. It is extremely strong and puncture resistant. Its magnetic properties may prove useful for special applications.

Flexible Package Forms

Wrappers.—Flexible packages are made from materials whose final shape conforms to and is governed by the product enclosed, as opposed to rigid packages which require the product to conform to the shape of the container. The simplest flexible package is a wrapper which may be loose or tight and may be sealed shut by using heat or various types of adhesives. Some wrappers are left unsealed as is done on stick chewing gum wrappers, others are partly sealed to hold them in place as is done with carton overwraps. The simplest wrappers are sheets of paper, metal foil, or plastic film with or without decoration. Many films can be heat sealed but paper and foil require coatings in order to make them heat sealable. Early types of paper wrappers were glued shut. The earliest types of heat-seal coating were waxes. Many advances have been made in wax formulations and hot melts so that today some wrappers can be sealed with pressure alone (cold tack adhesives). The advent of plastic resin coatings such as the vinyls, nitrocelluloses, Sarans, and polyolefins, made it possible not only to add heat seal properties but also to substantially improve strength and barrier properties of laminates of paper and plastic. Metal foil is in itself an excellent barrier. Coatings and other substrates serve only to add strength, decoration, and heat sealability.

A wrapper that comes in direct contact with a food such as candy or a loaf of bread is called an intimate wrap whereas if it wraps an inner package such as a carton it is called an overwrap. Wraps may be purchased as roll stock or as precut sheets. The latter are usually used for hand or semi-automatic production.

Other variations of wrappers include labels which wrap only part of a rigid package and serve primarily to identify the contents. Some completely wrapped around labels on paper containers contribute barrier protection also.

Another variation of a wrapper is a bundling overwrap which combines several smaller packages into one larger unit. Plastic film shrink wraps serve this function. The ultimate is extremely large pallet overwraps where a plastic film stabilizes an entire pallet load of smaller packages.

There are at least 13 distinct types of twists and folds by which wrappers may be closed neatly before sealing. Overwrapping machinery has been developed for each type of wrap and fold and for specific products. Speeds range as high as 1000 units/minute for twist wrap candies. Hand wrapped supermarket tray packs run about 16 per minute, whereas a pallet overwrap such as Reynolds Metals Co.'s "Reynolon POW" can be applied at the rate of about one per three minutes with substantial savings in time, material costs, and shipping costs.

Preformed Bags or Envelopes.—A bag is a tube with one sealed end and is made from a flexible material. After filling with product it may be closed by sealing, tying, or clipping shut. An envelope is a bag with a flap that can be folded over to close. A paper bag was one of the first converted packages. Paperbags may be made in four standard styles; flat, square, satchel, and self-opening. A flat bag has a longitudinal back seam and no gussets. The bottom is folded over and glued or sealed.

A square bag has side gussets which give it a much greater capacity. The satchel bag has no gussets but the bottom is folded into a hexagonal shape and sealed. When filled the bag will stand by itself. The self-opening bag has side gussets or they can be contour sealed in order to more closely fit irregular products.

Pouches (Form-Fill-Seal).—A pouch is similar to a bag except it is simpler in construction and is made from roll stock, filled, and sealed on automatic high speed machinery. There are many, many possible variations of pouch sizes, shapes, and compositions. A few general types can be described.

Pouches can be defined as vertical, or horizontal, depending on how they are formed. Vertical pouches are made by drawing one or two webs downward around a forming mandrel, sealing one or two seals to form a tube and adding a top and bottom seal for closure. When one side seal is used and centered, it is called a pillow type. If the side seal is inserted on one side and a fold on the other, it is called a three-side seal pouch. When two side seals are used, it is called a four-side seal pouch or fin-sealed pouch. Product is added while the pouch is in the vertical position and usually before it is severed from the web. Side seams can be lapped or butted, and gussets are optional.

Horizontal pouches are made by drawing one or two webs in a horizontal direction. Usually when two webs are fed they are positioned in the horizontal plane and formed into a four-side sealed pouch. Product is added before the top web is sealed in place. Frequently the bottom web is formed into a packet. When a single web is fed, it is usually drawn over a plough to fold it into the vertical plane. The bottom may be sealed or left unsealed. Side seals are made and the package is filled through the top, usually after being severed from the web. Again gussets are optional.

When packaging small items, frequently the individual pouches are not separated but left in a continuous strip. Perforations or notches are inserted so that individual packets can be torn off as desired.

Odd-shaped pouches can be formed by using contoured heat-seal jaws. Tetrahedral pouches are made by forming a tube on a vertical form and fill machine and inserting cross seals alternately at 90° angles to the preceding cross seal. The tetrahedron uses the least amount of material for a given pouch volume.

Some pouches or envelopes are sold and displayed as such, however many are given added protection by inserting them in folding cartons. This is nearly always done for small items or portion packs.

Collapsible Tubes.—The collapsible tube has traditionally been made of soft metal, tin, lead, tin-lead, and in recent years aluminum. One end of the tube is a threaded dispensing nozzle which can be closed by a screw cap closure. The entire tube is formed by impact extrusion from a slug of metal then it is trimmed, threaded, and annealed. Exteriors are decorated by roller coating and offset printing. Interiors may be lined by dipping, flushing, or spraying prior to exterior coatings. Linings may be waxes, or vinyl, phenolic, or epoxy resins. Product is filled through the open bottom which is then crimped or sealed shut. In use the closure is removed, the interior seal (if present) is punctured, and product is dispensed through the nozzle by squeezing the tube.

Tubes are marketed on cards, in blister packs, or in folded cartons.

Blow-molded plastic tubes are now being used for some products. Their main disadvantage is lack of dead-fold. They tend to spring back when squeezed making total dispensing of the product difficult. Tubes are best for packaging thick liquids or thin pasty solids.

FACTORS INVOLVED IN THE CREATION OF A FOOD PACKAGE

What is a Package?

A food package is a structure designed to contain a food product in order: (a) to make it easier and safer to transport; (b) to protect the product against contamination or loss; (c) to protect the product against damage or degradation; (d) to provide a convenient means for dispensing the product.

The addition of printing or other decoration to the exterior of the package serves: (a) to identify the contents as to type and quantity; (b) to identify the manufacturer's brand and quality grade; (c) to attract the buyer's attention; (d) to persuade the buyer to purchase; (e) to instruct the purchaser on how to use the product.

Ease and Safety of Transport.—The first packages were probably invented to make transportation easier. It is impossible to carry a liquid such as water or milk in one's hands, and very difficult to carry small solids such as grain, berries, or fruits. Open topped rigid containers such as cups, pails, bowls, urns, baskets and panniers or flexible containers such as sacks and wine-skins made transportation possible. Modern packages are carefully tested to insure they will survive the rigors of transportation.

Product Protection.—The development of wrappers and of closed containers was for the purpose of protecting products against spillage, evaporation, or pilferage losses and against contamination by dirt or vermin. Modern packages also protect against contamination from microorganisms and against product degradation from exposure to environmental factors such as heat, light, moisture, and oxygen.

Dispensing.—Easy opening and reclosure features make it easier and safer for the consumer to dispense desired quantities of a food from a package.

Selling the Package and Identification of Contents.—Since most bulk shippers and many individual packages are made from opaque materials, some means for identifying the contents is imperative. Over the years this has evolved from simple markings to highly

Courtesy of General Foods Technical Center

FIG. 17.   MODERN PACKAGING LABORATORY

Pilot packaging laboratory with machines for making many types of packages and tests for machine performance.

decorative labeling which not only instructs the purchaser as to the quantity, quality, and type of contents but also first attracts his attention and then persuades him to buy.

Birth of a Package

In designing, developing, and testing a food package, the manufacturer of the food product today can draw upon a host of new materials and new packaging designs that did not exist 5 to 10 years ago. He can rely on the talents of his own packaging development laboratories or he can call upon outside packaging laboratories for assistance. Most of the latter have been established by manufacturers or converters of packaging materials for the purpose of developing new markets for their products. In either case, there are certain steps that must be followed in bringing a food package to the market place with reasonable assurance that it will sell.

Idea Initiation

The idea for a new package may be initiated in any one of a number of ways, in the mind of a research scientist, a product or marketing man, a designer, a field salesman, or a production man on the line. But whatever its source it will be aimed at one of the following:

(1) To modify an existing package for an existing food product in order to improve sales through:
   (a) improved aesthetics of package design;
   (b) improved price due to lower cost materials, to improved manufacturing efficiencies, or to improved handling performance;
   (c) improved product quality due to improved package protection;
   (d) improved product utility due to improved package performance (such as an easier opening device or a better pouring spout) which provides greater convenience.

(2) To put a new food product in a well-tried and proven package in order to improve sales through expansion of the product line.

TABLE 7
HOW PACKAGE SIZE AFFECTS RATIO OF SURFACE AREA TO CONTENTS

	Package Sizes		
	1	$\frac{1}{2}$	$\frac{1}{4}$
Net weight, lb.	1	$\frac{1}{2}$	$\frac{1}{4}$
Dimensions, in.	$6\frac{5}{8} \times 3\frac{5}{8} \times 2\frac{1}{2}$	$5 \times 3 \times 2$	$4 \times 2\frac{1}{2} \times 1\frac{1}{2}$
Area, sq. in.	75.2	62	39.5
Ratio, sq. in./oz.	4.7	7.75	9.87
Ratio, sq. in./cu. in.	1.25	2.07	2.63

From Brickman (1957B).

(3) To develop an entirely new package concept for a proven food product to improve sales through:
   (a) revived customer interest;
   (b) improved product quality;
   (c) improved product utility (convenience).

(4) To develop a new package concept for a new and untried food product.

Typical examples of each of these types of endeavor would be:

(1) the substitution of a polyolefin film for a cellophane film in order to gain better performance (less breakage) when the product is frozen;

(2) the marketing of a new soft drink in a glass bottle with twist-off crown cap;

(3) the marketing of milk in a blown polyethylene gallon container;

(4) the marketing of soft margarines in metal or plastic tub containers.

Regardless of the source of a given idea, the market manager in charge of the food product must decide whether there is a need for such a package and whether it can be sold in sufficient quantity and at a price that will produce the desired return on investment or profitability. He is the one who pushes the "stop" and "go" buttons which get the program started, carry it through to completion, or at some point terminate it. In making these decisions there are a number of steps, which when taken more or less in sequence, will develop factual information and help him arrive at a sound judgement.

Initial Market Appraisal

Marketing personnel make an initial appraisal of the market potential for the proposed product and in so doing establish "ball-park" figures on optimum sizes and maximum costs for the product and the package.

Initial Design Concepts

Design concepts are collected from all available sources by the Packaging Engineer assigned to the project. "Blue sky" thinking and "brain storming" meetings are encouraged at this point. Ideas are then screened to rule out the impractical—that is those that are drastically out of line costwise or which can not be manufactured by existing technology. Two lists are prepared.

The first list includes all ideas which are feasible with relatively minor changes in materials or manufacturing techniques. The second list includes those which would be excellent packages but which require research to develop a presently not available material or manufacturing technique. In most instances it is decided to go with items from the first list, as these involve less investment in time and money. There may, however, be important reasons for selecting an approach from the second list. A successful long range development may be much less costly per unit or it may be unique and patentable.

Design Concept Testing

Where many ideas abound or where there is some question as to acceptability of a single concept, a consumer panel test can be arranged. Here one or several groups of consumers are asked to look at artists' sketches or at mock-up samples of each idea and to comment upon them. From this interplay of comments helpful conclusions often can be drawn. Care must be taken in interpreting results, however, and skilled professional market research people should be utilized, as it is all too easy to make incorrect decisions based on inadequate data.

Package Design and Performance Testing

The Packaging Engineer now takes over. He examines the criteria defined by Markets and makes a selection of packaging forms and raw materials. This selection involves many factors, to name but a few: (1) package sizes; (2) nature of product (perishable, fragile, liquid, solid, etc.); (3) package barrier requirements (air, moisture, grease, etc.); (4) temperature extremes to be encountered; (5) does package require easy opening device, reclosure? (6) will product be packaged before or after processing? (7) shelf-life; (8) stackability; (9) in use conditions; (10) ease of disposal; (11) competitive packages; (12) dealer pack; (13) shipper; (14) pallet size or other load unit; and (15) will package be reused?

Within each package form or category other more specific questions need to be answered; for example, suppose the package is to be a flexible lamination formed into a sealed pouch:

(1) How many plies should the laminate have?
(2) What is the function of each ply?
(3) Which ply should be closest to the product?
(4) What thickness should be used?
(5) How should the plies be joined together?
(6) How should the pouch be formed and sealed?
(7) What outer protection will be needed? (Box, shipper, etc.).

After making his selection, he prepares sample packages of each proposed design concept that has survived previous screenings. He may introduce several minor variations for each concept. He then tests the samples to see which ones best meet the performance criteria.

These may include tests for strength, product compatibility, heat resistance, permeability, water immersion, resistance to cold, and the like. This period of testing will eliminate some design concepts as technically or economically unfeasible and will narrow the choice to a relative few. By this time the selection of the basic concept can be fairly well established and unresolved differences will be relatively minor.

Selection of materials, preparation of samples, and performance testing may be a short term activity or it may involve several years of intensive work. This will depend on the type of package development. Minor modifications are short term problems. In some new package concepts total systems may be required. This may well involve learning new technologies, or developing new equipment for manufacture of the package, for filling and closing it, or for testing it. For example, in developing a flexible package for heat-sterilized foods, one of the most difficult problems was how to measure the change of temperature of the

Courtesy of General Foods Technical Center

FIG. 18.   CONTROLLED ATMOSPHERE TESTING ROOM

Interior of controlled atmosphere package testing room with
temperature and humidity set to simulate climatic conditions
in Jacksonville, Fla.

Courtesy of National Biscuit Company Research Center

FIG. 19.   CONTROLLED ATMOSPHERE TESTING CABINET

Large controlled atmosphere cabinet with glass doors to permit locat-
ing samples before opening to minimize time cabinet is open.

Courtesy of General Foods Technical Center

FIG. 20. CONTROLLED ATMOSPHERE SERIES OF TESTING ROOMS

Doors to a series of controlled atmosphere rooms provided with locks
to prevent access of unauthorized persons, thus preventing theft and
insuring validity of observations.

food in the center of the package, when the package was sealed and was
moving in a tortuous path through a pressurized system of steam and air
at 250°–260° F inside a closed heavy steel "cooker."

Manufacturing Capability

Now the packaging engineer concentrates his attention on establishing
manufacturing feasibility (although he has not ignored this facet in his
previous efforts). Trial manufacturing runs are made to get more accurate estimates of cost and to find and "debug" possible problems. Preliminary shipping and handling tests are conducted to help establish design
of shipping containers and package performance. Sufficient information
is generated at this point for manufacturing to start estimating production costs with some assurance. Plans for consumer placement tests can

Courtesy of General Foods Technical Center

FIG. 21. DETERMINATION OF OXYGEN PERMEABILITY

Apparatus used for determining oxygen in head space voids and intercellular spaces in packaged foods utilizing Beckman Oxygen Analyzer.

be begun and test markets studied. Design of the package artwork and copy can be brought to tentative completion.

Marketing and Advertising Aspects of Package Design

Not too long ago, the primary aim of a package was to provide protection. Marketing and advertising concepts were generally unsophisticated. Our rising standard of living and the advent of self-service groceries led to a bolder and freer approach to packaging. Packaging now is an integral part of merchandising, and every packaging decision has an important effect on sales. The package is now considered to be a top-management marketing tool.

Self-service stores and their impact on customer buying habits are of great significance in establishing package size. In the United States, supermarkets account for about 90% of all food sales. This means that the American housewife does her major shopping once a week. Bigger packages are usually demanded and large economy sizes are especially popular. The introduction of nylon film laminates into the cured meat market was in great part due to the demand for larger size packages. Polyester laminations were not capable of drawing to the depths required of one- and two-pound units. In Europe, the situation is different. In 1957, the number of self-service shops was estimated to be 11,000

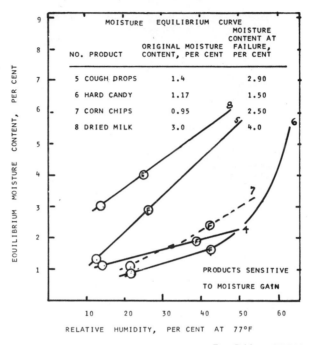

From Brickman (1957A)

FIG. 22.   MOISTURE EQUILIBRIUM CURVES FOR FOOD
PRODUCTS SUSCEPTIBLE TO MOISTURE GAIN

Open circles indicate initial moisture, circles with letter F indicate
moisture content at which products become unsalable.

and by 1965, it rose to over 100,000 stores. Nevertheless, most European
housewives still shop daily and therefore purchase smaller packages.
Individual nations differ as to the growth of supermarkets. They are
most highly developed in Germany, Norway, Holland, Sweden, and
Switzerland, and are increasing rapidly in Austria, Denmark, Finland,
and England. Only in France, Belgium, Spain and Italy are they relatively underdeveloped.

A housewife must choose from a wide range of packages, products,
and sizes every time she shops. In just one-fifth of a second the package
must be able to convey the message, "buy me!" No advertisement is
ever read as often as a package.

If the package is to sell in this brief period of time certain factors must
be considered: (1) does the package have a distinctive shape? (2) does it
feature a brand name in a standardized location? (3) does the manufacturer's trademark or symbol appear plainly? What about manufacturer's name? (4) is the product name readily identifiable? (5) what type of

decorative technique is to be used? (6) are colors appropriate and distinctive? (7) is the package attractive and pleasing? (8) does the package give adequate and clear information on quality and quantity of contents and how to use them?

In addition to these rather obvious questions there is an area which involves the psychology of design. This includes considerations of the implications of color and shape, for example, the connotation of cleanliness and coolness for blue colors, of tranquility for green, of warmth for red and orange, and of strength and cheerfulness for yellow. Strong hues are masculine; pastels are feminine. Ovals and circles are liked by all, but only men like triangles.

Symbolism enters into selection of design motifs, but one symbol may be meaningful to one culture and meaningless to another. Snow flakes have meaning in northern countries but mean little to a resident of tropical Africa. The toadstool-mushroom is good luck in Central Europe, but

TABLE 8

COLOR AND SYMBOLISM IN PACKAGING FOR ASIAN MARKETS

Country	Color	Connotation	Symbol	Connotation
China	white	mourning (avoid)	tigers, lions, & dragons	strength (use)
Hong Kong	blue	unpopular (avoid)	tigers, lions, & dragons	strength (use)
India	green & orange	good (use)	cows	sacred to Hindus (avoid)
Japan	gold, silver, white, & purple black	luxury & high quality (use) use for print only, prefer gay, bright colors.	cherry blossom chrysanthemum	beauty (use) royalty (avoid)
Malaysia (Population is mixed Malay, Indian, Chinese)	yellow gold green	royalty (avoid) longevity (use) Islamic religion (avoid)	cows pigs	sacred to Hindus (avoid) unclean to Moslems (avoid)
Pakistan	green & orange	good (use)	pigs	unclean to Moslems (avoid)
Singapore	red, red & gold, red & white red & yellow yellow	prosperity and happiness (use) Communist avoid	tortoises snakes pigs & cows	dirt, evil (avoid) poison (avoid) same as for India & Pakistan (avoid)
Taiwan Thailand	black	avoid	elephants elephants	strength (use) national emblem (avoid)
Tahiti	red, green, gold, silver, and other bright colors.	use		
Arab & Moslem States	white	avoid	animals pigs Star of David	avoid religious pollution (avoid) political (avoid)

Information courtesy of Hygrade Packaging Company (New Zealand).

is meaningless elsewhere. The shape of a design may be detrimental. Recently, a U.S. designed tunafish can was introduced in Japan. The design featured a tuna above the waves with its nose pointing down toward the water. The package sold poorly and questions were raised as to its failure. Market research pointed out that a down-pointing nose on a tuna signified that the fish was dead to the Japanese housewife. As soon as the design was modified, sales increased.

A purple package introduced in Hong Kong was a flop because purple is an unlucky color to Chinese people. On the other hand red signifies prosperity. When the package color was switched to red, the product was a success. Purple signifies death in some countries, luxury in others. White signifies purity to some and mourning to others. A green colored wrapper on bread connotes mold, while a brown wrapper connotes "oven-baked freshness." Brown glass containers for milk are popular in Northern Europe. Despite the better keeping quality of the milk, housewives in America rejected the brown glass, perhaps associating it with dirt.

In designing packages for export there are other problems relating to language, for example "gift" in German means "poison."

Consumer Placement Tests

While the initial design concept tests were conducted with panels made up of consumers, the number of people involved was relatively small. When the product has been brought to the point where technical and economic feasibility appear good and major design features can be fairly well established, it is nearly always wise to prepare a few hundred samples of each minor design variation and to conduct a consumer placement test. In this test, sample packages are coded and a set is given to randomly selected consumers in pre-determined communities. Questionnaires are filled out before and after the consumer has tried the product. The consumer comments on quality of the product, on convenience and attractiveness of the package, on how it performed, and on what she would be willing to pay for the product. Her answers serve as guidelines in further narrowing the number of design concepts and in establishing criteria for test markets.

The placement test may also reveal a glaring weakness hitherto overlooked. This may require further research or further design changes and a second consumer placement test. It may even represent total rejection. It is better to learn it at this point than at a later more costly point in the development program.

Consumer Reactions Are Questionable

It is important to note that even "concept testing" may lead to failure due to the unreliability of results. A new package should not be tested and developed solely via consumer conceptual testing. In our present day affluent society, people buy psychological pleasures and not biological necessities. It is important that psychological devices be integrated into a concept test. An additional problem inherent in concept testing is that only creative people deal with concepts. The majority of people must relate to discrete "things" and not broadly defined concepts. For these reasons, concept testing can be unreliable. For example, when consumers were asked whether they would want to purchase yellow margarine, over 90% said that yellow margarine was not needed. Yet, today it is still on the shelves and provides a profitable market. Again, when consumers were asked whether they would use instant coffee, they said it was an unnecessary product. Yet, it too is highly successful.

Test Market

When consumer placement test results have been analyzed, selection of the design or designs to be used in the test market is made. Final artwork and copy approval is obtained and manufacturing is asked to produce a sufficient quantity of the product in each package design to fill and sustain the test market areas. Specifications are provided by Packaging. Marketing Research plans the test as to what market areas, what type of stores and distribution, what type and quantity of advertising, etc. Manufacturing establishes the production capability, produces the necessary quantity of packaged product, and ships it to the test market areas. Packaging personnel remain in close touch with Manufacturing during the production runs and make quality checks on the product as produced. Packaging engineers participate in the market test by observing package performance in the field.

Marketing Research conducts the test market and reports and analyzes results. If these are favorable, final design, artwork, and package specifications are established and the program is turned over to Markets for final exploitation. Manufacturing and Markets work together in establishing full product capability and in developing a national distribution and selling program.

The test market is the final step in the development of a package. It is here that all the predictions and guesswork meet reality in the marketplace. Here the consumer must back her words with her money. A test market is expensive and should not be undertaken unless all foreseeable problems have been defined and solved. The package must have a very high level of probable success to justify putting it into the market area. Not the least danger is in allowing competition to become aware of a new development. Competitors have been known to enter the test market area and by heavy advertising and discount prices disrupt the accuracy of the test. It is best to conduct test markets in several areas at the same time to minimize this threat. Package failure during a test market may prove catastrophic. It can be disastrous to a program to have to withdraw a product for performance reasons. If all prior work has been done carefully and thoroughly, the risk can be minimized.

BIBLIOGRAPHY

ANON. 1967A. Modern bottling. Bottler and Packer *41*, No. 6, 68–74.

ANON. 1967B. Hygienic judgment of plastics in the framework of the food law. Wbl. Papfabr. *95*, No. 7, 262–266. (German)

BINSTED, R., DEVEY, J. D., and DAKIN, J. C. 1962. Pickle and Sauce Making. Food Trade Press, Ltd. London.

BRICKMAN, C. L. 1957A. Measuring gas transmission of flexible materials in pouch form. Package Eng. *2*, No. 1, 21–25, 54.

BRICKMAN, C. L. 1957B. Evaluating the packaging requirements of a product. Package Eng. *2*, No. 7, 19–25, 56.

CHESKIN, L. 1966. Selling psychological satisfactions. Commercial Chemical Development Association, Fall Meeting, 80–90 (Chicago).

DAVIS, A. 1967. Package and Print. Clarkson N. Potter, Inc., New York.

ELLIS, R. F. 1963. Metal containers for food. *In* Food Processing Operations, Vol. 2, M. A. Joslyn and J. L. Heid (Editors). Avi Publishing Co., Westport, Conn.

FOX, K. R. 1967. Plastics in the packaging of frozen foods. Conference on "Advances in Packaging with Plastics," Nov. 1967 (London).

FRIEDMAN, W. F., and KIPNES, J. J. 1960. Industrial Packaging. John Wiley & Sons, New York.

GIOLITTI, G. 1966. Plastics containers for sterile foods. Ernährungsforschung, *11*, No. 3, 375–383. (German)

GOULD, J. 1969. Packaging an image . . . his? hers? yours? Australian Packaging *17*, No. 5, 47–49.

GRIFFIN, R. C. 1960A. The role of aluminum foil on temperature moderation of frozen food packages. Frosted Food Field *30*, No. 3, 34.

GRIFFIN, R. C. 1960B. What are little boil-in-bags made of? Frosted Food Field *30*, No. 6, 30.

GRIFFIN, R. C. 1965. Geometry helps fix size of fin-sealed pouch. Package Eng. *10*, No. 2, 92–102.

GRIFFIN, R. C., and MATTHEWS, S. B. 1966. Find material needs for tetrahedral package with geometry. Package Eng. *11*, No. 3, 86–94.

HALL, L. P. 1967. Glass packing of fruit and vegetables—a review. Food Process. Marketing *36*, No. 434, 429–445.

HOLDSWORTH, S. D. 1968. Progress in the handling and packing of cans, etc. Food Trade Rev. *38*, No. 3, 54–58.

JOSLYN, M. A. and HEID, J. L. (Editors). 1963. Food Processing Operations, Vol. 2, Avi Publishing Co., Westport, Conn.

KANE, J. N. 1950. Famous First Facts. H. W. Wilson Co. New York.

KAREL, M. 1967. Use-tests only real way to determine effect of package on food quality. Food in Canada *27*, No. 4, 43–50.

KAUFMAN, M. 1963. The first century of plastics. The Plastics Inst., London.

KEFFORD, J. F. 1967. Science in food packaging. Food Technol. Australia *19*, No. 5, 204–213.

KNIGHT, E. H. 1876. Knight's American Mechanical Dictionary. Houghton, Mifflin, and Co., Boston.

LEONARD, E. A. 1967. Packaging to meet requirements in a changing world. Australian Packaging *15*, No. 7, 51–58.

LINDOP, R. 1967. Composite containers. *In* Packaging Materials and Containers, F. A. Paine (Editor). Blackie and Son, Ltd. London.

LONG, R. P. 1964. Package Printing. Graphic Magazines, Inc. Garden City, New York.

LOPEZ, A. 1969. A Complete Course in Canning, 9th Edition. Canning Trade, Baltimore.

MAY, E. C. 1938. The Canning Clan. MacMillan Co., New York.

O'BRIEN, R. *et al.* 1965. Machines. *In* Life Science Library. Time Inc., New York.

PAINE, F. A. 1962. Fundamentals of Packaging. Blackie and Son, Ltd. London.

SACHAROW, S. 1965. The permeability of flexible plastic films. Plastics Design Process. *5*, No. 9, 20–23.

SACHAROW, S. 1966. Testing of slippage of flexible film packages. Opakowanie, *71*, No. 6, 12–15. (Polish)

SACHAROW, S. 1967A. Boil-In-Bag in USA is very popular. Neue Verpackung, *20*, 70. (German)

SACHAROW, S. 1967B. Odor and taste in packaging. Israel Design Packaging *1*, No. 4, 13–14. (Hebrew)

SACHAROW, S. 1968A. Packaging materials. *In* Encyclopedia of Polymer Science and Technology, H. F. Mark and N. G. Gaylord (Editors), Vol. 9. Interscience Publishers, New York.

SACHAROW, S. 1968B. How to print paper, film and foil. Embalagem. *4*, No. 17, 24–27. (Portuguese)

SACHAROW, S. 1968C. Extrusion or laminating? Kunst. Rund. *15*, No. 12, 603–606. (German)

SACHAROW, S. 1969A. Flexible packaging. Packaging India, *1*, No. 2, 15–21.

SACHAROW, S. 1969B. Flexible packaging. Packaging India, *1*, No. 3, 13–18.

SACHAROW, S. 1969C. A survey of the new plastics for packaging. Plastics Design Process. *9*, No. 1, 14–17.

SELBY, J. W. 1961. Modern food packaging film technology, No. 39. British Food Manufacturing Industries Res. Assoc.

CHAPTER 2

Basic Food Processes

INTRODUCTION

There are some foods which can be eaten without any preparation—for example an apple can be plucked from the tree and eaten immediately. Some foods require a special kind of preparation—or processing—before they are edible at all, for example the bitter cassava, from which comes tapioca, is deadly poisonous due to the concentration of hydrocyanic acid in the root. The primitive peoples of the South Seas learned to heat the root to drive out the poison. Then they could make sauces from the juice and starch foods from the solids. Tapioca starch is widely used in puddings. Still other foods are essentially tasteless until processed—the vanilla bean and the mustard seed have little flavor until fermented.

Some foods are difficult to eat unless prepared, for example grain needs to be husked, crumbled, and boiled in water to be eaten easily, ordinary meats are tough if not cooked, and some such as abalone or octopus must be pounded thoroughly before cooking or they will have the consistency of vulcanized rubber.

Almost all foods will deteriorate or spoil after they have been harvested or killed. Some are more sensitive than others. Spoiled foods not only taste and smell bad but they can be extremely dangerous. Tainted meat or spoiled vegetables can make the eater extremely ill and even kill him. Many forms of microorganisms, including pathogenic bacteria, readily multiply in foods and produce dangerous toxins. *Claviceps purpurea* is a fungus which attacks grain such as rye or alfalfa. People eating the fungus ingest it and the by-products manufactured from it and are subject to a disease called ergotism. Sufferers from this disease encounter deterioration of blood pressure and the blood vessels. In extreme cases gangrene of the fingers and toes may set in.

Clostridium botulinum is a microorganism that inhabits the soil in every country in the world. It is also found at the bottom of seas and lakes. It therefore is likely to be present in anything grown in the soil and in fish. The clostridium produces a toxin which is one of the most deadly poisons known to man. One part of toxin in ten million parts of food will kill a mouse. It is many times more deadly than cobra venom. The toxin attacks nerves causing weakness, paralysis, and death. Fortunately, the toxin can be destroyed by about ten minutes of heating at 212°F. The microorganism however forms spores and will survive six

65

hours of boiling, then later may resume its vegetative state and manufacture of toxin.

By trial and error men have experimented for thousands of years and have perfected—without knowledge of the causes—methods for preparing foods to make them more palatable and also procedures for preserving foods to make them safer to eat.

In Table 9 there are listed a number of food processes and their purpose. In the first category are listed those processes which aim at extracting the edible portion of the food from the inedible portion. In the

TABLE 9

TYPES OF FOOD PROCESSING

I—Separation
    Picking
    Washing, dry cleaning
    Dehusking, peeling, shelling
    Destemming, deleafing
    Beheading, disemboweling
    Coring, deboning, pitting
    Bleeding, juicing
    Pressing
    Straining, filtering
    Distillation, steeping
II—Size or Shape Alteration
    Grading, sorting
    Cutting, slicing, dicing, flaking
    Shredding, chopping, comminuting
    Grinding, milling, powdering
    Pasting, mixing, beating, kneading
    Rolling, extruding, molding
III—Chemical, Physical, or Microbiological Changes
    Ageing, ripening, enzymatic action
    Smoking, curing, brining, pickling, syruping, spicing
    Malting, brewing, fermentation
    Steeping, extraction, straining, filtration
    Cooking, baking, distillation
    Evaporation, concentration, drying
    Condensation, coagulation, flocculation
    Pasteurization, sterilization
    Refrigeration, freezing
    Acidification, alcoholation, chemical preservation, carbonation

second category are listed those processes which aim at changing the shape of the food in order to make it easier to handle, to eat, to cook, or to process further. The third category lists processes that are performed on the food to institute chemical, physical, and microbiological changes which improve its flavor and aroma, its tenderness and texture, or its ability to resist deterioration or spoilage.

PRINCIPLES OF FOOD PRESERVATION

Food preservation can be accomplished in several ways: reduction in degree of spoilage by removal of contamination, delaying spoilage by making conditions unfavorable, and elimination of spoilage by destruction of contamination or the factors producing it.

Food Preservation by Removal of Contamination

Although most raw food products contain microorganisms, it is possible to reduce the degree of contamination by certain procedures in a food processing plant. All pieces of equipment used and the plant facility itself must be kept as clean and sanitary as possible to reduce the level of contamination that could be accidentally added to the food. Bulk packaging can contribute to this by protecting raw food from contamination or empty containers from air borne dirt. People working in the plant must be clean, wear sanitary clothing, and be free from disease. Containers must be washed before use.

During preparation of foods dirt is shaken or washed off using sanitary water. This reduces the population of bacteria that must later be dealt with. Many raw foods are peeled or skinned before using. Some bacteria are removed and discarded with the peelings. No bits of food that fall onto the floor or cling to machinery parts near the floor should ever be swept up and put back in the process; they should be discarded.

Some plants reuse part of their water for example in cooling canals for hot cans. This water must be monitored for contamination and chlorinated to maintain a suitable level of sanitation.

The newer techniques of ultrafiltration of beverages to remove yeast spores are examples of prevention of spoilage by removal of contamination. By removing or reducing the original population of microorganisms the number of descendants through vegetative multiplication are correspondingly reduced.

Food Preservation by Delay or Inhibition of Spoilage

Food spoilage can be reduced by inhibiting or preventing the multiple growths of microorganisms or by not giving them time to multiply. The time factor is used throughout food processing. From the moment foods are harvested they are promptly placed in cold storage or immediately processed. All processing steps are as rapid as possible.

Unfavorable conditions for growth of microorganisms vary, depending on the organism. All microorganisms require moisture in order to multiply and hence drying is an effective delayer of spoilage. Many microorganisms are inhibited in their growth by the presence or absence of certain chemicals. Thus salt and acids and certain chemical preservatives like sodium benzoate and sodium bisulfite are effective in slowing bacterial growth. Some prefer the dark, some the light. Some must have oxygen others must have none. In the process the original food may be altered such as for example lyophilization of a fat, hydrolysis of a protein, fermentation of a starch to a sugar, a sugar to an alcohol, or an alcohol to an acid.

For nearly every microorganism there is a temperature which is optimum for growth. Given food, moisture, the right conditions chemically and its optimum temperature, it will multiply at its maximum possible rate. A drastic lowering of this temperature results in a slower rate of microorganic growth; thus refrigeration slows spoilage, and freezing slows it even more. Frozen foods held at 20°F will deteriorate faster than foods held at −20°F largely because of chemical actions, but both will retain quality longer than foods held in a refrigerator at 40° to 45°F.

A moderate heat treatment can destroy or weaken microorganisms. Partial destruction, e.g., pasteurization, reduces the number that can multiply. Weakening appears to be due to thermal shock. The bacteria eventually multiply but there is a longer time delay before they begin to multiply.

Food spoilage is due principally to the action of enzymes, molds, yeasts, and many kinds of bacteria. While some of these can be destroyed by mild heat in a short time, certain spore forming bacteria, particularly the dangerous ones require high temperature and long times to be killed. The length of time required to kill a particular microorganism at a particular temperature can be determined experimentally. It is usually measured as the time to kill a specified large number of bacteria. Carefully prepared cultures of pure bacterial strains are inoculated into nutrient materials and incubated for growth after heat treatment for different time intervals at the desired test temperature. The number of inoculated tubes that show growth are recorded against the length of cook time. The time at which no spoilage occurs is a measure of the heat sensitivity of the microorganisms. Some of the most dangerous bacteria require high heats and longer times because they are spore formers. Under ideal conditions they act like any other bacteria and are readily destroyed by heat, but when conditions (such as temperature) become unfavorable they revert to a spore form which is very difficult to kill. Later when conditions become favorable again, the spores "vegetate" and become active bacteria. Most spore formers are unable to multiply in highly acid foods. For this reason so long as the food remains acid they are not a threat to health. Those organisms that can multiply in acid media are relatively sensitive to heat and can be destroyed at lower temperatures. This principle was utilized by Pasteur in improving the processing of wines. Heating of the acid fruit juices before fermentation destroyed wild strains of yeast or bacteria that might cause unfavorable results. Then desired fermentations were begun by using pure cultures of the wanted yeasts or bacteria. Pasteurization was later used on other foods such as milk to destroy most disease bacteria but not all bacteria. Thus milk was rendered safe to drink and spoilage was retarded, but refrigeration was still needed to further retard spoilage. Even under refrigeration pasteurized milk will eventually spoil.

Cooking is a form of pasteurization. It retards spoilage, though not preventing it.

Food Preservation by Destruction of Spoilage Organisms

Nicolas Appert discovered how to prevent food spoilage by heat pro cessing long before scientists found explanations as to why they spoiled (see Chapter 1). Appert's first experiments and his first commercial processes were far from efficient. There were unpredictable occasional high rates of spoilage. Imperfect closures leaked. Sometimes his pro cesses were inadequate. But, it was a beginning. Peter Durand's tin can was another step forward. All through the 19th century men struggled to contribute to this growing industry. There were many tries, many successes and many failures. As the development of the tinplate can reached closer and closer to an acceptable level of performance, the development of pressure cookers and the development of satisfactory processes became a commercial reality. One can find success stories in the middle and late 1800's, but the food canning industry really didn't become a strong factor until about the beginning of the 20th century. By then the work of Pasteur in France and Koch in Germany in identifying bacteria with food spoilage had left the laboratory bench and become a technology which could be used by industry. Canning became less and less an art and more and more a science. Today, the principles laid down by Appert and refined by those who followed him are used to process hundreds of different foods and total production is over 30 billion pounds. Pasteur found that microorganisms can be destroyed or weakened by moist heat. Later workers proved that different organisms in different foods required different amounts of heat to destroy them. They also found that many desirable properties of food were destroyed by the long cooks needed to destroy the microorganisms. The search for better methods led to higher temperatures and shorter cooking times. This search continues. Where Appert cooked for hours and later workers for tens of minutes, more recent workers using still higher temperatures are processing in minutes. HTST (high temperature short time) processes for some liquids use times that can be measured in seconds.

A number of different kinds of processing machinery have been developed to make the heat sterilization process more efficient. Others have concentrated on developing new types of packaging which would improve efficiency. Aseptic packing systems have been developed where

Table 10

GENERAL CLASSES OF FOOD PRODUCTS AND TYPES OF STEEL BASE REQUIRED

Class of Foods	Characteristics	Typical Examples	Steel Base Required
Most strongly corrosive	Highly or moderately acid products, including dark colored fruits and pickles	Apple juice Berries Cherries Prunes Pickles	Type L
. . . Moderately corrosive	Acidified vegetables Mildly acid fruit products	Sauerkraut Apricots Figs Grapefruit Peaches	Type MS Type MR
Mildly corrosive	Low acid products	Peas Corn Meats Fish	Type MR or MC
Noncorrosive	Mostly dry and non-processed products	Dehydrated soups Frozen foods Shortening Nuts	Type MR or MC

Source: Joslyn and Heid (1963).

package and food are separately sterilized then combined in a sterile environment.

Other forms of energy are being utilized. Radar ovens are being used to cook foods rapidly by high frequency radio waves. There may be more utilization of different energy forms used in the future in food processing. Ultrasonic energy, microwave energy, and laser energy, have not yet been fully explored.

Food Canning

Canning is defined as the processing of food in a hermetically sealed container with heat in order to reduce the population of bacterial contamination to a commercially safe level. Before a food can be so "sterilized" it first must be prepared and put in the container.

Preparation for Canning.—Before canning foods must be separated from their inedible portions, washed, graded for size, and inspected to remove defects. They also must be put into the shape desired by cutting, slicing, dicing or the like. Most vegetables have to be blanched using steam or hot water to soften tissues, eliminate air, and destroy enzymes. Containers must be washed prior to filling. Filling may be done by hand or automatically. This depends upon the product. For some products there is no automatic machinery available. The amount of solid product and the amount of liquid product added must comply with Food Laws governing the fill of containers.

Exhausting is the final step before closure. It is done to drive out air

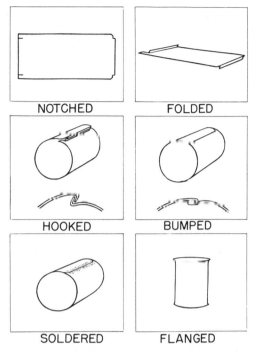

NOTCHED   FOLDED

HOOKED   BUMPED

SOLDERED   FLANGED

Courtesy of American Can Co.

FIG. 23. SEQUENCE OF CAN-MAKING OPERATIONS

and preheat the contents. It can be accomplished in "exhaust boxes" or it can be made part of the container closure operation. Proper exhausting creates a vacuum in the final container.

Container closure involves putting the lid in place and seaming a can or application of a screw-on or crimp-on lid to a glass jar or glass bottle.

Heat Processing.—For foods with acid pH of 4.5 or lower heat processing at 212° F is adequate. Most fruits but few vegetables meet this requirement. Pickled products or products that have been acidified with acids such as vinegar can be processed at this temperature. Low acid foods (those with pH greater than 4.5) must be processed at temperatures above 212° F. In early days, the temperature of boiling water was raised by adding salts. The pressure retort eliminated this procedure. In a retort water can be heated above 212°F or steam at temperatures substantially above 212°F can be introduced. Sealed containers are placed in the retort, the retort is closed, air is displaced by water and/or steam, and "cooking" begins when the retort reaches the desired "cook temperature." From previous experience, by making a trial inoculated test pack, or by calculation process times can be determined for any food in

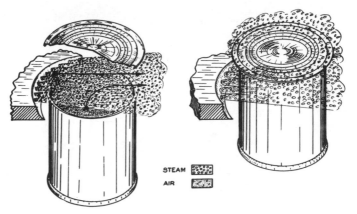

FIG. 24. OPERATING PRINCIPLES OF STEAM FLOW CLOSURE

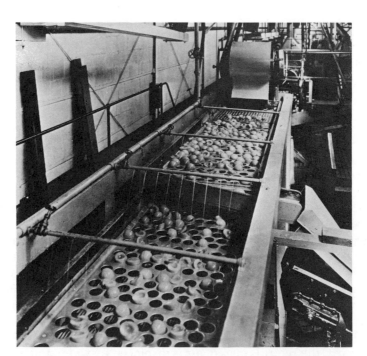

FIG. 25. PREPROCESSING OF FRUIT

Peach halves are rinsed with water prior to canning operation.

any container at any temperature. When the food has been given the necessary heat treatment, the retort is opened and the containers are removed to a cooling canal where temperature is reduced. With tinplate cans temperature is brought to about 100°–120°F average internal temperature in order that the remaining moisture on the outside will evaporate without causing rust. Additional air cooling may be used. (With aluminum cans rust is no problem.) If cooling is inadequate food continues to cook and quality suffers. This is known as "stack burn." Some cans and most glass containers must be cooled in the retort under pressure to avoid buckling, breaking, or leakage. Jar lids are not built to withstand internal pressures. Pressure cooling is done by maintaining pressure with air or with steam while cooling water is introduced. Air is less tricky for without air pressure steam condensation could cause lower pressures outside the containers than on the inside. This can cause container damage, cans can bulge or seams crack, bottles can lose their lids. Flexible packages can delaminate or have heat seals pulled apart.

Types of Processing.—*Still Retorts.*—A still retort is nothing but a pressure tank that can be filled with containers, closed, heated, cooled, and opened. Both horizontal and vertical types are available. Heating can be done in the presence of steam or water. Care must be taken to purge air by means of bleeders and to distribute steam uniformly (via headers). Packages must be so loaded as not to prevent free circulation of the heating medium. Traditionally, cans are jumble packed into perforated crates or baskets.

A newer type called the "Malo" system eliminates the basket and automatically loads cans through the top of a vertical retort and discharges them through the bottom. Steam enters through the top and forces out the hot water introduced at the start to cushion the impact of loading. This is still a batch system, but it is automated and drastically reduces labor required.

Agitating Retorts.—A few batch retorts have been designed wherein the atmosphere is circulated or the packages are moved or both. Such motion improves heat transfer, particularly when the heating medium is not pure steam.

Other agitating retorts have been designed as continuous processors. The FMC Sterilmatic conveys cans through locks and a spiralled conveyor system which rotates the cans. This rotating action improves heat transfer and substantially reduces the processing time needed. Latest models are capable of temperatures as high as 290°F and speeds of 550 #303 cans per minute. The same type of equipment can be operated at lower temperatures for canned beverages. Glass containers or more delicate metal containers can be loaded into carrier shells and then conveyed through the retort.

FIG. 26.   FILLING OF PEAS INTO CANS

Peas are filled into open ended prewashed cans and then sent on to next operation all automatic and high speed.

The Robins Hydrolock conveys containers through a warm water immersed lock into a pressurized chamber. The packages continue through the warm water which is now under pressure and then enter a circulating atmosphere hot chamber. Here steam, or steam air mixtures, or heated water sprays may be used to sterilize the food. On leaving the chamber the packages reenter the warm water which cools them under pressure sufficiently to permit their safe exit through the lock. Final cooling is done at atmospheric pressure. Carriers can be mounted on the conveyor chains which permit rotation of the packages.

Hydrostatic cookers are open ended permitting continuous conveyance of containers in and out. The pressure in the sterilizing section is counterbalanced by the head of water in the vertical legs leading in and out. Containers may be rotated as with the Robins Hydrolock.

Other Types of Cookers.—Fluidized bed cookers are continuous systems where the heating medium is particulate. For example sand or ceramic pellets may be heated by a flame and agitated from underneath by air. The agitation tends to float the particles in the air stream resulting in a "fluid mass" having the consistency of a thick liquid.

The "Steriflamme" process was developed in France by M. Beauvais, H. Cheftel, and G. Thomas. This uses gas flames to bring the preheated, rotating container to sterilization temperature. Heat penetration rates are extremely high and processing is very fast for convection heating foods.

Courtesy of National Canners Assoc.

Fig. 27.   Can Unscrambler

Retorted cans of food are removed from jumble packed retort baskets and unscrambled (foreground) and sent on to labelers and carton packers. Each can is coded for identification.

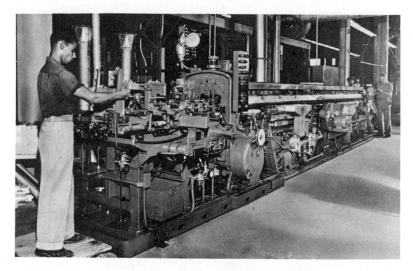

Courtesy of American Can Co.

FIG. 28.   AUTOMATIC BODY FORMING MACHINE WHICH FABRICATES CAN
CYLINDERS FROM FLAT METAL SHEETS

Aseptic Processes.—Aseptic processes differ from conventional
processes in that the heat of sterilization in the latter case is applied
after the product is sealed in the package, whereas in the aseptic system
the product and container are sterilized separately, then combined and
the container closed in such a way as to prevent recontamination. Hard
to heat conductive foods can be heated much faster in such systems
through the use of heat exchangers. These may be plate-type, tube-type,
steam injection-type, or moving film or scraped surface-type.

"Flash 18" Process.—This is a variation of an aseptic system. Here
containers are cleaned and filled with sterilized product at 255°F. The
hot product sterilizes the can during the "hold" period after closing. All
of this is accomplished at "atmospheric pressure" because the entire
operation is done within a pressurized chamber of 18 psig. Workers
enter and leave through locks. Cans are partially cooled inside the
chamber then pass outside for final cooling.

AUTOCLAVABLE FLEXIBLE CONTAINERS

A great deal of effort has been expended in the development of flexible
laminated products which will withstand retorting. Extensive studies
have shown that aluminum foil is an essential ingredient. Rigid or flexible plastic containers are available that will survive retorting but these
have not provided adequate barrier properties for long term storage of

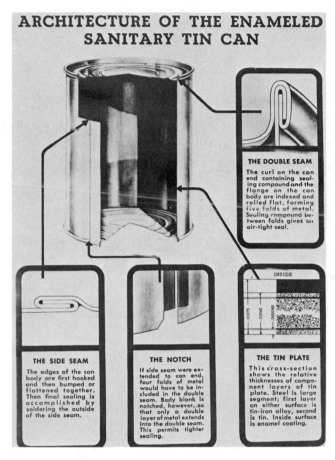

ARCHITECTURE OF THE ENAMELED SANITARY TIN CAN

THE DOUBLE SEAM

The curl on the can end containing sealing compound and the flange on the can body are indexed and rolled flat, forming five folds of metal. Sealing compound between folds gives an air-tight seal.

THE SIDE SEAM

The edges of the can body are first hooked and then bumped or flattened together. Then final sealing is accomplished by soldering the outside of the side seam.

THE NOTCH

If side seam were extended to can end, four folds of metal would have to be included in the double seam. Body blank is notched, however, so that only a double layer of metal extends into the double seam. This permits tighter sealing.

THE TIN PLATE

This cross-section shows the relative thicknesses of component layers of tin plate. Steel is large segment; first layer on either surface is tin-iron alloy, second is tin. Inside surface is enamel coating.

INSIDE

Courtesy of American Can Co.

FIG. 29.   CONSTRUCTION FEATURES OF THE
SANITARY TIN CAN

foods at room temperature. Properties of food that deteriorate on exposure to light and oxygen will do so much more rapidly at room temperatures than at refrigerated or frozen conditions. The barrier properties of aluminum foil prevent the entry of light and oxygen. Methods of filling and closing are available which will insure that initial air content is at a sufficiently low level.

Flexible laminations for autoclavable packages are comprised of an outer plastic film which contributes strength, is usually transparent, and can carry the decoration or labeling information, aluminum foil, and an inner plastic film which contributes some strength, inertness to product constituents, and heat sealability. Lamination of these compo-

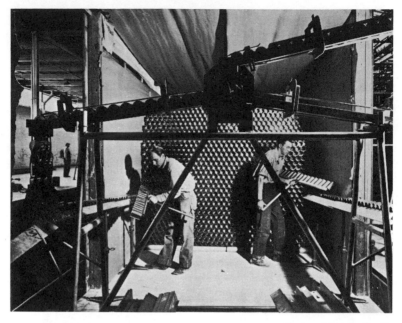

Courtesy of American Can Co.

FIG. 30.   BOXCAR LOADING OF SANITARY CANS FOR SHIPMENT
TO FOOD PROCESSOR

nents is accomplished by carefully compounded adhesives which must
develop a high level of adhesion and maintain it even under autoclave
conditions. Adhesives are usually two-part, curing-type adhesives that
develop maximum bond strength after a period of ageing. Primers are
often required on the aluminum to achieve desired bond strengths. A
typical laminate might be .0005 in. polyester film-primer-inks-adhes-
ive-primer-.00035 in. aluminum foil-primer-adhesive-.0015 in. polyamide film. For some foods the inner film may be .003 in. polyolefin film.

Most autoclavable pouches are fabricated as fin-sealed pouches with
filling being done through the top prior to making the final seal. Some
work has been done with acid foods in form and fill bags. Such a package for sauerkraut is being marketed successfully in Germany using a
Hassia type SK form and fill machine with a specially designed filler
and processing system.

In making packages, it is essential that care be taken to avoid contam-
inating final seals with large particles of food. Minor contamination can
be tolerated provided the type and thickness of interior film are carefully selected.

Since flexible pouches can expand if internal pressures are allowed to

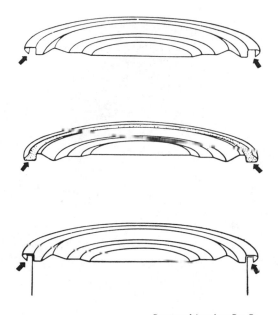

Courtesy of American Can Co.

FIG. 31.   INSIDE CURL OF THE CAN END IS LINED WITH
SEALING COMPOUND, A RUBBER GASKET MATERIAL

develop, processing must be done with overriding air pressures during heating and cooling. The less air (or other compressible gas) left in the container during filling the better.

Because flexible pouches are thinner in cross-sectional dimension than a can, the food mass can be heated and sterilized in much shorter times. This means food quality is substantially improved. On the other hand, the flexible package provides little protection against crushing. Soft products like peas can be damaged unless care is taken to protect against such damage by using outer paper board cartons, and by minimizing mechanical abuse during shipping.

Retorts require modification only to the extent that they must provide careful control of temperature and pressure. They must provide overriding pressure, they must hold the package shape (to avoid food mass dimensional changes), and they must insure adequate circulation of heating media around every package. Both batch-type and continuous-type retorts are being evaluated for this purpose.

COOK IN FILMS

Cook in films are the latest addition to the parade of convenience packages. By and large the broadest definition of a cook in film would be

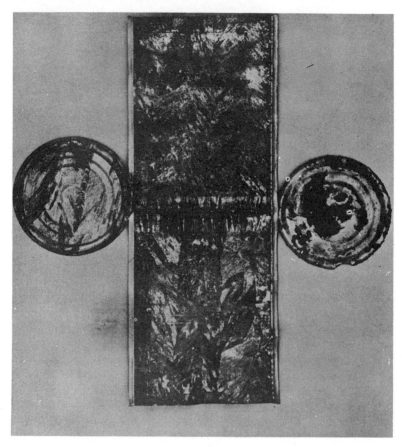

From Desrosier (1963)

FIG. 32.   DETINNING OF INNER SURFACE OF SANITARY CAN
AS MAY OCCUR WITH ACID FOODS

a package which takes the boil-in-bag concept out of the pot of water
and puts it into the heated skillet or into the oven, still retaining the
advantage that the pan and the oven are protected against spattering or
spillage.

The cook in film is able to withstand the temperatures encountered in
frying, baking, roasting, or oven simmering of a wide variety of foods. At
the same time it reduces losses due to excessive evaporation, it permits
normal browning where desired and it materially shortens the time
required to bring a food to the desired temperature.

Even more important the cook in film can be the total package used to
protect the food from the point of preparation through storage, handling,

Courtesy of U.S. Army, U.S. Army Natick Lab.

FIG. 33. AUTOCLAVABLE POUCH

U.S. Army sponsored research has been underway for a number of years to develop a flexible package which can survive retorting. Illustration shows an experimental package on the left and the comparable can on the right.

retailing and into the home. By doing the precook preparation before packaging the convenience to the housewife is even more enhanced.

Package Form and Use

There are at the present stage of development two basic approaches to the use of cook in films. One is the home use of a premade plastic film bag. The other is the commercial use of the cook in film as a fundamental part of the food package.

Consumer Bag.—In the home use the consumer purchases a number of plastic bags at her supermarket and keeps them in her kitchen as she now retains other bags for sandwiches, garbage disposal, refrigerator left overs and the like. She also purchases a supply of aluminum containers of suitable size. At the time she plans to cook a roast, a meat loaf, or other prepared dish she has the option of preparing only one or of preparing a multiple quantity reserving the extras for freezer storage and later consumption. She prepares her food in the usual way except that after placing it in the aluminum container, she puts the entire container into the plastic cook in bag and ties the end shut with a twist tie.

Or, if she desires, the aluminum container can be kept outside of the bag. Just before cooking, one or two small punctures are made in the top of the film to avoid build-up of pressure. No other precautions are necessary. Oven settings can be set a little higher than usual because evaporation losses are small. Foods which normally require browning will brown right through the film. Cooking times will be shortened 25–40% depending upon the food. Roasts will be self-basting and the juices can be collected and retained for subsequent use not burned onto the pan. Dry foods such as oven heat-and-serve rolls can be cooked without using a pan and will be softer and moister when served. Bagged foods like hashes, meat slices and gravy, or beef stews can be placed in a skillet and simmered without spatter leaving no pan to wash.

Commercial Packages.—Commercial packaging using cook in films can take one of several possible paths. The most obvious of course is to duplicate the consumer bag approach. Here the food packager merely makes certain the food is oven-ready then substitutes the cook in plastic film and foil container for whatever package he now uses. The food can be marketed from the refrigerator case—as with cut up poultry, or from the freezer as with frozen prepared dishes. The housewife merely puts the package directly into the oven without any further preparation. All the advantages previously mentioned apply. She doesn't have to buy a special bag and she doesn't have to remove and discard an old package, which in a frozen item usually means thawing first.

A second approach being given serious consideration is the use of foil containers designed to accept a lid made from cook in film. Here the entire container is not overwrapped, nor is the food bagged. The film can be applied as a loosely shrunk-on cap or it can be heat sealed. A foil laminated plug lid with tear out panel may be crimped onto the container in which case the film does not have to be heat sealed, but may also only be crimped in place. Or a film membrane can be sealed or crimped on and then a foil tuck wrap applied over all. The latter would be removed at a suitable time prior to or partially through the cooking cycle. This approach is suited primarily to foods which will be prepared, at least partially cooked, and frozen, such as spinach soufflés, egg omelettes with sauces, scalloped and au gratin potatoes, certain frozen baked goods. It is of particular interest to the vending machine concept.

The modern trend is to merchandise food that saves the housewife time and effort yet provides a high quality product. Cook in films accomplish this aim by: (1) saving preparation time; (2) saving cooking time; (3) eliminating spatter in ovens and dirty pots and pans; and (4) preserving excellent quality and freshness over desired storage periods.

Package Requirements.—A cook in film must be able to withstand the action of the heat in the oven and the hot liquids and fats which may spatter on it and yet, where frozen foods are used, it must be able to survive storage at freezer temperatures. Only a few films can handle these extremes. These include: polyesters, nylons (polyamides), and the more expensive polyfluoride and polyimide materials. There are some companies planning to market higher homologues of the polyolefins. Polybutylenes, polypentenes, or higher homologues may prove to have sufficient high temperature resistance for oven-cook-in films.

PRECOOKED FOODS FOR SPACE TRAVEL

The requirements for foods to be used in space travel are unique. They must be easily digested. They must be highly nutritious. They must be easy to prepare. They must be light weight. They most certainly must not spoil, and they must not produce crumbs. The latter requirement may seem odd, but in the weightlessness of space travel crumbs floating around can cause many difficulties. They could get into an astronauts eyes or lungs, or even worse—get into delicate equipment and cause malfunctions.

Most foods are therefore prepared under highly stringent conditions in laboratories that are dust-free, lint-free, and sterile. Even the air is filtered. Most of the foods are freeze-dried to reduce weight and liklihood of spoilage. Freeze-dried foods are also easy to prepare. They can be eaten dry or cold or hot water may be added. A typical menu provides a wide variety of food and drink items and a level of 2,520 cal per day.

Since cost is secondary to performance, foods are packaged for space travel in pouches fabricated from laminates of polyethylene-adhesive-polymonochlorotrifluoroethylene- adhesive-polyester-adhesive-polyethylene. Special closures include a sealed-in-tubing for beverages and paste-like products so the food can be consumed without spilling a droplet. In the absence of gravity the astronaut squeezes the package to expel the food through the tube into his mouth.

Packages are carefully sealed, labeled, and overpacked to prevent possible damage.

DEHYDRATION

Dehydrated foods are as old as civilization. Vases containing samples of dehydrated foods were found in Egyptian tombs. Marco Polo reported that dried milk was used by the Tartar tribes. In the United States, early settlers found that a staple food of the Indians was "pemmican." It was made of dried meats, such as venison, crushed to a powder and mixed with fat. Colonists adopted various Indian techniques and utilized dried corn ("samp"), dried fish, and dried beef ("jerky").

Original drying methods utilized the heat of the sun. In 1795, the first artificial drying process was introduced by Eisen. Vegetables were exposed to hot water before drying. Soon after, two Frenchmen, Masson and Chollet, dehydrated vegetables by slicing and drying them with a controlled warm air flow.

In the Civil War, General Sherman's soldiers used dried foods extensively. During the Boer war, dried compressed vegetables were used by the British Army in South Africa. In 1917, approximately 12 million pounds of dehydrated vegetables were produced in the United States for war time use.

Between World Wars I and II, the dehydrated food industry was at a low ebb. Although development continued in Europe, most US firms preferred to import whatever quantities of dehydrated foods they needed. The onset of World War II catalyzed further American development. Soldiers needed lightweight foods and traditional European sources of supply were terminated. After the War, the rise in consumer affluence coupled with the demand for convenience created new interest in dehydrated foods.

What Is Dehydration and Why Is It Needed?

Food dehydration involves the removal of water without destroying the cellular tissues of the food nor altering its caloric values. A variety of processes are used to effect water elimination. The presence of water in a food tends to accelerate: (1) mold and bacteria growth; (2) discoloration and off-flavor; (3) enzyme action; and (4) caking and other physical changes. By reducing the water content of a food below its critical value, using proper packaging, and careful storage increased shelf-life is possible.

The rapid rise in dehydrated food sales has been a direct result of improvements in processing techniques. Sales of instant coffee rose from 227 million two-ounce units in 1951 to more than a billion units in 1960. Dehydrated potatoes rose from 50 million pounds in 1956 to over 500 million pounds in 1965. In 1967, over $30 million of freeze-dried foods were sold.

Methods of Food Dehydration

Sun Drying.—Sun drying is still used for the dehydration of various fruits. Some varieties of fruits are sulfured to prevent darkening. They are then taken to a drying yard and exposed to the sun for 1 to 10 days, depending on the climate. Different fruits require different lengths of exposure. Apricots are exposed from 1 to 7 days, bananas 1 to 2 days, figs 2 to 4 days, etc. In general, the process is rather slow and used mainly for selected fruits.

Courtesy of Reynolds Metals Co.

FIG. 34. DRIED SOUP MIX POUCH

The complete moisture and vapor barrier qualities of aluminum foil enable high quality dehydrated convenience foods to be packed in convenient, inexpensive flexible pouches rather than metal cans. The aluminum foil is combined with plastic film and paper to gain the qualities of each material.

Spray Drying.—Spray drying is used for some liquids such as milk or dissolved solutions such as brewed coffee. The liquid is finely atomized into a gaseous heating medium such as air. As the moisture evaporates, solid particles are formed which float in the air stream. Evaporation occurs so rapidly on the surface of the particles that the temperature of the product never rises high enough to cause thermal damage. A variety of atomizers and driers are available from different manufacturers.

Atmospheric or Hot-Air Drying.—This consists of cutting the product into small pieces and spreading them out on trays which have slat or wire bottoms. Hot air is passed over and under the trays in order to effect dehydration. Drying takes place in tunnels about 35–40 ft long and 6 ft high by 6 ft wide.

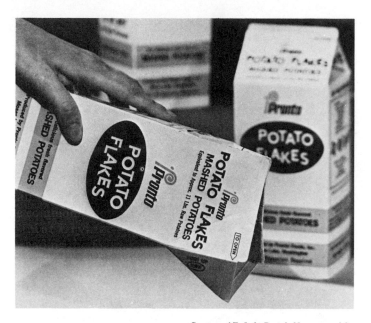

Courtesy of E. I. du Pont de Nemours and Co.

FIG. 35.   ADAPTATION OF MILK CARTONS TO DRIED FOODS

Dried potato flakes packaged in containers like the familiar polyethylene coated milk carton have been tried in the retail market. Two
years of success with this style of carton in institutional sizes made it
the choice for the 16-oz retail package. The carton, coated with Du
Pont's "Alathon" polyethylene resin, features easy opening, pouring,
and reclosing for the housewife, and offers economic and other
advantages for the packager.

Hot-air drying methods are used principally for vegetables and fruits
(which are not sun-dried). Instant potatoes, onion flakes and apple
slices are commonly hot-air dried.

Foam-Mat Drying.—This involves beating a liquid into a stabilized
foam, spreading the foam in a thin mat on a perforated belt, and passing
warm air through the foam mat. Air temperatures are lower than for
spray drying and contact times are longer.

Drum Drying.—Drum drying may be atmospheric or vacuum. In
atmospheric drying, the material is applied by feeding devices onto large
steam-heated drums two to six feet in diameter. As the drums revolve,
the material dries and is subsequently removed by a scraper.

Vacuum Drying.—Vacuum driers may be of any configuration, that
is drum driers, belt driers, kiln dryers, tunnel driers, etc. The difference
lies in the pressure being reduced below atmospheric pressure. The
higher the vacuum; the lower the temperature required to remove mois-

ture and volatiles. This permits drying of heat sensitive foods. Condensation of volatiles may be necessary. Capital investment is higher for vacuum equipment as opposed to atmospheric driers. The absence of oxygen helps avoid food quality deterioration. Some foods, when vacuum dried, are heated so that bubbles form. This makes the product lighter. It may be flaked as is or it may be crushed to reduce bulk. This is somewhat akin to foam drying.

Dehydrofreezing

Dehydrofreezing is a process which combines dehydration and freezing. A certain amount of water is extracted by heat and the product is then frozen. It must be stored under refrigeration and has a few advantages over conventional frozen foods. Since the food shrinks during dehydration, it requires less storage room than conventional frozen foods. Several vegetables, meats and fruits have been dehydro-frozen.

Freeze-Drying

The major interest currently in dehydrated foods revolves around a process first used in the early 1920's for medical applications. Blood plasma was preserved by a freeze-drying technique. Although the original process was slow and expensive, it has been developed into a rapid and economical procedure. Freeze-drying is an extension of vacuum drying. The product is first frozen and then moisture is removed by sublimation under high vacuum. Just enough heat is applied to accelerate drying without causing the product to decompose. Ice crystals are converted directly into water vapor without passing through a liquid phase. The freeze-dried product retains its original shape and is porous. This porosity allows for more complete and rapid rehydration than is possible with air-dried foods. In addition, the weight is reduced to 10–15% of the original weight. Moisture levels are as low as 2 per cent compared to a minimum five per cent with other methods. Large items such as steaks, fish and chicken can be freeze-dried without crushing or shredding the product. At present, freeze-drying is being applied on an increasing scale to a wide variety of products ranging from soup to vegetables. No refrigeration is required for freeze-dried foods.

Packaging Requirements

All dehydrated foods are subject to two prime causes of spoilage, i.e., moisture and oxygen. They are also fragile, sensitive to light, and if fatty, subject to contamination by foreign flavors. Dehydrated fruits and vegetables are especially subject to insect attack.

Moisture Sensitivity.—The problem of absorption of atmospheric moisture is more critical for some products than others, depending on the desired shelf-life. This is particularly so for freeze-dried foods

FIG. 36. MILITARY PACKAGING SHOWS ADAPTABILITY OF FLEXIBLE PACKAGING

Experimental ration utilizing flexible packaging for both dehydrated foods (potato pattie) and processed foods (chicken loaf).

because of their open structure. Even the slightest degree of moisture absorption by the product is harmful and undesirable. Even though the caking of a powdered dehydrated food may not constitute complete spoilage, it is unacceptable to a consumer and must be avoided.

Oxygen Sensitivity.—Although moisture is the common enemy of all dehydrated foods, many of them are highly susceptible to oxidation. Oxygen adversely affects products containing high levels of fats or oils and those having a high carotene content. Fats and oils turn rancid and carotene develops an odor like newly mown hay.

Freeze-drying, while offering a superior product in comparison to other dehydrated foods, yields a product highly susceptible to oxygen. Both vacuum and gas packaging are used to eliminate oxygen from a package.

Light Sensitivity.—Many dehydrated foods are sensitive to light. Bleaching occurs on some products and others darken due to the effect of light rays. Light also accelerates the development of rancidity by means of oxidative spoilage.

Mechanical Abrasion.—Although all dehydrated foods are subject to mechanical damage, this problem has been reduced with compressed foods. Powders, shreds and thin slices of air-dried foods cannot be damaged by mechanical handling to any great degree. In freeze-dried food, the product is extremely friable and brittle. Mechanical abrasion inside a loosely packed can containing freeze-dried foods can crumble and break up large items, such as shrimp, meat, or mushrooms, into unusable granules or powders. Freeze-dried shrimp or steak should be put in a package with a degree of rigidity with some cushioning or some means of immobilizing the product. Materials must be strong and flexible enough to resist being punctured by sharp granular or chunky items.

Flavor Contamination.—If dehydrated foods are fatty, they can be contaminated by foreign flavors extracted from packaging materials or from adjacent external sources. A good grease barrier and a good vapor barrier is needed to prevent this.

Insect Attack.—Insect attack is a problem with dried fruits and vegetables. Fumigation against insects is of no value after the material is in the final container. If insect eggs are present, they are unlikely to hatch in the dried food because of its low moisture content, but they will if the package is broken. Even a minute pin-hole in the package will attract insects and encourage entrance. If there is any doubt relative to insect infestation, it is advisable to heat the product and the empty package to 135° F for five minutes immediately before packaging.

Packaging Materials and Form

Three different types of packages are used for dehydrated foods: rigid containers, flexible pouches, and lined cartons. All fulfill the requirements imposed by the product and an eventual choice is based on economics and sales appeal.

Vacuum and Gas Packaging.—If it is desirable to remove most of the oxygen from a package to achieve prolonged shelf-life, the package can be evacuated and sealed under vacuum, or the air can be replaced by an inert gas.

Vacuum packaging presents certain problems with dehydrated foods in flexible pouches. Products with sharp corners tend to rupture the material causing vacuum loss. Many soft products are themselves crushed by the application of high vacuum. The unappealing shape of a vacuum pack makes it difficult to merchandise. In addition, the use of vacuum packaging demands heavy materials in order to resist the strains imposed by the process. This adds cost. These problems are not encountered with rigid containers.

Gas packaging is a more practical process for oxygen removal. The air is replaced by flushing or flooding the package with an inert gas. A vacuum may be pulled and then gas added or the interior of the package may be flushed prior to sealing or the package may be filled and sealed within an atmosphere of inert gas. All of these procedures reduce the oxygen level to between 1-2%. By back flushing with an inert gas, the product may be under sufficient pressure to prevent movement within the package and yet avoid the damaging stresses of a full vacuum package.

Three gases can be used for the gas packaging of dehydrated foods: nitrogen; carbon dioxide; and "Vitagen."

Nitrogen is more commonly used and is truly inert. Carbon dioxide may be absorbed by certain foods and create a partial vacuum after a period of time. "Vitagen" is fairly new and is a product produced by the burning of natural gas in generators.

FREEZING

Drying, smoking, pickling, or canning of foods alter the physical appearance and the flavor and aroma of the product. Eskimo tribes invented a method of food preservation which avoided these changes. They caught fish in sub-zero weather and the fish froze immediately upon exposure to air. Many months later, upon thawing, the fish were just as edible as when first caught. These procedures were not recognized or discovered by others until relatively recent times. The use of low temperatures for food preservation goes back to the mid 19th century. The first patent issued for a freezing technique was awarded in 1842. In 1876, frozen meat, using ice and salt, was shipped from the United States to England. In 1912, berries were slow frozen in barrels in Puyallup, Washington. These early products were not quick frozen and suffered from structural alteration.

On freezing, the water contained in food is converted into ice crystals at a temperature between 23°-32°F. With slow freezing processes large crystals are formed which may distort and rupture the tiny cells. After thawing, valuable nutrients are lost through the punctured cell walls. In 1912-1915, Clarence Birdseye, while in Labrador, observed what the old Eskimos had long known that animal products which had been killed many months earlier were edible after being exposed to the Arctic weather. He decided that the secret lay in the rate at which the product was frozen. With quick freezing, small ice crystals are formed which have less effect on the structure, flavor, and texture of the product.

In 1928, one million pounds of fruits and vegetables were frozen in the United States. Today, sales of frozen foods account for over $7 billion annually and the growth rate is rapidly increasing. In 1968, the US per

Courtesy of E. I. du Pont de Nemours and Co.

FIG. 37. ANOTHER FROZEN CONVENIENCE FOOD

Here a complete pizza item is prepared for the consumer and frozen ready to bake in the oven. The "Elvax" hot melt coated carton provides adequate protection against moisture loss.

capita consumption of frozen foods amounted to 68 lbs. Per capita consumption in Europe ranges from 22.7 (Sweden) to 0.6 (Italy). Many areas of the world have not yet been exposed to frozen foods. In Portugal and Spain, sales are neglible. The Eastern European countries are just emerging as large scale users of quick frozen foods.

Method of Quick-Freezing.—Four basic methods are used to quick-freeze food products. In all four methods, temperatures below − 30° F are utilized. The product may be packaged prior to freezing for convenience and protection against dehydration.

Blast freezing employs a current of very cold air directed against the product. An alternate method utilizes a tunnel in which the product is moving in the opposite direction. This process is well suited for individual fruits and berries and irregular shaped items.

Plate freezing consists of refrigerated plates on which the product is placed. By compressing the plates together in order to make close contact coupled with the use of an internal circulating refrigerant, at a low temperature, rapid freezing is obtained. Rectangular shaped items in thin cartons are rapidly frozen.

In immersion freezing, the product is immersed in a liquid refrigerant such as glycerol, alcohol, brine, or sugar syrup. It is useful for large products such as turkeys and other types of poultry.

Cryogenic freezing is usually done before packaging using liquid nitrogen or powdered dry ice at very low temperatures.

After freezing, foods should be kept at a storage temperature of 0°F or lower until ready for distribution to the retailer. By the controlled use of packaging and retail storage temperatures of 0°F or lower, frozen foods can be stored for long periods of time. Small size foods such as peas and lima beans may be quick frozen while moving thus preventing the individual products from freezing to one another. These individually quick frozen (IQF) foods will pour readily from a package if not allowed to thaw and refreeze.

Package Requirements.—Over $500 million of packaging materials are used by the frozen food industry. A wide range of materials account for this volume.

Frozen foods are specifically designed for long term storage. A suitable packaging material must protect the product from dehydration and oxidation.

Under the dry conditions maintained for freezing and cold storage, foods tend to loose moisture. Quick frozen meat, poultry, and baked goods are particularly sensitive. A material having a low moisture vapor transmission rate is needed.

The development of rancidity in fatty and oily foods is due to oxidative deterioration. Flavor losses occur coupled with the fading and discoloration of various pigments. In order to prevent oxidation, a suitable oxygen barrier is needed. Oxidation occurs at a slower rate at low temperature ranges; however, it is still a factor in the packaging of frozen foods.

If dehydration and oxidation are allowed to occur, "freezer-burn" results. A discoloration of the surface of the food coupled with texture loss results. In addition a suitable package must not permit the loss of organic volatiles important in the maintenance of flavor. The intermingling of different food odors also needs to be prevented. A gas barrier is necessary to prevent the transfer or pickup of odors.

A final package must be able to withstand the condensation of atmospheric moisture on the exterior. If water vapor from the atmosphere permeates the package, excessive frost formation will occur on the interior. In order to prevent such condensation, air cavities within the package must be minimized.

All materials used in packaging frozen foods should maintain their integrity at temperatures as low as −40°F, or as low as −320°F for cryogenic freezing. They must be capable of being properly printed for retail sales and easily stacked in either a vertical or horizontal frozen

Courtesy of Chemische Werke Hüls Aktiengesellschaft

FIG. 38.   BOIL-IN-BAG PACKAGE

This frozen Hungarian Goulash meal is quick frozen in a film pouch
and can be reconstituted in boiling water.

food cabinet. In addition, a suitable material may have to withstand any
abrasion or puncturing caused by the sharp edges or points present in
many frozen foods.

BIBLIOGRAPHY

ANON.   1966.   Aseptic bottling by four methods. Dairy Ind. *31*, 997.
ANON.   1967.   Metal box air sealing helps Eskimo maintain packaging quality. Food Trade Rev. *37*, No. 10, 108–109.

BETZ, W. 1967. The aseptic packaging of foods in plastics materials. Verpackungs Rdsch. *18*, No. 9, 1016–1020. (German)

BILLING, O. 1967. Food preservation: new methods of treatment—new packaging requirements. Packing *4*, No. 6, 17–19. (Swedish)

CLARKSON, H. 1966. The packaging of prepared meats for microwave catering. Food Trade Rev. *36*, No. 12, 49–50.

COOK, G. G., and MARTON, S. 1968. Thaw pack bids to eliminate paper can for frozen fruits. Quick Frozen Foods *30*, No. 11, 35–40.

DAVIS, R. B., and MAUNDER, D. T. 1967. New system for aseptic pouch packing. Modern Packaging *40*, No. 14, 157–163.

DESROSIER, N. W. 1963. The Technology of Food Preservation, 2nd Edition. Avi Publishing Co., Westport, Conn.

FODA, YOLT et al. 1967. Effect of dehydration, freeze drying, and packaging on the quality of green beans. Food Technol. *21*, No. 7, 83–86.

GHOSH, K. G. et al. 1968. Lightweight flexible packaging for accelerated freeze dried (AFD) meat. J. Food Sci. Technol. *5*, No. 1, 12–16.

GRIFFIN, R. C., HERNDON, D. H., and BALL, C. O. 1969. Use of computer derived tables to calculate sterilizing processes for packaged foods. II. Application to broken-line heating curves. Food Technol. *23*, No. 4, 121–126.

HERNDON, D. H., GRIFFIN, R. C., and BALL, C. O. 1968. Use of computer derived tables to calculate sterilizing processes for packaged foods. I. Application to straight-line heating curves. Food Technol. *22*, No. 4, 129–140.

JASPER, W. 1966. Evaluation of the heat to be removed during the partial freezing of poultry by immersion. Kalte Technik *18*, No. 8, 294–297. (German)

JOSLYN, M. A., and HEID, J. L. (Editors). 1963. Food Processing Operations. Vol. 2. Avi Publishing Co., Westport, Conn.

LOPEZ, A. 1969. A Complete Course in Canning, 9th Edition. Canning Trade, Baltimore.

MARG, D. F., and ANDERSON, D. N. 1966. Heat penetration of bagged foods. Mod. Packaging *40*, No. 4, 149–150.

PRESTON, L. N. 1967A. Quick frozen foods. Food Packaging Design Technol. June, 33–37.

PRESTON, L. N. 1967B. AFD foods. Food Packaging Design Technol. July, 21–28.

SACHAROW, S. 1966A. Food processing trends thrust packaging materials into the space age. Paper, Film Foil Converter *40*, No. 5, 58–61.

SACHAROW, S. 1966B. Taking the mystery out of plastics for frozen food packaging. Quick Frozen Foods *29*, No. 2, 50–51.

SACHAROW, S. 1967A. Deep-freeze storage is tough test for converted packages. Paper, Film, Foil Converter *41*, No. 1, 54–56.

SACHAROW, S. 1968. Frozen food packaging. Food in Canada *28*, No. 12, 32–34, 47.

SACHAROW, S. 1969. Packaging materials for frozen foods. Frozen Foods *22*, No. 5, 29–40.

SACHAROW, S., and GIBLIN, J. P. 1967. Plastic films for frozen food packaging. Frozen Foods, *20*, No. 4, 18–20.

STREAT, C. 1967. Packs with impact. Food Manuf. *42*, No. 9, 49–53.

TRESSLER, D. K., VAN ARSDEL, W. B., and COPLEY, M. J. 1968. The Freezing Preservation of Foods, 4 Vols. Avi Publishing Co., Westport, Conn.

CHAPTER 3

Red Meats

INTRODUCTION

Meat was one of the earliest of man's foods. When he learned to make fire (about 750,000 years ago), he learned to roast meat.

Up until comparatively modern times meat was eaten promptly after it was killed. The seller of meat purchased live animals and kept them near his shop to be killed when a buyer was at hand. There were problems in keeping live animals through the winter as grain supplies might be needed for human life. In cold climates animals were slaughtered at the approach of winter and great feasts were held. Much experimentation was done to find ways of keeping meat longer, and even in Biblical days spices and salt were used to pickle or brine the surplus. Other techniques included drying, smoking, and dry salting. The resulting tough, salty, stringy meat encouraged the development of stews, meat pies, and sauces or gravies.

In the seventeenth century, pirates in the West Indies were called buccaneers since they sold and used dried or "boucaned" beef, making them one of the earliest meat processing industries in the New World. Other early meat packers in the United States were New England farmers, who preserved meat with salt.

Commercial meat packing developed with the concurrent development of direct-expansion ammonia refrigeration in the late 1800's. This replaced natural ice refrigeration methods and enabled meat packing to become a year-round operation. Scientific research into meat science began in the 1890's when many large meat packers organized research programs to study processing and packaging.

In the days of immediate consumption or alternative preservation of meat, packaging was limited to that necessary for protection against contamination and for ease of transportation and storage. The development of the canning process in the early 1800's was quickly adapted to meat and meat products. Wm. Underwood Co. of Boston was one of the first to can meats in 1845. Tinned canned beef was introduced in 1868 by the Libbey brothers, and the canning of meat soups and stews by Campbell came in the 1870's. Despite the popularity of these canned products, fresh meat and some cured meat products were still purchased from the local butcher and carried home in any suitable wrapper. Later, special juice-resistant butcher's wrapping papers were developed.

Not until the development of supermarkets in the 1930's and 1940's did any significant change occur. They introduced the prepackaged retail cut of meat, already weighed and priced, in order to reduce labor costs. Without cellophane, which enabled the buyer to see the meat, this would not have been possible. In more recent years the supermarkets have demanded packages which would provide increased shelf-life and reduce their losses due to spoilage. This has required increased attention to the development of packages and packaging materials for this end use.

Different packaging principles, dependent on inherent biological requirements of the product, are involved in packaging fresh, cured, and processed meat. In the latter category the various methods of processing each have their own peculiar packaging requirements.

FRESH MEAT

In the United States, over seven billion pounds of fresh meat are packaged annually. In 1967, this amounted to a per capita consumption of 177.5 lb. In most supermarkets fresh meat is packaged manually as an in-store operation with automatic labeling equipment, cut-off devices, and film dispensers serving to increase speed. The most commonly used package is a film-tray combination.

Characteristics

Fresh meat is highly perishable and biologically active. Even with packaging to reduce moisture loss and refrigeration to reduce the attack of microorganisms, a shelf-life of only 2 to 3 days can be expected. Change of color is the first cause of reduced sales appeal.

Color.—The pigments present in fresh meat are the conjugated proteins, hemoglobin and myoglobin which form complexes with the oxygen inspired by the animal. Hemoglobin transports oxygen in the blood; myoglobin is a storage mechanism for oxygen in the cells. The purple-red color of a freshly cut surface is due to the concentration of myoglobin. Upon combination with oxygen oxymyoglobin is formed, and a bright red color results. The conversion of myoglobin to oxymyoglobin is completely reversible and involves an oxygenation process. When the bright red surface is exposed to air for several days, an additional reaction forms the brown pigment, metmyoglobin. Except for the presence of reducing coenzymes which convert it to myoglobin, metmyoglobin is highly stable and its formation is essentially nonreversible.

Both oxygen reactions occur at a rate proportional to the partial pressure of oxygen; however, the rate of oxidation of myoglobin to metmyoglobin reaches a maximum at a partial pressure of oxygen of only 1–1.4 mm Hg. The rate falls as the oxygen pressure is raised and reaches a constant value over the range of 30 mm to 760 mm Hg. If a freshly cut

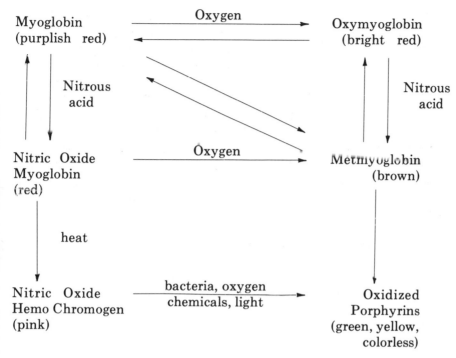

Courtesy of the American Meat Institute Foundation

FIG. 39. COLOR CHANGES IN FRESH AND CURED RED MEATS

piece of meat is wrapped in a highly oxygen-permeable film, a sufficient amount of metmyoglobin is formed to cause the color to deteriorate in 2 to 4 days.

The formation of metmyoglobin is also dependent on other factors. Oxygen utilization by meat occurs first through a solubilization of oxygen in the surface fluid and subsequently by a diffusion process into the interior. Maintaining low temperature storage conditions enhances oxygen solubility. Under poor sanitation conditions microorganisms utilize the available oxygen, causing the conversion of oxymyoglobin to metmyoglobin. Bacterial spoilage is also accelerated in the higher pH range, which is generally between 5.3 and 6.0 depending on the feeding and handling procedure prior to slaughter.

Another type of surface discoloration, a dark reddish-brown color, occurs with loss of moisture. As dehydration takes place, the concentration of pigment increases on the surface of the meat. Interior moisture, containing dissolved pigments, then migrates to the surface and evaporates, causing pigment to concentrate further.

Literature contains conflicting evidence about the effect of light on fresh meat color. Some say ultraviolet light accelerates desiccation and oxidizing of myoglobin, others that light is almost without effect. Tests run under controlled conditions on beefsteaks prepackaged in films and stored for various lengths of time at 30°F in a dark environment and then in a lighted display cabinet support the former contention. The color score of the steaks remained stable during storage in the dark but display under light caused color degradation.

Flavor, Odor, and Texture.—Undesirable flavors, odors, or textures can occur in fresh meat if the actions of enzymes, molds, bacteria, and oxygen are not properly controlled. Tenderness in texture is achieved by ageing the freshly slaughtered carcasses of beef and lamb for up to two weeks at 35°F, giving them an "aged flavor." (Pork is not aged because the meat is already tender and because of the rancidity problem.) The better grades of carcasses do not lose much weight during ageing because the good fat covering minimizes moisture loss and surface contamination. Lower grades do not have this protection and therefore are seldom aged. Ageing at higher temperatures can be accomplished if special methods such as ultraviolet radiation or bacteriostatic agents are used to prevent surface spoilage from molds and bateria.

Packaging Principles

The package is not intended to prevent bacterial contamination of fresh meats. They are handled so frequently prior to retail packaging that contamination is inevitable. The development of flavors and odors due to spoilage from bacteria or molds is prevented by control of temperature. Except during ageing, fresh meats should be stored as close to 32°F as possible and never above 50°F. The principal role of the package is prevention of moisture loss, the exclusion of foreign odors and flavors, and the moderation of oxygen transfer. A relative humidity of 85%–90% is needed to prevent desiccation. Unless the package material is a perfect barrier, the atmosphere outside the package should be kept at this level. A good barrier material will prevent absorption of odors and flavors from external sources. Control of oxygen permeation requires a compromise between development of ideal color and prevention of oxidative degradation reactions. Oxygen is needed for "bloom" in red meats, but it will promote rancidity in fats, particularly in pork fat. Usually refrigeration is used to control rancidity as much as possible and a moderate oxygen permeability is sought. The package also should be capable of resisting tearing and puncturing from normal handling in a retail outlet, and it should offer a pleasing appearance to the purchaser.

Package Types and Materials Used

In supermarket packaging, most fresh meat products are placed in rigid trays and overwrapped with a transparent film. The nonshrink films are now being challenged by heat-shrinkable films which produce a tighter package.

Trays.—Until recently, most fresh meat trays were made from molded pulp and paperboard. These trays were absorbent, economical, and fairly rigid. Disadvantages were loss of strength due to decomposi-

Courtesy of Chemische Werke Hüls Aktiengesellschaft

FIG. 40.   FOAM POLYSTYRENE TRAY

Red meats are packaged in overwrapped expanded polystyrene foam trays.

tion by excessive moisture absorption, adherence to the meat when frozen, and poor visibility. Polystyrene foam trays offer an aesthetically appealing white background for red meat and have a "quality" feel. The use of blotters underneath the meat has served to absorb meat juices not absorbed by the tray.

Recent years have witnessed a growing concern by the consumer for full visibility in fresh meat packages. Molded pulp, paperboard, and foam trays are still used; however, clear formed-polystyrene trays are being introduced. The recent "Peek-A-Boo" bill introduced in New York City provided legislation for clear, fully transparent meat packages in supermarket retailing. Although the law requiring these packages is operative only in New York City, many other localities are reacting to its regulations. Clear plastic trays offer improved visibility but are fairly expensive, and the sealing of the film overwrap on the back of the tray reduces visibility. Meat juices may also create a poor sales effect. The use of blotters eliminates excessive juice accumulation.

Transparent Films.—The earliest transparent film used for wrapping fresh meat was a special grade of cellophane. In order to retain the red "bloom" of a freshly cut piece of meat, a film with an oxygen permeability equal to 5,000 ml O_2/sq m/24 hr/atmosphere at 75° F with 100% rh inside and 52% rh outside the package is required. The cellophane used was one-side nitrocellulose coated in order to be permeable to oxygen but impermeable to water vapor. The uncoated side was placed next to the surface of the meat. The moisture in the meat caused the one-side coated film to become saturated, and since wet cellophane is more permeable than dry cellophane, its oxygen permeability increased, while the nitrocellulose coating on the outside prevented excessive moisture loss to the atmosphere. One-side coated cellophane is still widely used for fresh meat packaging. Two-side coated film cannot be used because of its low oxygen permeability which accelerates the formation of metmyoglobin. An additional grade of cellophane used for packaging heavy and irregular-shaped cuts of meat is one-side polyethylene coated film, which is extremely durable, punctureproof, and permits tight contour wrapping.

The introduction of synthetic plastic films was widely hailed by the fresh meat industry. Softer films, capable of packaging meat, have captured a large segment of cellophane's market. The first non-cellulosic plastic film used by meat packagers was rubber hydrochloride. Originally introduced in 1934, the self-service meat packaging industry became one of its major markets. While cellophane wrapping generally is done with sheets, most plastic film systems utilize rolls and sealing wires. Rubber hydrochloride is more durable than cellophane and may be pulled tightly around the package.

Low density polyethylene film may also be used for fresh meat packaging. At gauges lower than 0.001 in., it is sufficiently permeable to oxygen and provides a suitable moisture vapor barrier. One problem, however, has been the condensation of moisture on the inner surface of the polyethylene. In order to avoid this, one variety has been produced with minute holes. Other types are available whose surfaces have been treated to prevent condensation. Polyethylene, however, is not widely used because of its extensibility, low strength in thin gauges, and marginal clarity and gloss. By modifying polyethylene with vinyl acetate a good clear film can be made, which has suitable oxygen permeability.

Another widely used transparent plastic film for fresh meat is polyvinyl chloride. Gauges as low as 0.0007 in. may be used and a "soft" package results. The film is highly plasticized in order to provide oxygen permeation and contour wrapping. Advantages include excellent gloss, clarity, and package "oling." Extruded polyvinyl chloride films are making a strong bid for in-store wrapping of fresh meat.

Biaxially oriented polystyrene film has been used for some meat packages. Its clarity and permeability provide an aesthetically pleasing and functional package, but it suffers from low durability and sealing problems.

Shrink Films.—The larger and more irregular cuts of fresh meat are usually wrapped in shrink films. The advantages of shrink film include neat appearance, ease in handling, and a contour fit which requires less film. At present, shrink wrapping accounts for about ten per cent of the fresh meat market.

The trend toward the use of heat shrinkable polyvinyl chloride has been increasing, because the average cost on automatic lines is about 0.75 cents per tray. Cellophane wrapped packages are about 25% more expensive. Other shrink films used include rubber hydrochloride polypropylene, irradiated polyethylene, and polyvinylidene chloride.

Package Forming and Filling.—Most supermarket packages are wrapped manually in a film-tray combination. The overturned tray is positioned centrally on the film with its edges diagonal to those of the film. The two pairs of opposite corners of the film are folded over the bottom of the tray and sealed. The entire wrapping operation is fairly simple and 10 to 15 packs may be produced per minute. The film is generally unprinted because of the many different sizes of cuts. Sealing is accomplished by heat or by means of pressure-sensitive labels. In a large volume operation, semi-automatic and automatic machinery become a realistic necessity.

Several systems are based on semi-automatic machinery involving sleeving the film around the tray manually. The package is then fed into the machine for weighing, labeling, and pricing, with speeds up to 20 packs per minute possible. Fully automatic machinery involves a roll fed mechanism. The filled tray is inserted on an in-feed conveyor system. Cut-off length, weighing, and subsequent operations are fully automatic, with speeds up to 35 packs per minute possible.

If the cut of meat is large, irregularly shaped, and requires a shrink film wrap, the package may not contain a tray. Hot air tunnels are used to effect a tight wrap.

Centralized Pre-Packaging

Most supermarkets have many problems in packaging fresh meat. Standardization is difficult and product turn-over serves to complicate the overall operation. Re-wraps for torn packages, leakers, or discolored products account for about ten per cent of all packages prepared. The average store recovers only about 50% of the retail price on rewrapped meat. These problems plus the shortage of skilled butchers, overhead, and special services orders have focused considerable attention on the packaging of fresh meat in central locations.

The principal factor involved in a centralized pre-packaging operation is proper shelf-life. If the conventional two to four day shelf-life is retained, the area over which the product can be shipped is severely limited. By increasing the final shelf-life to seven days or more, centralized meat cutting and packaging becomes practical. Close temperature control, excellent sanitation, and color maintenance through oxygen control increase the shelf-life of fresh red meat to seven or more days with advantages to the supermarket of better distribution, more uniform cutting methods, quantity discounts, and reduced overhead. Pork has been distributed in wholesale cuts for several years and many other meats are now centrally packaged. Primal cuts of fresh meat are supplied to institutions and retailers. They are then cut into consumer portions as an in-store operation.

A successful centralized operation should be situated fairly close to the retail store so that meats can be cut specifically for certain geographical areas and shipping costs reduced. After hanging the carcass at a temperature of 30° to 34°F for several days, the meat is prepared. The cutting room is maintained at 60°F and "bloom" occurs within 15 to 20 min. Control of shipping temperatures is necessary for the operation to be successful.

Many US firms have examined central consumer packaging systems, but few have adopted it, and most have not been successful because of poor shelf-life. In European supermarkets, however, centralized packaging is a distinct reality with nearly 100 firms preparing meat in central locations. As the success of centralized packaging of primal cuts becomes established, it is anticipated that central packaging of fresh retail cuts will become more widespread.

Vacuum Packaging of Fresh Meat

An alternate method of packaging fresh red meat is vacuum packaging, but it is difficult to introduce on the retail level because the purple-red color is not aesthetically appealing. Since vacuum packaging provides a barrier to the product from oxygen and moisture, it is suitable for periods up to three weeks. During this holding period, the enzymes in the meat continue the total tenderizing process called ageing, and meat "blooms" satisfactorily at the end of the storage period. Major retail cuts of beef now vacuum packed by the packer are boxed for ageing, storage, and handling on pallets and are used by hotels and restaurants. Wholesale cuts vacuum packed can be opened in the supermarket, re-cut, and re-packaged. Large size briskets and roasts are now vacuum packed on automatic machinery using deep draw films.

Vacuum packaging might well be integrated into a centralized packaging operation. Economic advantages include more efficient use of labor, reduced freight and storage, and holding of the trimmings at a centralized location. Quality advantages include more uniformity, better ageing, and reduced bacterial contamination.

The most commonly used film for fresh meat vacuum packages is polyvinylidene chloride. It offers low oxygen permeability and shrink characteristics, so that large cuts may be held for up to 21 days with minimum loss of moisture. Other materials useful include polyesters, polyamides, rubber hydrochloride, and cellophane-rubber hydrochloride, cellophane-polyethylene, or polyester-polyethylene laminations.

FRESH SAUSAGES

Fresh meat sausages are made from comminuted and seasoned meat blended with various cereal binders, spices, and preservatives. Their packaging requirements are similiar to fresh meat; however, they contain a comparatively heavy bacterial population. They are highly perishable, and the use of a preservative does not appreciably reduce perishability.

Fresh sausages are subject to oxidative color changes, bacterial spoilage, dehydration, and on occasion sensitivity to high intensity illumination.

Prior to the use of packaging materials, fresh sausages were sold unwrapped or bound together in multiples. The first transparent film used to wrap fresh sausages was coated, heat-sealable cellophane. The film was designed to maintain a proper balance of moisture. Protection from moisture loss is necessary; however, a truly moistureproof wrap would maintain a too high relative humidity within the package. This would encourage mold growth and slime development.

Other transparent films are used for pre-packaging fresh sausages including polystyrene, polyethylene, and cellulose acetate. Polystyrene does not heat-seal readily and does not have a high enough permeability to water vapor to offer maximum shelf-life. Cellulose acetate is expensive and also is not easily heat-sealed. Polyethylene in the perforated form has been used to some extent. Since it is a limp film, it does not machine well on high speed packaging machines.

A small proportion of fresh sausages are wrapped by hand in pre-cut sheets of printed film. For automatic packaging, two types of equipment are used. A direct wrap may be applied to the collated sausages and the ends folded and heat-sealed around the pack or a film may be over-wrapped on a paperboard or rigid polystyrene or polyethylene tray. Speeds of up to 75 per minute are possible on this equipment.

Fresh sausages are mostly sold in natural casings. Cellulose based casings and alginate formulations are used to a lesser extent. In the skinless variety, the sausages are linked in a natural casing and the meat is coagulated by a heat treatment. The casing is removed prior to packaging.

NEW DEVELOPMENTS

In past years, most meat was delivered to retail stores without any package protection. It is now customary to put the wholesale cuts into polyethylene or polyvinylidene chloride bags (0.002 in. thick for weights up to 10 lb) which are heat-sealed and then shrunk. Total shelf-life is increased to about 21 days and spoilage, weight loss, and contamination are reduced. Corrugated boxes, usually 350 lb test board with a wax coating or wax imprugnation, are used for wholesale shipments.

In the convenience food area, a novel package has been developed whereby entire joints are packed in cotton netting ready for oven use. The cotton withstands oven temperatures and has no affect on meat flavor.

An interesting package has been developed comprising an inner sealed container of oxygen-permeable material and an outer sealed container of oxygen-impermeable material. The entire package is vacuumized and the outer container is removed prior to sale to admit oxygen to the inner container and allow the meat to bloom. Nitrogen packaging may also be used.

A significant new application is the packaging of hamburger in oxygen-impermeable tubes made of two plies of 0.0006 in. to 0.00075 in. polyvinylidene chloride with metal clips used to secure the ends. The material is opaque with printing between both plies. The interesting point is that beef trimmings are ground and are immediately put into the tube before the surface blood oxidizes, giving the meat a purple-red color which is masked by the opaque package. Upon opening the package, the hamburger turns red with oxygen contact, and shelf-life averages ten days, in contrast to the 2 to 4 days for conventional fresh meat packages.

Several suppliers are working on methods of coating fresh meat. These include the use of edible coatings based on gelatine and high amylose starch and various easily strippable coatings based on acety-

Courtesy of Cryovac Division, W. R. Grace, Co.

FIG. 41. A CHUB PACK USED FOR A GROUND BEEF PRODUCT

lated monoglycerides. The meat is curtain-coated without the use of conventional equipment.

PRESERVATION OF RED MEATS

Frozen Meats

Although frozen cuts of red meat have found little consumer acceptance, practically all women freeze some meat in the home. The advantage of the frozen product is that each item can be prepared and frozen under the ideal conditions for preserving fresh texture and flavor. Because frozen meat is prone to dehydration and loss of surface texture, it requires high protection against moisture loss and temperature fluctuation. A barrier against oxygen and light is also desirable on fatty products that are subject to rancidification. Freezing with liquid nitrogen is the preferred method (see Chapter 2).

A frozen meat film must accommodate expansions and contractions that may occur during freezing and thawing. Low density polyethylene (.0015 in. to .0020 in.) is used for this purpose as it provides adequate

TABLE 11

RANCIDITY DEVELOPMENT IN PACKAGED FROZEN PORK CHOPS DURING STORAGE
AT 5°F.[1]

Packaging Material	Storage Time to Beginning of Rancidity,[2] months
Vinylidene copolymer (Cryovac)	Over 14[3]
Cellophane (moistureproof)	3–4
Vinyl chloride-acetate copolymer and nitrile rubber base film	3–4
Rubber hydrochloride base film	2–3
Locker paper (wax controlled)	2–3

[1] Lowry and Nebesky (1953).
[2] Criterion for rancidity not given.
[3] End of test.

clarity and good low temperature toughness at low cost. Other suitable materials are polyamide-coated polyethylene and polyethylene-coated polyester.

If cost savings can be realized, the resistance to consumer acceptance may change. Meanwhile packers are supplying frozen meats to institutions such as hospitals, motels, and restaurants.

Courtesy of Reynolds Metals Co.

FIG. 42.   ALUMINUM FOIL CARTON OVERWRAPS

Frozen chopped beef steaks are packaged in a paperboard carton overwrapped with a printed aluminum foil-paper lamination.

Irradiated Meats

The irradiation of fresh meat is still highly experimental. Recently the FDA withheld approval for irradiated hams and withdrew prior approval of irradiated bacon. The problem of off-odors and off-flavors that occur in radiation sterilization of fresh meats has now been largely overcome by irradiating the product at $-112°F$ because such low temperature inhibits the associated chemical reactions. It is best to use metal cans or an impervious flexible laminate to protect the product from re-contamination.

Dehydrated Meats

Many meats are successfully preserved by dehydration. They are very well suited to military rations because of their lighter weight and their not requiring refrigerated storage. All dehydrated meats require protection from moisture, oxygen and mechanical damage.

Hot-air drying has been attempted on cooked ground meat and pork fat. It is not suitable for uncooked meat or cooked chops and steak. The process is too slow and causes surface hardening.

Freeze-drying is a more acceptable method. All types of meats can be freeze-dried; however, their stability heavily depends on the method of packaging. Many commercial firms are using freeze-dried meats in the production of soup mixes and as intermediates for other food formulations. Freeze-dried meats and products containing freeze-dried meats can be packaged in tinplate cans or heat-sealable laminates containing one or more layers of aluminum foil. Representative laminates are polyester-polyethylene-foil-polyethylene and cellophane-polyethylene-foil-polyethylene. Since freeze-dried products are friable, the use of an intermediate layer of polyethylene serves to give the pouch better resistance to crushing. Also, many freeze-dried products are gas packaged rather than vacuum packaged, as vacuum packaging tends to crush the fragile product. Freeze-dried beef, gas packed in carbon dioxide appears to give a more stable red color than when nitrogen is used.

Cured Meat

At present, about 60% of the cured meats are vacuum packed, and the market is growing. Sales of vacuum packaging materials are expected to reach $70 million by 1970.

Meat Curing.—Meat curing was developed to prevent spoilage and reduce the need for refrigeration. The original curing process consisted of treatment with salt and saltpeter. The carcass was immersed in brine, and the resultant product yielded an attractive pink color. Cured meat may be ground, comminuted, or consist of specific cuts. Most meat curing is confined to beef or pork.

Present day curing practices utilize salt, sugar, sodium nitrate, and sodium nitrite. Salt is generally used as a flavor enhancer and secondly as a preservative. "Mild cures" involve a reduced concentration of salt. Sugar is added for flavor and both sodium nitrate and sodium nitrite are color-fixation agents. The product may be immersed in the curing solution or the curing agents may be added in the dry state. For sausage, dry curing agents are added in the dry mix. Hams and bacon may be immersed in the solution or the solution may be pumped through the vascular system of the meat.

Frankfurters.—Frankfurters are made from a wide variety of ingredients. The meat formulation is usually 40–60% beef and pork. Filler meats such as tongues and snouts may also be used (up to 20%). Non-meat ingredients such as cereals, milk solids and corn syrup solids are also commonly used. Flavor is obtained by various spices and seasonings, i.e., pepper, nutmeg and mustard. Curing is obtained by salt, sugar and curing salts.

The meat cuts are reduced in size by grinding. Comminution of the mixture is then necessary. In this process, the beef components are chopped first with the salt, spices, sugar, cure and seasoning plus water and ice. Pork is then added and the entire mixture is chopped to final texture. A colloid mill is usually used to emulsify the mixture. One passage is normally enough; however, in order to prepare a fine emulsion, two machines may be used in-line.

A vacuum is applied to remove large bubbles of trapped air. This is done either by placing the truck containing the emulsion in a vacuum chamber or using a vacuum mixer with an agitator.

The sausage is then stuffed into casings by a hydraulic or air driven piston machine. This machine extrudes the emulsion from a stuffing horn at a specified pressure.

Linking is performed by hand or machine. If ascorbic acid is added, processing follows immediately after stuffing. In the absence of ascorbic acid, the sausage is held for a period of time in order to cause cure to proceed.

Three phases of the heating and smoking operations are usually employed. First, the sausage is tempered by mild heating (90°–100°F). Then a smoke is applied and the temperature is raised (165°F). This allows the cure to develop and the protein to denature which yields a firm product. Finally the product is cooked with a hot water spray (170° –180°F).

The product is then cooled by cold water showering. The entire processing time may take between 65 min to 2-1/2 hr, depending on the curing reaction and overall formulation.

Coloring may be added by colored cellulosic casings or by a dye added to the hot water shower.

For frankfurters processed in cellulosic casings, peeling is done by hand or machine. Machines are usually employed.

The final product is then commonly vacuum packaged in plastic films in two layers on automatic form-fill machinery. Standard Packaging's 6-14 machine employs a thermoformed bottom web of nylon-polyethylene with either a polyester-polyethylene or a nylon-polyethylene top web. Vacuum is drawn up to 23 in. prior to sealing the top web in place. Other packages are made from cellophane overwrapped paperboard trays, or paper banded wrappers. Institutional units consist of paperboard boxes with the layers separated by wax paper or parchment inserts.

Characteristics —Unlike fresh meat, which has a shelf-life of only 2 or 3 days, cured meats will survive for several weeks depending upon the level of cure. Bacon, frankfurters, and smoked links will last about two weeks; luncheon meats and Canadian bacon may last up to four weeks. The general trend is a demand for longer shelf-life, consequently more attention is being paid to the use of vacuum and gas packaging and to more expensive barrier-type packaging.

Color.—The attractive pink color present in cured meats is due to a modified pigment called nitroso-myoglobin. This pigment is formed by

Courtesy of Nixon-Baldwin Div., Tenneco Chemicals, Inc.

FIG. 43. STRUCTURED FILM LUNCHEON MEAT PACKAGE

Luncheon meats are packaged in structured films. The formed web is a combination of ionomer and nylon; the flat web a combination of ionomer and polyester.

the reaction of myoglobin with the nitrites present in the curing solution. Although nitroso-myoglobin is more stable than oxy-myoglobin, it is readily oxidized to metmyoglobin. The stability of the nitroso pigment is dependent on the partial pressure of oxygen in the atmosphere to which the pigment is exposed. The reaction is also subject to an increase in rate with a rise in temperature. Raising the pH inhibits the tendency of nitroso-myoglobin to undergo oxidation.

All of the green and blue visible spectrum as well as ultraviolet light catalyzes the breakdown of nitroso-myoglobin by oxygen into met-myoglobin and nitrates. The use of colored filters in supermarket display cabinets is unattractive and not completely effective. It has been reported that fading was reduced in products containing free nitrite and sulfhydryl groups, but if the product lacked one of these components, reduction of fade was not noticeable. Most meats contain free sulfhydryl groups and two per cent nitrite dip is an effective way of retarding or restoring light fade. Even if faded, cured meats recover some of their original color upon exposure to darkness. It is good practice for supermarkets to turn off their display lights when not in use to permit color restoration.

The intensity of light in display cabinets for cured meats should not exceed 25 foot candles, and for ham the intensity should not exceed 15 foot candles. Since the color fade reaction is related to the presence of oxygen, prepared or cured meat products can be vacuum packaged to prolong color life. The presence of oxidizing chemicals will, of course, also tend to fade color.

The development of a brownish color on the outside surfaces of a cured meat is usually due to dehydration; however, excess nitrite can cause the same off-color. If insufficient nitrite is present, meats will have a weak color that readily fades to gray. Green streaks may be due to excess nitrite (nitrite burn) or to the action of bacteria such as *Lactobacillus viridescens*. Certain fats have high concentrations of organic peroxides (pork fat is particularly high). These fats can become rancid even under frozen storage, and the result is not only a bad flavor and odor, but also a color fading on the surface.

Bacterial Spoilage.—The development of a slimy surface is related directly to heavy growth of bacteria or of yeasts. This can be prevented by lower temperature storage, but once encountered the meat is beyond recovery. Some bacteria produce carbon dioxide; others produce sour flavors or odors. Since molds are strictly aerobic, vacuum packaging is an effective means of control.

The bacteria responsible for spoilage in cured meats can largely be determined by the canning method. Excessive cures or reuse of cover pickle cause the growth of halophilic flora. Bacterial counts are sharply reduced by curing and smoking. Lactic acid bacteria may be present in commercial curing brines and may survive inadequate heat processing. These tend to grow in the interior of the product during refrigeration. Green cores may be caused by *Lactobacillus viridescens.* Souring is usually caused by *Streptococcus faecium.* Gas development is caused by heterofermentative strains. If the meat has been properly cured, recontamination is still possible during handling and packaging. Since the surface contains a fairly high salt concentration, a barrier exists to bacterial penetration even though large numbers of lactobacilli and microbacteria may have been deposited on it.

There is conflicting evidence in literature on the effect of vacuum packaging on shelf-life extension. Most researchers feel that vacuum packaging does prolong the shelf-life of cured meats. However, a study of sliced cured meat on the Swedish market discovered no difference in total bacterial count between vacuum packed and nonvacuum packed products after storage.

Vacuum packaging helps to establish anaerobic conditions, thus encouraging the growth of anaerobes and facultative spores. The possible existence of *Cl. botulinum* has been the source of much research, but this is unlikely to grow in cured meats at normal refrigeration temperatures. Also, the concentration of salts in the meat tends to inhibit its growth. Cured meat products contain few putrefactive anaerobic sporeforming bacteria. Even if they are added during slicing and packaging, they would not grow in competition with the lactic acid bacteria that are developed during normal storage.

Packaging Principles.—As with fresh meat, the package for cured meats is not expected to perform as an absolute barrier to bacterial contamination; however, some protection is inherent in any barrier. The principal role should be to minimize light fading by preventing the entry of oxygen and the loss of moisture. Spoilage is expected to be reduced by refrigeration or freezing, therefore the package must be able to survive these conditions as well as present an attractive appearance.

Package Types

Some special cuts of cured meats are sold in the same case as fresh meat cuts and are packaged accordingly. For example, smoked pork chops are packaged in a film overwrapped tray in the same display as fresh pork chops. Chub packs, which are tubes stuffed with a soft product, twisted, tied, or clipped at each end, are used for some ground meats and sausage preparations. Other smoked sausages may be vacuum packaged in film envelopes with or without trays or cardboard inserts. Sliced meats are sometimes loosely packed in a film pouch, but more frequently they are shingled or stacked and vacuum packed. Some of the newer plastic films can be drawn into cavities which make a neater package for stacked slices or other shapes. Most hams are overwrapped and shrink wrapped. Some are boned and sold in cans (see cooked meats).

Package Materials

Overwraps may be plain film or foil laminations. Film overwraps can be selected from a variety of materials. Plain or coated cellophane, polyethylene, polypropylene, polyvinyl chloride, polyvinylidene chloride (PVDC), rubber hydrochloride, or various combinations. Foil overwraps are usually combinations of aluminum foil and paper. The selection of a suitable overwrap is based on economics, shelf-life and machine availability.

Overwraps do not afford long-term shelf-life with cured meats. Tightly fitting overwraps are obtained with PVC, rubber hydrochloride and polyvinylidene chloride. Also, these films are available in reel form for ease in in-store handling. Cellophanes are used in sheets and may be more difficult to apply. Foil laminates are useful for light protection where the supermarket lights are not turned off at night. They also offer consumer aesthetics. Chub packs can be made from polyethylene alone. Sometimes they are a combination of two films where one is a cellophane and the other a polyethylene. Shrink films are used for packaging trayed items and large cuts of cured meats. Those meats that are packaged in bags or pouches may be nitrogen gas flushed or vacuum packed.

Courtesy of Cryovac Div., W. R. Grace Co.

FIG. 44.   SHRINK-FILM TIGHT WRAP

A cut of corned beef is wrapped in a PVDC-PVC copolymer film. After closure the package is immersed in hot water to shrink the film tight.

The use of vacuum or gas packaging is necessary for longer shelf-life. Under these conditions the film is likely to be a high barrier combination such as cellophane-PVDC-polyethylene, or polyester-PVDC-polyethylene. Drawn cavity packages are usually made from combinations of polyethylene, PVDC and either polyester or polyamide films. The polyamides are softer and will give better draw, thus are capable of packaging large size units such as one-lb bologna. At present, over 70% of packers use a polyamide laminate as the drawn web and 30% use a polyester combination.

Package Making, Filling, and Closing

When automatic pouch making machinery was initially developed, most packers preferred to use preformed pouches. Cellophane-PVDC-polyethylene and cellophane-foil-polyethylene were the most frequently chosen materials. Hand fed, manually operated filling equipment was generally used. At present most large packers still use preformed pouch type units for small volume products, and many smaller packers rely on preformed pouch machines for almost all their production; but the trend is toward automatic, vacuum- or gas-packaging machinery which form, fill, and close a drawn cavity package from roll fed stock at higher speeds and with less material consumption.

The basic concept for pouch units incorporates an air pressure equilization principle. By means of a chamber, air pressure is equalized so that no collapsing or distortion occurs when a product is vacuum packed, and a stretching mechanism prevents top seal wrinkles during evacuation. Hand-loaded models are available with single or multiple chambers, and pouch machines can vacuum and/or gas flush packages. For some sliced meats, such as chipped beef and liverwurst, only gas flushing is feasible because vacuum packaging causes adhesion of the slices to one another. Nitrogen or carbon dioxide can be used to create an oxygen-free environment. The packaging materials are the same as for vacuum packaging. The meat absorbs some of the gas and creates a partial vacuum in the package. Gas packaging is not widely used by the meat industry due to poor efficiency and lack of specifically designed meat packaging equipment.

Most vacuum machines evacuate a package to an absolute pressure of 25 in. Hg using a rotary piston pump. The pump should be closely situated to the machine and additional outlets eliminated. It is very important to obtain the highest degree of evacuation possible. If gas-flushing is desired to render the package oxygen-free, it can be accomplished in three ways. One concept vacuumizes the package and then gas-flushing is conducted. Another flows nitrogen through a tube to displace air from the package. Still another conducts the total operation of filling and sealing in a nitrogen atmosphere.

The basic concept for drawn cavity packages is the use of two webs of material. The "formed web" is thermoformed into varying size cavities while the second "flat web" is used for sealing. The package is evacuated or flushed in a chamber or through a small opening in the final seal, which is then sealed shut. Automatic integrated lines operate at about 40 units per minute per cavity. Approximately 300 meat packing and processing plants in the United States are using this type of packaging line with about two-thirds making a shallow drawn cavity and one-third a deep cavity. Typical package sizes are 12 and 16 oz for frankfurters, 8 and 16 oz for bacon, 6, 8, and 12 oz for sliced luncheon meats. The trend is toward larger packages and deeper draw materials.

Shrinkable vinylidene chloride copolymer films are used for packaging cooked hams. The package is evacuated prior to shrinking. A hot water immersion or hot air tunnel is used to effect shrink. In addition, the product is pasteurized by exposure to heat. This process is used for other large cuts of cured meats such as bacon, corned beef and various sausage mixtures such as the Polish sausage (Kielbasa). It is also possible to stretch wrap large cuts of cured meats with tight contour films.

Cured sausage meats are usually marketed in natural or synthetic casings. A newer trend is the use of a vinylidene chloride copolymer as the casing. The meat is stuffed into the casing automatically and sold as an overwrapped film unit.

New Developments

Most bacon is now packed in recloseable cartons with transparent windows. Since slices have to be weighed and check-weighed before packing, the use of a paperboard insert is mandatory. The board must resist "wicking" due to the extremely high fat content of bacon. The tuck carton usually consists of a bleached white sulfite board coated with wax or polyethylene. The paperboard insert is also coated. The window is typically made of polystyrene. A new concept is vacuum-packed bacon. Although only about ten per cent of bacon sold in the United States is presently being vacuum-packed, many packers are considering it. Vacuum-packed bacon may be packaged in a wide variety of plastic and foil laminations. A preformed pouch may be evacuated or a package formed from roll stock on several commercially available machines. Polyester-PVDC-polyethylene and polyamide-PVDC-polyethylene are the two film combinations most widely applicable for a vacuum packed bacon. The shallow draw requirements of bacon make the film choice not too difficult; however, individual evaluations are necessary before a decision is reached. The use of a foil

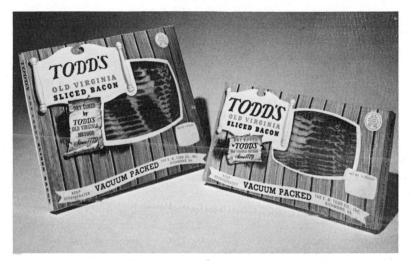

Courtesy of E. M. Todd Co., Inc.

FIG. 45.   VACUUM PACKAGED BACON IN A PRINTED CARTON

composite "flat web" in combination with a transparent film composite "formed web" may be necessary due to aesthetics. There are many variations possible, but the need for low oxygen permeability and for color retention does dictate the use of polyvinylidene chloride.

A new material offered to the cured meat industry is rigid polyvinyl chloride. The final package consists of a rigid formed web of 10–14 mil PVC and an unformed covering web of different type laminates. Some packers use PVDC coated PVC for better protection. This may be necessary at the cavity corners which tend to thin out at the lower gauges. Originally introduced in Germany, the formed PVC material is now produced in the United States. Scientific-Atlanta offers a high speed machine for the gas-packaging of products in semi-rigid PVC containers. Speeds range up to 38 per minute. The unit uses preformed containers and has a revolving turret. A paper-foil composite may be used as a covering material. Advantages of the rigid package include better stacking and increased product protection.

A recent award was given to the "Flavor-Tite" package developed jointly by Standard Packaging Corporation and Swift & Company. The package consists of a rigid polyvinyl chloride base and a flexible polyester top with a heat-seal coating. As the housewife pulls back the flexible cover, the heat-seal coating breaks, exposing a pressure sensitive adhesive. The adhesive is then used for reclosure features. Advantages to this package include rigidity and peelable features.

Another interesting bacon concept is "Saran-Pac," in which plastic film is extruded in line at the packing plant. The final package is composed of PVC-PVDC-meatboard-PVDC-PVC. Limiting factors to acceptance of the "Saran-Pac" system are its high cost compared to a conventional roll-stock machine, and its limitation to only one size package, which requires a large volume operation.

COOKED MEAT

Cooked meats are intended for consumption without any prior heating although they can be heated if desired. Many of these products have also been cured to provide the desired flavor. Tinplate or aluminum cans frequently are used as the package, and unless the product is sterilized, the package must be handled and stored under refrigeration. At present tinplate is the standard material; however, aluminum is making inroads. Aluminum cans offer light weight, corrosion resistance, variety of shapes and easy opening. Both types of cans require coatings to prevent the action of salts on the metal as well as other effects. The coatings used are called meat enamels and consist of modified epons with aluminum pigment. If unprocessed cooked meat is thinly sliced, short term shelf-life is possible with a cellophane or PVDC overwrap or pouch under refrigerated conditions. Meat pies require a breathing film to allow for moisture escape because the ingredients present tend to deteriorate in a damp atmosphere. An equilibrium humidity of less than 75% insures adequate ventilation. Cellophane, cellulose acetate and ventilated polyethylene may be used as overwraps.

Autoclavable Meat Package

A recent trend in cooked foods has been the development of flexible packages for autoclaved meat products. If a food product has a pH below 4.5 the very dangerous spoilage bacteria, such as *Cl. botulinum* and other bacteria such as the putrefactive anaerobes will not multiply. Food poisoning from eating improperly processed acid foods is, therefore, extremely unlikely. The growth of harmless yeasts and molds can be prevented by pasteurizing the product at temperatures up to 212°F. Foods with pH above 4.5, on the other hand, must be processed at high temperatures (240°F to 270°F) to insure destruction of the dangerous bacteria. A satisfactory packaging material must be capable of withstanding the processing conditions for the food it is to contain and must also provide both an oxygen barrier and protection against subsequent recontamination by microorganisms.

Most canned meats have unlimited shelf-life; however, two years is considered to be the norm. A flexible package should provide a minimum of one year storage. To obtain this shelf-life or longer, a composite

Courtesy of The Metal Box Co., Ltd.

FIG. 46. CURED MEAT SPREADS IN ALUMINUM CANS

All-aluminum cans with easy open tear out lids are used for cured meat spreads in Great Britain.

lamination must be used. The most functional material consists of a film-foil-film laminate. The outer layer of plastic film provides support and physical strength to the composite. Aluminum foil contributes excellent barrier properties and a heat-sealable inner ply completes the structure. Polyester, polyamide and/or oriented polypropylene are possible outer ply candidates. The sealing media may be polyethylene, polypropylene, polyvinyl chloride, or some of the newer copolymer films. An all plastic laminate may also be used for heat-sterilized meats, but the life of such packages is limited by the permeability of the packaging material. Cooked hams have been marketed in nylon-11 (Rilsan) film after storage for periods up to three weeks.

Pork luncheon meat, ham fingers, and chopped pork are being marketed in Europe in an autoclaved pouch. The meats are vacuum packed in a polyester film-foil-polyolefin film laminate pouch, autoclaved, and then placed in a printed board carton. The packaged meat products are processed in batch retorts using a pressurized water cook. No refrigeration is needed and ample shelf-life is reported. A Norwegian company is currently heat-sterilizing a full range of meat products in clear all plastic pouches. Polyester-polyethylene laminates are used for a maximum shelf-life of three months. Since oxidative deterioration is rapid at room temperature without the foil ply, these products are sold from chilled display cabinets at 35°–40°F.

SOLID FATS

Lard

Over 3 billion pounds of lard are produced annually in the United States. The lard is "rendered" by heating the fat tissue of swine to melt and separate the pure fat from the tissue. The best lard comes from caul and kidney fat. Fuller's earth, activated carbon, or other suitable clarifying media are used to bleach the product, after which it is filtered. Lard may be sold plasticized (smooth) by precooling to 110°F, then rapidly chilling, or it may be sold "grainy" by allowing it to cool slowly. Lard flakes from 3 to 10% may be added to raise the melting point. (Lard flakes are hydrogenated lard.) With the addition of an antioxidant higher melting lard can be stored without refrigeration in moderate climates. There are a number of antioxidants approved by the former Federal Meat Inspection Branch of the Department of Agriculture (now the Technical Services Division of the Consumer and Marketing Service of the USDA). The most effective however are BHA (butylated hydroxyanisole) and BHT (butylated hydroxytoluene). BHA together with propyl gallate and citric acid is the most commonly used stabilizer for lard.

When a more bland flavor and odor are required, further processing is necessary, that is additional bleaching, deodorizing and refining.

When properly refined and stabilized, lard needs only protection against soilage; consequently packaging has been principally in paper wrappings or waxed cartons. Nitrogen purging of lard helps further to improve shelf-life. Newer hot melt coated boards may prove superior to wax coated boards for this end use.

Other Solid Fats

Various tallows are obtained by rendering fatty tissue of beef cattle or sheep. Some of these tallows are useful in food products. A great quantity is consumed in the manufacture of oleomargarine. Little if any tallow is sold at retail. Bulk packaging may be in large tins, lined paperboard drums, or other convenient package. The main property requirement is grease resistance.

BIBLIOGRAPHY

ALIKONIS, J. J., and ZIEMBA, J. V. 1967. Edible coatings. Food Eng. *39,* No. 12, 78–80.

ANON. 1960. The Science of Meat and Meat Products, Am. Meat Inst. Found. W. H. Freeman and Co., San Francisco, London.

ANON. 1966. Yellow film lamination for meat packaging. Mod. Packaging *40,* No. 3, 66.

ANON. 1967A. New flexible vacuum pack. European Packaging Dig. *46,* 7.

ANON. 1967B. 'Blotters' absorb meat liquids on transparent polystyrene tray. Quick Frozen Foods. *29*, No. 12, 59.

ANON. 1967C. Close-up meat. Package Eng. *40*, No. 16, 140–141.

ANON. 1967D. Gas flush for semi-rigid pack. Mod. Packaging *40*, No. 16, 140 141.

ANON. 1968A. 'Revolutionary' wrap for bacon sides gives better protection. Packaging News *15*, No. 1, 15.

ANON. 1968B. All polystyrene packs increase sausage sales by 100 percent. Packaging News *15*, No. 4, 28.

BROWN, W. D., and DOLEV, A. 1963. Autooxidation of beef and tuna myoglobins. J. Food Science *28*, 207.

HOFMANN, L. 1967. Storage testing with meat tins. Verpackung *8*, No. 2, 49–51. (German)

MARRIOTT, N. G. *et al.* 1967. Color stability of prepackaged fresh beef as influenced by pre-display environments. Food Technol. *21*, No. 11, 104–106.

PELIZZERO, O., and AUBERT, S. D. 1968. Research on the packaging of sausage in plastics film. Imballaggio *19*, No. 144, 15–18. (Italian)

PRESTON, L. N. 1966A. Flexible packaging. I. Fresh meat. Food Packaging Design Technol Nov., 19–23.

PRESTON, L. N. 1966B. Flexible packaging. II. Fresh sausages and meat pies. Food Packaging Design Technol. Dec., 25–27.

PRESTON, L. N. 1967. Flexible packaging. III. Cured meats. Food Packaging Design Technol. Jan., 14–17.

PURR, A. 1967. The influence of technological processes on the quality of tinned lard during long storage. Verpackungs Rdsch. *18*, No. 4., 25–31. (German)

RADINGER, H. 1968. Packaging and selling of fresh meat. Verpackungs Wirtschaft *16*, No. 4, 16–17. *Ibid. 16*, No. 5, 19–20. (German)

RAMSBOTTOM, J. M. 1966. Packaging luncheon meats. I. Meat *32*, No. 10, 46–50. *Ibid.* II. *32*, No. 11, 38–40.

SACHAROW, S. 1965. Packers need to know many things about packaging films. Natl. Provisioner *153*, 81–84.

SACHAROW, S. 1966A. Packaging of food. IV. Laminates move to wrap up huge processed meat market. Paper, Film, Foil Converter *40*, No. 12, 60–64.

SACHAROW, S. 1966B. Vacuum packed bacon. Fleischwirt *46*, 1217–1234. (German)

SACHAROW, S. 1966C. The vacuum packaging of cured meats. Fleischwirt *46*, 1110–1113. (German)

SACHAROW, S. 1967A. Meat packaging with flexible films. Svensk Emballagetidskift *33*, No. 1, 6–9. (Swedish)

SACHAROW, S. 1967B. Meat packaging in the USA. Fleischwirt *47*, 575–576. (German)

SACHAROW, S. 1968. Supermarket shelf-life puts burden on meat packaging. Mod. Converter *12*, No. 12, 20–22.

TÄNDLER, K. 1966. Packaging films for meat and meat products. Einkaufsberäter für das Fleischgewerbe *2*, No. 2, 15–16. (German)

TÄNDLER, K. 1968. What must be observed during vacuum packaging of meat? Neue Verpackung *21*, No. 6, 756–758. (German)

WALLENBERG, E. 1967. The effect of light and discoloration. Scandinavian Packaging No. 1, 8–10.

CHAPTER 4

Poultry and Eggs

POULTRY

The raising of domesticated flocks of geese, chickens and other fowl was known to many early civilizations including the Chinese, Egyptians, Greeks, and Romans. Seventeenth century English feasts loaded tables down with roasted meats and fowl. "Four and twenty blackbirds baked into a pie" was not just an imaginative nursery rhyme. As late as the 1830's lark pie was an English favorite.

Few birds have created a sensation to match that of the turkey, which was known to the Indians of the Americas but first discovered for the Europeans by the Spanish explorers. In 1511, the Bishop of Valencia ordered Miguel de Passamonte to fetch ten birds to Spain on his next voyage. Within a century the people of France, Italy, and Spain were driving flocks of turkeys to market. The Portuguese brought them to Goa and to India. From there they spread throughout the Middle East.

The market for poultry meat is growing rapidly all over the world. Consumption is increasing both over all and per capita, particularly in the United States and Europe. In the United States alone, eight billion pounds of poultry meat (live weight) were produced in 1965 and nearly twenty billion pounds are projected by 1985. Most of this growth has been in broiler chickens and in turkeys. The per capita consumption of turkey in the United States has doubled since 1948. World production is expanding steadily with the United States the largest producer followed by the USSR, Canada, and Great Britain.

The growing of poultry has become a large industry and its very largeness has been the main reason why prices have been held relatively stable. Special breeds have been developed for special purposes. The Rock Cornish game hen, for example, is a small roasting chicken produced by cross breeding the Cornish with the White Plymouth Rock. Birds that are best suited for producing eggs are not necessarily good meat producers and vice versa. The broiler industry in particular has concentrated on development of fast growing birds who attain market weight in the least possible time with the most efficient usage of feed. That they have met with remarkable success is evidenced by the decrease in time to bring a broiler to 3.5 lb from 78 days in 1950 to about 37 days in 1962, and by the decrease in pounds of feed required from 3.3 lb to 1.9 lb. Cockerels are no longer permitted to survive as their meat cannot be sold profitably. Likewise, laying hens used to produce hatching stock are no longer a primary meat source, but rather a by-product. Research on the role of vitamins and minerals in the diet of poultry conducted during the 1920's and 1930's made it possible to formulate feed scientifically and to raise broilers the year round in confined hatcheries and brooders. Successful close herding was only possible with the introduction of disease resistant stock and new medicines. Automation of equipment brought labor savings, but ultimate cost reduction was made possible only by vertical integration of all operations. In 1934, the USDA reported 11,405 chicken hatcheries. By 1963, there were only 2900. Modern chicken farms are capable of producing as many as one to one and a half million broilers per day.

CHARACTERISTICS OF POULTRY MEAT

Poultry meat is high in protein, low in calories, and easy to chew and digest. It is a particularly good food for the young, the old, convalescents, and weight-watchers. The quality of poultry meat is primarily a function of the heredity of the flock, however freedom from disease and proper nutrition are critically important. From the moment the bird is killed the meat is subject to bacterial growth, off-odors, and discoloration. Since most of the fat is located under the skin, very little runs throughout the muscle tissue. Cooked chicken breast contains only 1.3% fat while a cut of beef contains 13-30%. Poultry fat is higher in unsaturated fatty acids, although not so high as some vegetable fats. Fresh poultry fat is almost tasteless and odor free. Rancidity occurs more readily because of the unsaturation and is accelerated by oxygen, heat, and ultraviolet light. Turkey fat is less stable in this respect than chicken fat. Even refrigerated poultry carcasses are susceptible to absorption of odors and can develop off-odor from either bacterial decomposition or fat rancidification. Shelf-life varies from as little as a few days to as much as seven days for top grade depending upon type of processing. Frozen poultry is less susceptible to the foregoing but more vulnerable to desiccation or "freezer burn." Surface desiccation of skin tissue or underlying meat causes discoloration, toughness and loss of flavor.

PREPACKAGE PROCESSING OF WHOLE POULTRY

Birds are weighed live, shackled, stunned, killed, bled, scalded, defeathered, singed, and washed. After removal of shanks and oil glands, they are eviscerated, inspected, and decapitated. Giblets are removed, cleaned, wrapped in parchment paper or polyethylene film, and replaced in the body cavity. Carcasses are sorted according to weight and chilled, usually in a mixture of ice and water. USDA regulations specify carcasses must be chilled to at least 40°F internal temperature or lower. Chillers may or may not be agitated. Agitated chillers cause more damage and more moisture pickup (up to 12%). Also, they require giblets to be withheld and added later. After chilling, birds are diverted to be cut up, left whole, refrigerated or frozen. Each type of end product requires a different handling procedure and a different packaging system.

REFRIGERATED POULTRY

Shipment to Retail Market

Wet Shipment.—Birds are packed in containers, covered with about 30 to 60 lb of crushed ice and shipped. One trip containers may be wire-bound wooden crates, wax coated corrugated boxes, or polyethylene coated paperboard containers. Metal containers or wire baskets can be reused if decontaminated. Wire-bound crates are popular, easy to handle, and stackable, but they are not easy to open, to unload, or to dispose of. Paperboard containers eliminate drip and are easier to unload. They are smaller but not so easy to handle and not as stackable. They do however reduce evaporation losses. Metal containers are less costly because of the multi trip feature, but they require an initial investment and sometimes are lost or stolen. Converters of corrugated box board containers supply about 20% of the poulty shipper market. While there has been a general decrease in the use of wirebound containers, these are still widely used for most shipments of whole ice-packed poultry. Corrugated containers are used for 30% of the dry shipments of raw, frozen, or chill packed birds. New trends are toward greater use of these boxes for wet iced shipments due to improved coatings and improvements in insulative properties.

Dry Shipment.—If chickens are well chilled and contain a good level of moisture, they can be shipped "dry" in refrigerated trucks. No wet ice is added to the container. For long shipments sometimes dry ice snow is added. There are differences of opinion on the virtues of wet and dry shipment. Some feel dry shipment causes excessive losses of moisture and is not properly controlled for temperature. The Forest Products Div. of Owens-Illinois Corp. claim otherwise. Their shipper includes a blanket of porous cellulosic material with a polyethylene backing. The moisture draining from the birds is absorbed by the cellulosic material and wicked upward. Since the polyethylene prevents its escape, it recirculates and helps maintain moisture equilibrium. The saving in weight through elimination of up to 60 lb of ice offsets added costs and permits 38% more chickens to be shipped per container.

In Store Handling and Packaging

Whole Birds.—Small whole fryers, broilers, or roasting chickens, may be trayed and overwrapped as with cut up parts or they may be bagged. Bags are made from low density polyethylene and twist-tied or clipped shut. Their main disadvantage is the collection of bloody fluid which is unsightly and may leak out. With trayed birds blotters can be used to soak up the liquid.

Pressure sensitive adhesive backed labels may be applied to the outside of the package, or plain labels slipped onto the bird before overwrapping. A recent development is the use of transparent printed polystyrene film labels. These are used under clear film overwraps. Use of a very soft, highly plasticized, stretchy film such as polyvinyl chloride, can be used to produce a package that resembles a shrink film package. The whole bird is wrapped tightly in the film by rolling the bird end over end and sealing the film in place. An absorbent pad is placed on the back of the bird prior to the first wrap. Next the sides of the film are grasped, pulled tight, and sealed to the bottom. This yields a very tight and attractive package. For extra durability a band of film about four inches wide can be stretched transversely around the bird and sealed in place. Soft stretch films may cause tray distortion or may press blood

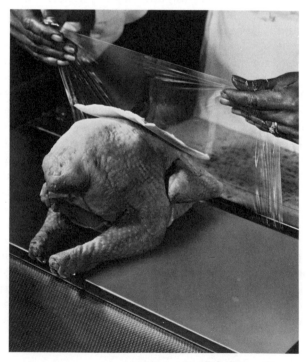

Courtesy of American Viscose Div., FMC Corp.

FIG. 47.   SOFT FILM POULTRY WRAP

A soft, stretchy PVC film is wrapped tightly around a roasting chicken. An absorbent pad is included to soak up juices.

Fig. 48.   Soft Film Poultry Wrap

An extra band of soft, stretchy PVC film is wrapped tightly around the "waist"
of the bird to insure an attractive package.

onto the film surface. Unsightly packages have to be rewrapped to
added cost.

Cut Up Birds.—A large proportion of all poultry sold is cut up ready
to cook fryers or broilers. These are cut up in the store into ten or more
pieces, put in trays, and overwrapped with film. As with red meats,
trays used to be paperboard or pulp, now they are made from molded
expanded polystyrene foam or from clear polystyrene. Advantages of
pulp trays are low cost, fast film wrapping, and moisture absorption.
Foam trays offer a clean, bright appearance coupled with increased
shelf-life. Plastic trays preserve color and quality for up to 48 hours
longer. This is attributed to their low moisture absorption.

In some states allowable moisture absorption is controlled by regula-
tory agencies. Plastic trays can mean up to five per cent savings in product, because no extra meat needs to be added to compensate for moisture loss into the tray. Other plus factors for plastic trays include a
fresher product, a more sanitary package, and when frozen the meat
does not stick on the tray. Different sizes are used depending upon the
pack; for example a $5 \times 10 \times 1$ in. tray is suitable for a baking hen while a
$5\text{-}1/2 \times 8 \times 1$ in. tray is recommended for a cut up fryer. Although shrink
films make a more attractive package initially, they are not essential for
overwrapping fresh poultry. Like soft stretch wraps they can press more

fluid out of the bird. Earliest materials were cellophanes, rubber hydrochloride, and non-fogging polyethylene. Until recently 98% of all packages used these films. The introduction of polyester and polyvinylidene chloride films provided better barriers and better puncture resistance, however their higher costs have held back their use. In addition to the typical tray-pack, other chicken parts such as livers, gizzards, hearts, necks, etc. are often marketed in lined folded cartons or in thermoformed polystyrene tubs with snap-on transparent covers.

Centrally Prepackaged Whole and Cut Up Poultry

By far the most important development in this field is one which was started about 4 or 5 years ago and which is gaining adherents—that is the prepackaging of chicken at the processing plant. Pioneers in this development were Holly Farms Poultry Industries, Inc. This is a totally integrated operation starting with the purchase of grains for feed and ending with chilled, cut, packaged, weighed and price marked packages of chicken going direct to the retail store in controlled refrigerated trucks. Over 60,000 broilers per hour are produced by this operation. Whole birds are chilled in 33°F water and then drained in a 34°F cooler. After insertion into polyethylene bags with a soaker pad over the neck they are further chilled in a −40°F freezer to 28°F. After a second holding period at 28°F they are weighed, price-labeled, boxed, and palletized on returnable 40 × 48 in. wooden pallets. During warehousing and shipping they are carefully held at 28°F. Special nitrogen cooled trucks are used for transportation. Birds that are to be cut up are packaged in white polystyrene foam trays containing a 15-ply soaker pad with a polyethylene coating on the side which contacts the meat. The trays are put into two mil thick stretchable polyethylene bags and heat-sealed on the bottom. These packages also pass through a freezer and into 28°F storage. There are several advantages claimed for this system:

(1) The meat is not frozen but water and juices are. This avoids drip and promotes ease of handling.

(2) The meat has a lower moisture content (but also loses less in storage). Ice pack chickens contain 12% moisture and frozen chickens have 8%. Chill packs average 6–7%, effectively eliminating "tray weep."

(3) Because of careful control of sanitation and of temperature the processor guarantees a minimum five day shelf-life.

(4) Despite the higher cost of the chill pack chicken, the system saves the retailer money on an overall cost basis. He is getting more meat and less moisture. He doesn't have to rewrap. His highly paid meat cutters can be assigned work on higher priced meat cuts. He has less waste and more efficient use of space.

FROZEN POULTRY

Cut Up Parts.—Some cut up chicken parts, when in surplus supply, are frozen by the retailer and sold from the freezer case. This is not prevalent practice however because he is not equipped for rapid freezing which is necessary to preserve quality. Cut up parts of chicken and turkey in this country and some cut up duckling parts are being prepackaged at central processing plants, frozen, and shipped frozen to market. Frozen cut up poultry suffers from bone discoloration. While overall quality is unaffected, eye appeal is inferior, hence it does not command a high level of buyer acceptance as yet

As with fresh parts the product is trayed and overwrapped. Shrink films make a more attractive package and should be resistant to breakage in order to prevent moisture loss and freezer-burn. Useful shrink films include irradiated polyethylene, oriented polypropylene, and oriented PVC, PVDC, and PVDC coated polyesters. Some work has been done with EVA modified polyethylenes (polyethylene vinyl acetate).

Whole Birds.—Larger chickens, especially those that are stuffed, capons, ducks, geese, turkeys, and most small specialty birds such as pigeon, game hen, guinea fowl and the like are sold frozen whole. Some birds are given special treatment such as injection of butter under the skin to give a special flavor during roasting. Some are prestuffed with special stuffings containing giblets, wild rice and spices. Regardless of the type of bird the method of handling is generally the same as for whole chickens through the chillers. Birds are then packaged and frozen. Freezing is usually done in blast freezing tunnels with air at $-20°F$ circulating at 1300–1500 ft per min velocity. Another method is immersion in a supercooled brine for a brief period followed by storage in a noncirculating freezer. Brines may be calcium chloride solutions, or other supercooled liquids such as glycols may be substituted. Still another method is immersion in liquid nitrogen, nitrous oxide, or carbon dioxide. The freezing by liquid gases takes only about 2-1/2 to 5 min for cut up fryers packaged in aluminum trays as opposed to 80 min in a brine solution.

Packaging of large whole frozen poultry is usually done by using a special PVC-PVDC copolymer film having a very high puncture resistance and ability to be heat shrunk by immersion in hot water. Smaller birds that need less puncture resistance or birds which will not be stored long can get by with a cheaper shrinkable polyethylene film. Preformed printed film bags are used. The bird is placed onto a mandrel. The bag is drawn over the bird and gathered around the mandrel. A vacuum is drawn through the mandrel, and the neck of the bag is clipped shut.

Courtesy of Armour and Co.

FIG. 49.   FROZEN WHOLE TURKEY PACKAGE

A heat shrink film wrapped frozen turkey is labeled "self-basting"
because butter has been injected subcutaneously.

During subsequent heat shrinking of the film the neck is sealed tighter. As with cut up frozen parts other shrinkable films used for frozen birds include shrinkable polyesters and cross-linked polyethylenes. Contour wrappers of whole birds have been attempted with ethylene vinyl acetate films. The packaged bird then goes to the freezing line. Depending upon size of the bird one or more may be placed in corrugated shippers. The timing of the latter action depends upon the method of freezing. In blast freezers and still air freezers birds can be cartoned before freezing. Brine, liquid freezing, or liquid nitrogen freezing is done prior to putting the birds in the shippers.

Frozen Turkey.—In preparing turkeys for freezing, live large turkeys are usually electrically stunned and bled. The bleeding process must be complete in order to produce desirable skin color. It takes between 1–3 min to bleed the bird. Other methods of killing include the "axe and chopping block" method and the method used by Jews in which the birds bleed freely. In the latter method, a "Shohet" cuts through the windpipe and gullet with a sharp implement.

After bleeding, the turkeys are put through a scalding tank. This provides for easy removal of the feathers. The water temperature is between 138°–140°F and the bird is immersed for 30–75 sec. This process, called, sub-scalding is widely used on birds that are to be sold frozen.

Defeathering is then conducted by either a machine or manually. The machine method involves rotating rubber fingers which removes all but a few pinfeathers. The latter are commonly removed by hand picking. After plucking, the bird is eviscerated in a separate cool room. The viscera are inspected by a veterinarian or someone under his supervision and the birds passing inspection are thoroughly washed. Difficult to remove organs are dislodged by a suction tube prior to washing of the birds.

The washed birds are then rapidly chilled to 35°F with ice slush. They are then drained and sized and graded for quality.

Frozen whole turkeys are usually sold in vacuum packaged shrink plastic bags. Shrink films commonly used are Cryovac "L", shrinkable polyester (Mylar) and Saran. The bird is inserted into the bag. A vac-

Courtesy of E. I. du Pont de Nemours and Co., Inc.

FIG. 50.   FROZEN WHOLE TURKEY PACKAGES

Young turkeys are put in "Mylar" polyester film bags with clip closures, the film is tightly heat shrunk, and then the birds (shown here) are given a 15 min liquid immersion quick freeze.

uum is pulled. Then the bag is secured by twisting and tying a knot or sealing with a metal clip. The sealed evacuated package is immersed in hot water (195° F) in order to shrink the plastic film. The net result is a skin tight wrap with no air pockets. Also, no voids are present where water vapor can migrate to the package surface and cause freezer burn to the skin. Turkey parts are packaged in film overwrapped trays.

The birds are placed in telescoping solid wall shipper cartons and then into freezers where temperatures from −50° to −30° F with or without fan circulation are used to freeze birds. Obviously the colder the air and the more circulation, the faster the birds will be frozen. Freezing should be accomplished within two to six hours for optimum quality.

Another method of freezing consists of immersing the film wrapped birds into a liquid freezing medium such as brine, glycerol, or other alcohols, then packing them into shipping cases.

Frozen turkeys are also being packaged in unplasticized polyester and polyamide bags which do not have to be removed for cooking. It is claimed this method permits browning, reduces weight losses, shortens cooking time and improves flavor and succulence.

CURED OR SMOKED POULTRY PRODUCTS

Smoked poultry meat has a unique taste and is treated as a gourmet item usually with a premium price. The cure must be lighter in salt and spices to maintain a delicate and bland flavor and not to make the meat taste identical to ham. Humidity should be controlled during smoking to avoid excessive drying and shrinkage. The light meat turns light pink and the dark meat a mahogany color. As with other meats, hardwood smoke is preferred and the type of wood imparts a characteristic flavor. Birds should be of top grade and quality and should be young to insure tenderness. Poultry sausages made from minced meat and giblets can be smoked also.

Curing solutions are made from salt, sugar, sodium nitrite, and spices. Birds are soaked in the pickle at about 35°–40° F for about 1-1/2 days per pound dressed weight and then soaked in cold water for an hour to remove excess salt. Smoking is done for about 5 to 12 hours at temperatures of 110°–130° F. Smoking at higher temperatures (170°–185° F) until the bird reaches 160° F internal temperature produces a ready to eat product. Smoked poultry must be kept refrigerated or frozen. Whole birds are wrapped in printed film to keep down desiccation. Polyethylene, polypropylene, polyvinyl chloride, and polyvinylidene chloride films are used. Because of the premium price, printed foil laminations can be used as overwraps.

COOKED POULTRY PRODUCTS

Cured or uncured poultry, usually chicken or turkey can be cooked and then processed into a wide variety of products, such as sausages, patties, sticks, croquettes, frankfurters, bologna, luncheon meats, pies, hashes, stews, soups, potted meats, patés, chow mein, slices and gravy, or the familiar "a-la-king." These products may be frozen or canned and are packaged accordingly. Paste or sausage like products may be packaged in film chubs made from cellophane laminations, polypropylene or polyvinylidene chloride. Canning in tins or jars is done by traditional methods. Canned whole chickens used to be popular, but now that frozen chicken is available the canned product is on the wane. The canned product tends to fall off the bones and to contain an excess of gelatin. This makes it suitable for casserole dishes but not as a whole roast. Frozen products are packed in sealed plastic bags (polyethylene coated polyester film) with printed paperboard cartons for protection. The bag can be used for reheating the product in boiling water. Lidded foil containers have also been used for frozen cooked products. Chicken luncheon meat items are packaged like their red meat counterparts.

An increasingly popular item is a boneless roast made by partially precooking the bird, removing the flesh, and forming it into a loaf-shaped roast. The meat is usually enclosed in skin and the roast is packaged in a lidded aluminum foil container. Some similar roasts are precooked and need only be thawed to use. Cooking is done in fibrous netting or in plastic film bags (such as polyamide). It takes several hours immersion in 160°F water to effect a proper cook. Turkey rolls, roasts, and parts have been marketed in like manner in paperboard cartons and shrunk films. Foil containers may be replaced by foil-paperboard composites. Cooked, diced chicken meat has been sold in polyethylene bags (two five-ounce bags in a waxed carton) as a frozen convenience item, which can be thawed and added to salads, cream dishes, and the like. Frozen partially cooked poultry meat is found in the familiar TV dinner tray and in chicken pot pies both packages being formed from aluminum foil.

For fine particle meat users, deboning machines have been developed, which debrade the meat from the bone. These accept raw or partly cooked meat, even the poorer cuts such as necks and backs. Brine flotation is used to separate the bone particles from the meat.

DRIED POULTRY PRODUCTS

There are several ways to remove moisture from poultry meat including freeze-drying, drum drying, spray drying, oven drying, or heating in

FIG. 51. ALUMINUM FOIL CONTAINERS

Boneless turkey roasts are becoming extremely popular. The foil
container serves as an oven roasting pan.

oil. Each has advantages and produces a somewhat different product.
Freeze-drying retains flavor and original form, while keeping bacterial
growth and enzymatic changes to a minimum. Little or no oxidative
rancidification can occur because the process involves a vacuum. The
products may be stored without refrigeration and contain less than one
per cent moisture. They are fragile and porous somewhat like a solid
sponge. They must be packaged to protect against moisture pickup,
oxygen, and crushing or other mechanical damage. Oxygen must be held
under two per cent and moisture under three percent. This dictates
hermetically sealed cans or pouches either vacuum packed or inert gas
flushed. A suitable pouch material for freeze-dried items would be a
cellophane-polyethylene-foil-polyethylene; however poultry fats pick
up or extract off-flavor from polyethylenes. This dictates a different
interior component, for example an ionomer film, polyvinylidene chloride, rubber hydrochloride, polyamide, or polyvinyl chloride.

The principal market for freeze-dried poultry meat is for military rations, campers, and ingredients for prepared dishes such as soups or stews.

Spray drying consists of taking finely comminuted cooked poultry meat and spraying it into a superheated atmosphere. The product is suitable for soup mixes or other preparations where a fine particle can be utilized. Poultry meat in various sizes can be put in trays and oven heated to dry. Vacuum is usually used to reduce heat and oxygen damage. Heated drums or rollers have been used for the same purpose.

POULTRY BY-PRODUCTS

Poultry fat is used as an ingredient in other prepared poultry dishes such as soups. Care must be taken not to use mesenteric fat, which may be contaminated and which is certain to have off-flavor. Heating to remove moisture must be held to below 220° F to avoid discoloration and the onset of rancidification. Best results are obtained by centrifuging to remove moisture. Antioxidants, such as BHA and BHT, up to 0.02% by weight may be added for stability. Refrigeration or freezing is highly desirable to avoid deterioration. Since most fat is used commercially, various types of bulk packaging are used. Several consumer packages of chicken fat have been designed employing screw top glass jars. They are marketed in large cities to Jewish people who use it in cooking.

Finely chopped poultry skin is highly nourishing, has excellent flavor and is added to many commercially prepared recipes. Some work has been done to use deep fried chicken skin as a snack food, however it is not marketed on a large scale—if at all.

Frozen duck feet are considered a delicacy in the Far East and are exported from Midwest processors for that purpose.

FUTURE TRENDS

More progress can be expected in genetic development of improved poultry strains. For example work is being done to develop 100 lb turkeys. At this weight they could be cut into retail parts like small red meat animals. The all white meat chicken is not far off.

Improvements in automation through machinery or through handling as is done at Holly Farms can be expected to hold prices down and quality up.

Irradiation sterilization may be perfected.

Further convenience foods and snack foods are a certainty.

EGGS

Like the flesh of the bird, its egg also varies in character and taste depending upon the species. Duck and goose eggs tend to have a strong flavor, but turkey, pheasant and partridge eggs are almost indistinguishable from chicken eggs.

Eggs are prized for their nutritive value and flavor and have been eaten raw or cooked in many different ways from egg-nog to omelette. They are an important ingredient of many other foods including baked goods, candy, and ice cream.

Although eggs have been traditionally a very popular food and although the total consumption continues to rise each year, the per capita consumption in the United States has declined over the past 15 years from about 390 to about 308. The production of over 65 billion eggs requires a large amount of packaging material, yet until comparatively recent years the egg package has remained singularly unattractive and unimaginative. There is good reason to believe this has contributed to the decline in consumption.

Much research has gone into the breeding of poultry for meat or egg production and great strides have been taken in the introduction of automation, integration, and amalgamation in order to lower costs. Small operations have been amalgamated into larger organizations with over a million laying hens. Farms have been vertically integrated from the production of feed and brood stock to the disposition and sale of the main product as well as by-products. Production operations have been made more efficient by the use of automatic machinery and new equipment such as more easily cleaned plastic chicken coops. All of this has helped to hold costs down. Even more effort is required however to stimulate buying through more imaginative packaging and marketing. Little or no effort has been spent on genetic alteration of the egg itself other than recent work aimed at producing a thicker and more uniform shell. One could ask, why not larger eggs, or larger yolks, or consistently double or triple yolked eggs?

Characteristics

Nature has given the egg a natural package—the shell, and an inner liner—the membrane.

Despite its relative strength the egg is a fragile object and must be protected against breakage. In addition the egg is alive and must be given a chance to breathe or it will die. Respiration may be slowed by refrigeration, or by treating the shell (usually by an oil dip) but if it is completely halted, the flavor and texture of the egg will deteriorate more rapidly. Eggs are also susceptible to deterioration from ageing and from attack by bacteria and fungi. Bacterial contamination through microfissures in the shell is minimized by washing the egg. Approved bactericides may be used in the wash water. Careful drying is absolutely essential to avoid mold. Best conditions for storage of eggs in the shell are 85 to 92% relative humidity and 29° to 31°F temperature. Under these conditions undamaged eggs can be stored for 4 or 5 months with little or no change in flavor or quality and for up to 9 months with only slight changes. After 11 months, differences can be detected but whites and yolks still can be separated, a very important factor in some end uses.

Eggs are susceptible to loss of moisture and particularly sensitive to absorption of foreign odors. Quality loss is evidenced through loss in weight (as much as five per cent in nine months), thinner and runnier white with loss of opalescence, and flatter yolk due to absorption of moisture from the white. Very old eggs develop a "straw like" taste probably due to enzymatic changes.

Broken out eggs are sold frozen or dried and are extremely sensitive to bacterial contamination. Eggs are used by pharmaceutical houses for deliberate inoculation and growth of viruses because they are such good hosts.

PACKAGING PROCEDURES AND MATERIALS

Shell Eggs

Eggs traditionally have been shipped to market in flats molded from paper or pulp. The flats may be stacked in wooden crates, in baskets, or in corrugated shippers. Over 260 million pulp filler flats and 52 million corrugated shippers were used in the US in 1965. Although most shell eggs are sold to the consumer pre-boxed by the dozen, a few retailers merely display the flats and allow the purchaser to select and box his own. Eggs may be sold graded or ungraded, if so identified, but higher grades command premium prices. Grading is usually done at the hatchery or at a central distributing point such as may be maintained by a farmers' cooperative. Grading includes candling, weighing, and occasional "breaking out." Top grade eggs have thick opalescent whites, tall domed yolks, and are free of defects such as blood, bits of shell or other foreign matter. In the United States fertilized eggs may not be sold for retail food consumption. In earlier times when eggs were produced in barn-yards rather than in modern factories this was more of a problem and candling was the means of detecting the fertilized eggs.

After sizing and grading, eggs are packed into dozen containers and these in turn into larger shippers. They may be held in storage for reasonable periods of time, but most consumers expect and demand fresh eggs. Prompt delivery to retail markets is an inherent part of the system and most egg packages are marked "strictly fresh."

The traditional package for one dozen eggs has been a folded paperboard box with dividers. This was replaced several decades ago by more rigid paperboard cartons with interlocked dividers and by boxes fashioned from molded pulp. The pulpboard box could be shaped to hold three rows of four eggs, as some of the earlier paperboard cartons were shaped, or in the now more popular two rows of six eggs. The interlocked carton is limited by its construction to the latter configuration. A wide variety of closures has been employed including metal/clips, wire staples, gummed tapes or labels, glued spots, or simple locking tabs.

Decoration of these older packages was particularly uninspired. Paper or pulp was supplied in off whites or dull greys. Inks were prosaic reds, blues, greens, blacks, and browns; and copy was simple line lettering to name the vendor, his brand if any, and the quality of the contents. They offered almost no visibility and little protection. All too often a cracked or broken egg was to be found. Breakage in one carton contaminated neighboring cartons, with resulting loss. Repacking in the store was an added cost as well.

Over the more recent past years attempts have been made to improve egg carton design. Paperboards have been upgraded to make them whiter, stronger, and more printable. Carton designs have been altered to give more support and protection to the egg, particularly during stacking of the boxes. Brighter and more attractive colors have been added. On some folded boxes lithographed process printing has been used to add interest and appeal. Party scenes convey the idea of a deluxe product. Appetizing food dishes suggest tempting uses. Color has also been used to identify grade or size differences.

New Trends

Folding Box.—The latest developments in the folded paperboard egg carton are designs which enable the buyer to see the egg. One approach is to form the box with an open top and apply a shrink film overwrap such as PVC, PVDC, or PE. A 27-point cylinder kraft back board is used with triple end panels and a honey comb-like construction for added stacking strength.

Molded Pulp.—The molded pulp egg carton is fighting for its life against stiff competition from molded plastic foams. Its chief advantage is its price. Its chief disadvantages are lack of strength and unattractiveness. It is possible to increase strength by using denser pulps or by adding fillers or binders with some added cost.

Attractiveness is limited by the inherent color and surface roughness of the material itself. It is also not conducive to fancy printing. One attempt to solve this has been to combine a deep celled molded pulp tray with a printed paperboard lid which can be glued or otherwise fastened on. Another attempt substituted a shrinkable polypropylene or polyvinyl chloride film overwrap and a printed paper label for the aforementioned paperboard lid. This gave strength, visibility, and decoration.

Courtesy of Reynolds Metals Co.

FIG. 52.   SEE THROUGH PACKAGING FOR EGGS

New trends in egg packaging stress the see through advantages of a
heat shrink PVC film overwrap. Here the package is formed from
folded and interlocked printed paperboard.

When tried in the Los Angeles area it outsold the conventional packages
five to one. Another packager put a pulpboard tray of eggs into a preformed, printed, shrinkable polyethylene bag to achieve the same general purpose. Still greater strength was attained with a transparent
"crash helmet" lid made from thermoformed oriented polystyrene. The
lid separated and protected each individual egg and was locked by tabs
to the pulpboard tray.

Molded Plastic Foam Egg Cartons.—A radically new approach in
the packaging of shell eggs came with the adoption of cartons and trays
made from expanded polystyrene foam. Two basic types of material are
used. Molded shapes are created by filling polystyrene foam beads into
a suitable mold and then applying heat and/or steam to expand the
beads into a coalesced mass. The resultant block shape has strength,
provides excellent cushioning, is very light weight, and is a good insula-
tor. Other shapes are thermoformed from a thin sheet of expanded
polystyrene foam. The latter material has a smooth satiny appearance
almost as if it had a thin transparent skin. It is very attractive and is
presently available in snowy white and pastel colors.

Egg cartons can be molded in almost any shape from these materials.
The present chief competitor looks much like the 2 by 6 molded pulpboard container, except for being more attractive. It can be printed with
high quality line graphics in simple designs and patterns. It can be filled

Courtesy of Container Corp. of America

FIG. 53.   FOAMED POLYSTYRENE IS LATEST INNOVATION FOR EGGS

Molded foamed polystyrene egg cartons provide superior cushioning
and protection against odors and moisture. The package is resistant
to fungus and mold growth. Consumers like its attractive luster and
the soft pastel colors.

and closed on fully automatic machinery. Although early versions had
some difficulties with variations in smoothness, stiffness, strength, and
dimensions, most of these have since been overcome. The cushioning
against shock, and the ability to be formed to any desired shape make
this an extremely functional as well as attractive package. Many new
ideas are being tested.

One manufacturer produced a block molded package in a boat shape.
After use of the eggs the carton became a child's toy. To achieve a simple
closure and better graphics, another company has tested a foam tray
with an A-flute printed corrugated sleeve. Others have tried a rigid plastic film sleeve in similar manner.

European designers have experimented with smaller unit packages.
One package featured two molded blocks with matched cavities for six

eggs. The blocks were designed to close with an adhesive but male and female locking grooves or tabs would probably serve as well. Openings permitted ventilation and visibility from all sides, yet the package was stackable and the flat top permitted decoration.

Another European package called the "Econopak" also featured a two part foam box and top. This one held two eggs and two rashers of bacon. A printed plastic outer sleeve completed the package.

Hermetically Sealed Plastic Packages.—Experimentation has been reported with a two-piece rigid polyvinyl chloride carton fashioned in such a way that it could be flushed with nitrogen and then sealed hermetically by pressure. Grade AA eggs were said to have kept fresh and in good quality in this package for up to three weeks. Further research on ideal gas combinations might yield still further advantages.

FROZEN BROKEN OUT EGGS

Although about 89% of eggs produced in the United States for consumption as food are sold in the shell about 6–10% are broken out and sold frozen, liquid, or dried. Broken out eggs are processed in different ways depending upon ultimate end use. They may be frozen whole and later thawed, or they may be homogenized and then frozen. They may be separated into whites and yolks and the parts each separately frozen. Eggs which are to be frozen are usually treated to prevent coagulation or "rubberiness." Ten per cent by weight of salt or sugar or five per cent by weight of glycerol may be added for this purpose. Salt is used where the ultimate use will be in salad dressings, or nonsweet cooked items; sugar or glycerol when the end use is in sweet items such as candy or some baked goods. Whole eggs or yolks are sometimes pasteurized before freezing. Eggs for freezing are put in 30-lb bulk containers. Freezing at $0°F$ may take up to 44 hr. Use of $- 20°F$ blast freezers can drastically reduce this time.

Bulk containers can be made of a variety of materials. Large cans, small drums, spiral wound paper canisters, and the newer molded plastic pails are all available for the purpose. Flexible polyethylene plastic bags in corrugated shippers would also be satisfactory.

DRIED BROKEN OUT EGGS

Dried eggs, again, may be processed as whole eggs, yolks, or egg whites. Drying is done with steam-heated drum driers or spray driers with and without vacuum. Foam spray-drying is a somewhat more efficient method of spray-drying but properties of final products made from the differently dried eggs may vary. The properties of whole eggs and yolks are sufficiently similar that they may be considered together. Both deteriorate rapidly at room temperature if glucose, which is naturally

FIG. 54.   LIQUID EGG PACKAGING

Broken out eggs can be packaged in polyethylene film bags which
will hold 10 lb—the equivalent of 100 medium sized eggs. The automatic filling device shown here measures the liquid eggs into the bag.

present in the egg (about 1.1% of egg solids) is not first removed by
fermentation or enzymatic oxidation to gluconic acid. Oxygen causes
off-flavors to develop and these will vary depending upon whether or not
sucrose was added (20–30% on dry solids) prior to drying. Addition of
sugar (sucrose) also prevents loss of leavening power of dried egg whole
or yolks. This power is what makes a sponge cake "spongy." Dried egg
white also is susceptible to glucose darkening of color and must therefore
be treated in similar manner. Dried whites can lose their whipping or

foaming properties if small amounts of glucose or of egg yolk are retained.

Packaging of dried egg white is principally concerned with protection against moisture absorption, whereas whole egg or yolks must be protected also against oxygen. In the latter case, vacuum or gas packaging is highly desirable. Since very little, if any, dried raw egg is sold at retail, we are again concerned with bulk packaging.

Economics and protection are the prime consideration. Vacuum- or gas-packed cans, cannisters or cartons with hermetically sealed liners, or metal drums are suitable for the purpose.

COOKED EGG PRODUCTS

In most cooked egg products the identity of the egg is lost. For example the egg used in a sponge cake is not identifiable. Consequently no attempt is made to cover such products in this chapter. They will be discussed in other appropriate chapters. There are some cooked egg dishes however where the egg remains identifiable, for example an omelette, or an egg foo yung. Some egg dishes of this nature have been prepared and then frozen. Packaged in foil laminates or foil containers they are readily thawed and reconstituted as convenience meals. The US Army has experimented with considerable success in freeze-drying of many egg dishes.

BIBLIOGRAPHY

ANON. 1966A. Broilers: ships—without ice. Food Eng. *38*, No. 7, 157.

ANON. 1966B. Showcasing the egg. Mod. Packaging. *40*, No. 3, 128–255.

ANON. 1967. Edible packaging offers pluses for frozen meat, poultry. Quick Frozen Foods. *30*, No. 4, 165–214.

ANON. 1968. Increased security for extra-fresh eggs. Emballages. *38*, No. 254, 134–135. (French)

CASH, D. B., and CARLIN, A. F. 1968. Quality of frozen boneless turkey roasts pre-cooked to different internal temperatures. Food Technol. *22*, No. 11, 143–146.

MAY, K. N. *et al.* 1966. A comparison of quality of fresh chicken packed in various containers. Georgia Agr. Expt. Sta. Bull. *N.S.168.*

RANKEN, M. D. 1967. Plastics in poultry packaging. Conference on Advances in Packaging with Plastics. Paper *14*. London

RUSSO, J. 1967. Cartons save 160 man-hours per week. Food Process. Marketing *28*, No. 9, 82–83.

SACHAROW, S. 1967A. Packaging of poultry in plastics-foil. Anyagmozgatás Csomagolás *12*, No. 1, 15–16. (Hungarian)

SACHAROW, S. 1967B. Plastics Laminations for packaging poultry. Industria Avicola *14*, No. 4, 44–48. (Spanish)

SACHAROW S. 1967C. Packaging of food. XIII. Prepacks take over poultry market. Paper, Film Foil Converter *41*, No. 4, 102–106.

TAYLOR, M. H. *et al.* 1966. Evacuated packaging of fresh broiler chickens. Poultry Sci. *45*, No. 6, 1207–1211.

WELLS, F. E. *et al.* 1958. Effect of packaging materials and techniques on shelf-life of fresh poultry meat. Food Technol. *12,* No. 8, 425–427.

WILSON, G. D. 1960. Sausage products. *In* The Science of Meat and Meat Products. Am. Meat Inst. Found. W. H. Freeman and Co., San Francisco, London.

ZEHNER, M. D. 1967. Consumer attitudes toward egg cartons. Michigan State Univ. Agr. Expt. Sta. Quarterly Bull. *50,* No. 1, 84–92.

CHAPTER 5

Milk and Dairy Products

INTRODUCTION

Milk is one of the basic foodstuffs and from it have been derived a variety of processed foods which are known as dairy products. There exists today a frieze that was carved in the third millenium BC which shows priests of Sumer milking, straining, and presumably butter-making at a dairy farm sacred to the goddess Ninhursaga. The Brahman Hindu *"Rig-Veda"* about 1400 BC describes a pastoral society which raised cattle, feasted on beef, and regarded milk as a highly valued food. The Promised Land of the Israelites was to be "flowing with milk and honey." Because milk soured quickly in the warm climates of early civilizations, it is not surprising that butter and cheese making were early discoveries.

MILK

Production

Milk and cream comprise nearly one sixth of the weight of food consumed annually by the average American family. Although more than 1-1/4 billion pounds of milk are produced in the United States, over one third goes to nonfood uses. Of the remainder about 60% is sold as liquid milk and cream and the rest is processed into various products such as butter, cheese, candy, and baked goods. In this country nearly all milk sold is from cows, although some goats' milk is produced. In other countries, goats' milk is more popular and still elsewhere milk is obtained from other animals—such as camels, donkeys, llamas, reindeer, and yaks.

Composition

Whole milk is a complex physico-chemical system consisting of a water solution of salts, lactose, and lactoalbumin. Proteins are colloidally dispersed in the water solution and milkfat solids are present in a partially emulsified suspension. The proportion of milkfat varies with breed of cow, individual, and period of lactation. Fat content according to breed is approximately as follows:

Jersey	5.3%
Guernsey	4.9%
Ayrshire	4.0%
Holstein	3.4%

Protein content remains relatively constant at about 26.5% of total solids and includes casein and the whey proteins. Casein is present as a calcium caseinate: calcium phosphate complex and can be precipitated at pH 4.7 by addition of acid. It is used in the manufacture of industrial adhesives and plastics. The whey proteins are albuminous in character. Both they and casein are primarily amino acids high in nitrogen.

Another major constituent of the watery portion is lactose or milk sugar. Lactose is easily fermented by lactobacilli into lactic acid, a principal ingredient of sour milk.

The fatty portion of milk is a mixture of triglycerides of saturated and unsaturated fatty acids. It also contains small amounts of cholesterol and lecithin type phospholipids. The latter have high emulsifying properties.

Characteristics

The glycerides of the lower fatty acids (butyric, caproic, capric, caprylic, lauric, and myristic) are mainly responsible for the characteristic flavor of milk fat, and the deterioration of these fats causes most off-flavor problems. Rancidity is caused by the lipase enzymes which hydrolyze the glycerides and release the strong flavored and strong smelling fatty acids. Pasteurization inhibits the lipase action and sterilization destroys it. Oxidation of pure milk fat produces a tallowy flavor probably from the formation of hydroperoxides. The reaction is accelerated by heat, acid, and metal catalysts—particularly copper.

The distinctive "oxidized" flavor of milk appears to be caused by oxidation of phospholipids. Another fishy off-flavor results from hydrolysis of lecithin to produce trimethylamine.

The vitamin content of milk is affected by exposure to light, heat and oxygen. Ascorbic acid (vitamin C) is deteriorated by heat and oxygen. Thiamine (vitamin B_1) is partially destroyed by heat. Vitamins B_2 and B_6 are destroyed by light. Vitamins A, D, and E are fat soluble and heat stable.

Since milk is an animal product and an excellent growing medium for bacteria, it is extremely important that it be produced under sanitary conditions. Prior to the legal enforcement of sanitary codes, diseases such as typhoid fever and tuberculosis were transmitted through distribution and sale of milk. Milk must now be produced from healthy animals under strict sanitary conditions.

In most markets pasteurization is compulsory. This consists of heating the milk to 145°F for 30 min or to 162°F for fifteen sec. Homogenization is done after the temperature reaches 135°F but before it reaches pasteurization temperature. Pasteurization destroys all pathogenic bacteria in the milk without much effect on its nutritive value.

Homogenization reduces the fat globules in size to about one tenth of their original diameter. This slows the rate of flotation of the fat in milk and cream.

Packaging of milk therefore should be designed to protect the product from contamination by dirt or bacteria and from the effects of light and oxygen.

Packaging Materials

Glass.—Until 1950, almost all milk was packaged in glass bottles. Milk was sold to consumers mainly by home delivery or through supermarkets in one-quart bottles. The increase in supermarket shopping and the decrease in home delivered milk have served to lessen the usage of glass milk bottles. In 1961, home delivery of milk in the United States accounted for about 40% of all milk sales. In 1965, home delivery accounted for less than 30% of sales. Glass bottle usage correspondingly decreased from 50% to 40%. Housewives favor a lighter package for the supermarket, while in home delivery, the heavier glass bottle reigns supreme. By 1970, the use of glass milk bottles is expected to decline to about 10%. In spite of decreasing US usage, glass milk bottles are still the basic package in many other lands. The glass bottle holds about 95% of the British market, 93% in Australia, 60% in Finland and other European countries where home delivery is still widely used.

Glass milk bottles originally were round, tapering to a rather wide mouth with a thick flange. The move to a squared body saved considerable space in the home refrigerator. Quarts, pints, and half pints were once the standard sizes, however there is now an increasing trend toward half gallons and gallons. Glass bottles average about 50 trips and are packed at rates up to 24,000 per hour. In Finland milk is packed in brown glass to filter out harmful light. US housewives will not accept the brown color despite the added protection it offers.

Bottle closures are formed from aluminum foil, high density polyethylene, polypropylene, and paperboard. Most closures are applied by automatic machinery at high speeds. Aluminum foil closures are supplied in reels about 2 inches wide and weighing 12 to 17 pounds. The foil is 2 mils thick and 1 lb is enough to cap 1300 bottles. Printing is applied prior to use.

Paperboard.—The introduction of wax coated paperboard captured the milk market in the late 1940's. Although consumers were attracted to the concept of a disposable milk container, some problems existed. Wax particles were common in the milk. The outer surface had a cloudy unattractive appearance. Although a few polyvinyl chloride coated cartons were tried, the introduction of polyethylene coated paperboard, in the early 1960's, solved nearly all the problems inherent in milk packaging. It was disposable, clean looking, and functional. In the United States about two-thirds of all milk is packaged in polyethylene coated paperboard cartons. Most retail sizes sold are quarts and half-gallons. Gallon paperboard cartons have a tendency to leak. Cartons of various types account for 85% of the German market, 75% of the Swedish and small percentages of the British and Australian market. Several basic types of cartons are marketed.

Plastics.—An all plastic milk bottle is light weight and tougher than its two competitors. In addition, it creates high impulse appeal and allows the milk to be seen. Disadvantages include difficulties in connection with printing, labeling, and various decorating techniques. The basic material in general use for an all plastic milk bottle is polyethylene, although several polystyrene bottles are also available.

American development of the all-plastic milk bottle has greatly accelerated during the last three years. Research is currently under way in the area of improved manufacturing technique. Blowmolding machines are being investigated for shorter cooling cycles. Filling systems are under study in order to meet present day speeds, the current range for plastic being only 100–200/min.

Package Forms
Milk Cartons.—Milk is produced by the lacteal glands in the udder of the cow. It is drawn from the cows teat canals by automatic milking machines using a pulsating vacuum principle. The milk goes directly into a bulk holding tank which is refrigerated at 40°–50°F. After a maximum storage of two days, it is transported to the milk processing plant. In some instances, the milk goes to a central receiving station where it is blended with high fat and low fat milk. It then goes to a processing plant.

Smaller farms ship the milk in ten-gallon milk cans. Most larger farms utilize bulk road tankers. These tankers are made of insulated, stainless steel which can hold up to 5,000 gallons.

At the processing plant, tests are conducted to control the milk quality. Representative tests run are fat and total solids, sediment, bacterial counts, freezing point, and evaluation of milk flavor. In some cases, antibiotic and pesticide residues are tested. The milk is held in refrigerated (40°F) vats and may be blended to a specified fat content.

The next step consists of classification. A centrifugal classifier is used. This machine removes sediment, body cells from the cows udder and some bacteria. The milk is then ready for pasteurization.

Pasteurization is necessary to kill any disease producing organisms such as *Mycobacterium tuberculosis*. Two methods are used to effect pasteurization: the batch or holding method (145°F for 30 min) and the high temperature-short time method (HTST) (161°F for 15 sec). Batch pasteurization is conducted in heated covered vats equipped with an agitator and thermometer. HTST methods employ a more complex system including heating plates, holding tube, flow diversion valve and time-temperature recording charts.

In order to check the effectiveness of the pasteurization method, the phosphatase test is run. Milk is incubated with disodium phenyl phosphate and 2,6 dichloroquinone-chlorimide is added. A blue color indicates improper pasteurization or contamination.

Homogenization generally follows pasteurization. This disperses the fat globules and insures an even-bodied product. Large fat globules are sheared in the homogenizer and reduced in size by about ten times. They become evenly distributed and no longer rise to the top of the milk.

The milk is then cooled and packaged in wax or polyethylene coated paperboard cartons. Many different cartoning lines exist and varying carton designs are used. A unique system is the Perga carton patented by the Jagenberg Company in Germany. The Perga carton consists of a two-piece container which is supplied to the dairies preformed, erected and nested in columns. Specially designed machines are used to fill and seal these cartons at rates up to 2,400 per hour. Since the filling and sealing operations occupy a small area and the machines are economical, this package offers definite cost advantages especially when there are fluctuations in consumption. The Perga carton is available in 1/3, 1/2, and 1 pint sizes. It is widely used in Australia.

In the United States, one of the major packages used is the "Pure-Pak." This is used by the dairies as pre-cut blanks which are formed, filled and sealed on one machine. The package is formed on automatic machinery, milk is filled and the familiar gable top is used as the seal.

Pre-formed.—Preformed cartons are supplied to the dairy in a fully erect form and ready for filling. The "Perga" carton is an example of a preformed carton. Both waxed and polyethylene lined paperboard are used. Paperboard caliper is 0.009 in. for the third and half-pint sizes and 0.011 in. for the one pint "Perga" size.

Pre-cut.—In a pre-cut carton system, printed coated paperboard blanks are supplied in a knocked down shape. The final carton is set up, formed, filled, and sealed on one machine. "Pure-Pak," "Seal-Right," "Blocpak," and "Tetra-Rex" cartons represent variations of pre-cut cartons. Both rectangular and gable tops are used.

Post-formed.—Post-forming uses rollstock, and forms, fills and seals in one continuous operation. The system may use polyethylene or foil laminated paper. The "Zupack" is a rectangular block while "Tetra-Paks" are tetrahedron shaped. If an aseptic unit is to be prepared with a "Tetra-Pak" shape, aluminum foil is included in the lamination.

Courtesy of Imballaggio, Milan, Italy (from an article by Dr. O. Pasquarelli)

FIG. 55.   MILK CARTONS

A tetrahedral package for milk is shown here beside a rectangular
carton. They are named "Tetra-Pak" and "Tetra-Brick"
in Europe.

Courtesy of Imballaggio, Milan, Italy (from an article by Dr. O. Pasquarelli)

FIG. 56.   ANOTHER EUROPEAN MILK PACKAGE

The "Zu-Pak" in 1/4-liter and 1-liter sizes is used commercially
in several different European countries. This package is formed
from polyethylene coated paperboard and is capable of standing
unsupported.

Rigid Plastics.—The recent increase in half-gallon and gallon milk purchases has accelerated the trend toward plastics. Housewives now prefer larger sizes and the number of quart sizes sold is steadily decreasing. Glass bottles in sizes larger than quarts are cumbersome and heavy and, as previously noted, gallon paperboard cartons tend to leak. Because the all-plastic bottle is more expensive than paperboard or glass

Courtesy of U.S.I. Chemicals Div. National Distillers and Chemicals Corp.

FIG. 57.   BLOWN RIGID PLASTIC CONTAINERS FOR MILK

Light weight plastic gallon jugs for milk feature molded-in handles making them easy to grasp.

bottles, it is impractical in quart sizes or smaller. In order to give the consumer even more convenience, several dairies are now offering 6- and 8-quart all plastic bottles. Haskon has introduced a six-quart bottle which closely resembles a conventional "Tapper."[1] A spring-activated device enables the consumer to obtain milk by pressing a button. The

[1]Registered trademark, Reynolds Metals Company, Richmond, Virginia.

container lies on a refrigerator shelf readily accessible. Several larger size bottles are designed in a conventional manner, but incorporate graduations, helping the consumer to judge the amount of milk remaining in the container.

When the all plastic bottle was initially introduced in the United States, closures and design were the basic problems. Caps did not fit satisfactorily and consumers felt that the handles were too small for regular usage. Handles have been improved by the introduction of superior manufacturing techniques. They have been made wider and smoother, permitting easy holding.

Capping and filling procedures must be modified in order to introduce a plastic bottle. The first bottles were closed with simple paperboard plugs. This was the closure method used on glass bottles; however, leakage occurred and closures were faulty. Metal screwcaps then appeared and were capable of preventing milk spillage. The cost of the metal cap was high and its use was restricted in several US states. A recent innovation in closure devices is a plastic top made of polyethylene with a diaphragm. The cap is applied to the bottle and the diaphragm is in the center of the cap. When the diaphragm is depressed, the cap expands, causing a tight fit. It is tamperproof and may be produced at high speeds. In addition, a paper label can be used for brand identification. Aluminum foil screwcaps and reclosable, heat-sealable caps are being developed and will soon be available.

Distribution.—A dairy considering the introduction of a plastic bottle can choose from several methods. The simplest approach is by direct purchase of ready-made bottles from a local blow-molder. No capital investment by the dairy is necessary and deliveries can be fairly rapid. Disadvantages include expense, and dependence on the supplier for functional bottles. The dairy may produce bottles as an in-plant operation. Equipment is purchased and the capital investment is rather high. In addition, personnel must be hired to run the machinery. The obvious advantages are low raw material cost and the capacity to supply a sufficient demand for daily milk production. The dairy may contract the services of a bottle manufacturer. An agreement is signed as to specific quantities and the bottle manufacturer installs the machinery in the dairy or in an adjacent plant. The entire program is managed by the supplier, but the total production is captive to the dairy. The dairy may purchase partially fabricated bottles and complete the manufacture in its own plant. Not all bottle designs are conducive to this approach. Freight costs are reduced since only partially formed containers are shipped. Polyethylene at present costs about 19 cents per pound in the American market. In general figures, a gallon milk bottle will cost about 5–7 cents while half-gallons average 3-1/2–4-1/2 cents. The obvious factors involved in determining final bottle price are the method of supply and the volume desired. Since most bottles are produced by a blow-molding technique, it is economically important to maintain at least four molds per machine. A greater number of molds will serve to reduce the final price. Thus, a dairy must carefully investigate the source of supply—whether in-plant or by an outside concern.

Returnable Plastic Bottle.—The higher competitive cost of an all plastic bottle compared to paperboard and glass has led to the development of a returnable plastic bottle. The problems inherent in a returnable polyethylene bottle are odor absorption and hydrocarbon contamination. One patented system incorporates a contamination detector. A highly sensitive electronic unit detects volatile organic contaminants immediately after the washing station. If contamination is found, the washed bottle is rejected and destroyed. This system is in use with several dairies and has received US Public Health clearance. It has been reported that 150–200 return trips are attainable. Even then, most rejects are based on physical damage rather than odor or organic contamination.

New Trends

All the above approaches are in current use by US dairies. The final choice depends on a thorough study of economics and production. A dairy must consider the most expedient approach and maintain a steady supply of bottles. A standardized design is also under consideration by many dairies. Most American dairymen feel that the 1 gal. bottle will be all plastic in the next two years, while plastics will penetrate at least 50% of the 1/2 gal. market. If the costs are reduced, a functional quart size can become a commercial reality in the near future. Since consumer demand often brings reduction in cost, the future looks promising. In France, an in-line blow molding, filling, closing and crating packaging system called "Sidel" is in use.

Flexible Plastic Pouches.—Liquid milk also may be packaged in plastic film or laminated pouches. The packaging of milk in pouch form was pioneered in France and has subsequently been introduced to several other European countries. Pouches offer economy, compact storage, and ease of disposal. Disadvantages include the need for support and an unconventional appearance. Since clear plastic does not offer adequate shelf-life, an opaque laminate is required. Most all plastic film pouches for milk are prepared from two-ply, low density polyethylene lay flat tubing. The outer ply is white and the inner ply black to protect the milk from ultraviolet degradation. The lay flat tubing is made by extruding two polyethylene resins through a coaxial die and then passing the two films through a second die to produce a .004 in.

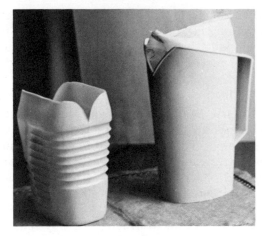

Courtesy of Imballaggio, Milan, Italy (from an article by Dr. O. Pasquarelli)

Fig. 58.   Flexible Plastic Milk Packages

An example of a flexible polyethylene bag used for milk. The molded rigid plastic pitcher makes it possible for the package to stand up and facilitates pouring.

laminate. On the packaging machine the tubing is sterilized by ultraviolet irradiation, cut to the desired length, sealed on the bottom to form a pouch, filled, and heat sealed at the top to close. Four filling systems are currently available—the "Polipack," "Prepac Polymaid," "Pintapac," and "Doypac." The pouches are fairly popular in Holland, France, and Finland. A specially designed holder must be used to support the bag upright after opening and during storage.

Paperboard-Plastic Liner.—In recent years, a "Bag-In-Box" concept utilizing a low density polythylene bag in a corrugated container has become popular. Both the bag and box are completely disposable. The normal size used is five gallons. The inner ply consists of either single or double ply .002 in. polyethylene. A spout is heat sealed into the bag and a plastic valve may be added for dispensing purposes. The bag collapses in use as the level of liquid falls. The corrugated fiberboard container is printed and used as the outer package. Systems available include "Pergall," "Polygal," "Liqui-Box" and "Cubitainer."

Aseptic Milk Packaging

Aseptic or "long-life" milk was originally introduced in Sweden in the early 1960's. Called the "Tetra-Pak" system it utilizes a laminate presterilizer and a filling environment heater. Aluminum foil is an integral part of the flexible laminate in order to provide a barrier against light and gas. Cow's milk is pre-heated to 165°–185°F then rapidly raised to

Courtesy of E. I. du Pont de Nemours and Co., Inc.

FIG. 59. BAG-IN-A-BOX PACKAGE

An example of a bulk container for milk to be used in the home. The polyethylene film bag with molded spout is supported by a rigid corrugated carton. Exterior of the carton is coated with an "Elvax"-wax blend to help prevent moisture weakening. Tear out panels permit user to see level of milk remaining.

285°–300°F. It is held at the latter temperature for 2 to 4 sec and then suddenly cooled by flashing into a vacuum chamber. It must be packaged under completely sterile conditions. No refrigeration is necessary for at least three months. If kept under refrigeration, a shelf-life of up to one year is possible.

Canned Milk Packaging

Sterilized canned milk is produced by using lacquered tin-plate cans which have a special seam stronger than a conventional can. They must be able to withstand the stresses imposed by high temperatures used to sterilize the contents.

Condensed and Evaporated Milk

Condensed milk is prepared by evaporating milk in a vacuum at 60° F. Sugar is added and the total water content is reduced by 40–50%.

Evaporated milk is essentially an extension of the same process, but contains no sugar.

Both the sweetened and unsweetened types are packed in tin plate cans of varying size. After canning, they are heated like all other canned foods. Sweetened condensed milk holds for a longer time after the can has been opened and has the consistency of thick cream. Both types require refrigeration and may be reconstituted by the addition of water.

CREAM

Cream consists of the intermediate stage between milk and butter. The aqueous phase has been reduced but the emulsion has not been broken.

Since cream contains a high percentage of butterfat, it is very susceptible to spoilage. In addition, it must be protected from water loss. The shelf-life of refrigerated creams ranges between one and two days without proper protection.

Packaging consists of units similiar to those used for milk, i.e., polyethylene coated paperboard. Since cream is used in lesser amounts, sizes sold are 1 pt and smaller.

Newer concepts include portion-control thermoformed packs made from linear polyethylene, polystyrene or polypropylene. These may be closed with a peelable lid or snap-on cover. Tinplate containers have also been used for larger sizes.

Whipped creams and synthetic formulations are sold in aerosol cans and polyethylene tubs with snap on lids. The recent introduction of many synthetic formulations may spur on the introduction of new packaging concepts. Imitation "cream" made from soybeans and vegetable oils is often marketed in wax-coated paperboard cartons. These products were originally introduced for the Jewish market and for people suffering from allergies to cow's milk.

Powdered coffee whiteners made from vegetable fat, protein, sucrose and other ingredients are packaged in amber-colored glass jars with a heat-sealed membrane and screw-top or in small individual sealed packets of coated paper. Powdered whiteners are usually quite hygroscopic especially if they contain a high level of corn syrup solids. A suitable moisture barrier is essential for proper packaging.

Frozen coffee whiteners are packaged in PE coated "Pure-Pack" type cartons.

ICE CREAM

Wealthy Romans transported snow from the mountains and served it flavored to their guests. Sherbet was an Arab and Persian favorite and may have originated in China. The Chinese are believed to have taught the art of ice cream making to the Indians, Persians, and Arabs, and they in turn brought it to Italy where it remained a favorite of the wealthier classes. In 1533, Catherine de'Medici journeyed to France at the age of 14 to marry the heir to the throne. Her Florentine cooks made "iced cream" popular in the French court and by 1666 ice cream making was an industry. It was first served in the United States in 1700 by Governor Bladen of Maryland.

Properties and Characteristics

Ice cream is prepared by mixing milk, cream solids, sugar and a stabi lizer. It is carefully pasteurized and homogenized for sanitary reasons as well as for texture. Flavorings are added and then the mix is frozen. The frozen mixture is whipped to incorporate air and develop a smooth texture. The amount of whipping is carefully controlled to achieve a desired overrun (volume increase). Bulk ice cream is then packaged and hardened at a low temperature (e.g., −50°F). Shaped bars are hardened prior to packaging. Rapid hardening is essential to prevent formation of large ice crystals. Off flavors or poor textures will occur in ice cream only if improper procedures are followed during its manufacture, storage, and handling.

A large amount of ice cream is consumed in the United States. Annual consumption is more than four gallons per capita or nearly 800 million gallons per year. Sherbet ices represent only about five per cent of the total.

The widespread growth of supermarkets, coupled with larger home freezers, has resulted in a steady increase in retail ice cream sales. Many retail packs, however, are deficient in consumer convenience and product protection.

The chief requirements of packages for ice cream are protection against contamination, attractiveness, ease of opening and reclosure, and ease of disposal. Protection against moisture loss and temperature fluctuations is desirable.

Package Form

Most bulk ice cream is packaged in a linerless, bleached sulphite board carton, coated with wax or polyethylene-wax blends for protection from moisture and oxygen. Once the carton is opened it is difficult to reclose and the paperboard tends to warp. Home freezers do not maintain adequate temperature stability for most packs, and condensation and ice accumulation cause deformation. Although economic considerations favor the simple rectangular paperboard carton, improved packaging often leads to higher sales, and a marginal price differential may be offset by higher throughput.

Several distinct new packaging concepts have appeared during the last few years. For example, multi-packs are on the increase for sticks, bars and cups. In 1965, they were used for 37.6% of novelties, and a year later for 46%, package innovations being partly responsible for the expanding market. Six-packs in polyethylene bags, bars in foil cartons and parfaits in plastic cups appeal to the housewife.

The huge half-gallon and gallon market, on the other hand, demands superior packaging. Consumers complain of poor recloseability, desiccation and oxidative darkening, and shippers are concerned with maintaining constant temperature in order to prevent heat shock.

Aluminum Foil Cartons.—Aluminum foil is a very effective barrier to light and radiant heat. As a result of surface heat reflection, bulk ice cream in foil cartons is kept at lower temperatures than that in paperboard cartons, and it has also been reported that the ice cream takes longer to thaw. An additional advantage is the attractive appearance provided by printed foil.

Cylindrical Containers.—Many pint and 1/2-pint packages are constructed of cylindrical paperboard materials with a variety of covers and coatings and all offering an easily opened and recloseable package with superior storage properties. Cylindrical containers cost more than rectangular cartons, but consumers want a convenient pack.

Spirally-wound containers are made by pulling materials over a cylindrical mandrel at an angle to the winding mandrel. A container can be tailor-made for a product by incorporating various different plies. Ice cream packs usually consist of plastic or wax coated containers with window-lid covers and metal reinforced edges.

Plastic Containers.—The most recent trend in bulk ice cream units has been all-plastic cylindrical containers. Most of these are made of polyethylene and have recloseable lids. The inherent properties of polyethylene restrict striking graphical design, and the cloudy appearance makes printing difficult, but the recent introduction of electrostatic printing may lead to design improvement in the future.

Other plastics that can be used are polystyrene, PVC and polypropylene. Rigid PVC containers offer clarity and excellent graphical design, but require FDA food approval. However, PVC is widely used in Europe and may soon become acceptable in the United States.

Convenience Concepts.—It has been proved that consumers will pay for package convenience. One ice cream manufacturer is offering a pail-type container for a gallon pack, and zip tapes and easy tear features have been incorporated in half-gallon packs. Other ideas may include tack down lids, folding top improvements and foil overwraps. The ultimate aim is to prevent product deterioration at a reasonable cost.

BUTTER

The origins of butter date to the earliest civilizations such as Sumer in Mesopotamia and the Aryans of India. It was long deemed a sacred food to be prepared only by priests and consumed by the priesthood and royalty, but inevitably became a staple food for all. Monarchs such as the Persian King Xerxes received it as tribute. Wealthy Egyptians had it buried with them in their tombs.

Over a million pounds of creamery butter are produced annually in the United States despite inroads made by butter substitutes.

Compositions and Characteristics

Butter consists of the fat which has been separated from the aqueous phase of milk. About 15% by weight of water is trapped in the butter. Because of this high moisture content butter, unlike solid fats, is susceptible to mold growth. Flavor and odor are easily affected by absorption from other materials or through deterioration of the butter by rancidification. The package must provide protection against these factors. Mold growth is inhibited by exclusion of air, and addition of salt during manufacture. Tight intimate wraps are used to help exclude air from the product. Materials and manufacturing areas must be kept sanitized to reduce the population of mold spores.

The package should also protect against moisture loss, as too much will result in color darkening as well as short weight.

Flavor and odor preservation is a critical requirement. Refrigeration is imperative in order to avoid rancidification. Light and oxygen promote photochemical oxidation. It is therefore desirable that the package be opaque and preferrably a high barrier against oxygen and foreign odors.

Long Term Storage Conditions

For long term warehouse storage, butter must be wrapped in impermeable materials in order to eliminate odor pickup and must be stored under refrigerated conditions. In paperboard cartons odor transference may occur. The only satisfactory method for eliminating odor damage is lining the carton with an odor barrier coating, or wrapping in an impermeable lamination. Superior materials to parchment are needed. Foil laminations are essential for this purpose.

Packaging Materials

Parchment.—The most commonly used butter wrap is probably vegetable parchment, which is sterile as produced and free from mold spores. Most butter manufacturers use a plasticized grade in order to meet the high speed packaging machine requirements. If parchment is used, it is essential that it not contain excessive numbers of microscopic pinholes. Also, it must be stored properly. It should contain not more than nine per cent moisture and must be stored in clean, dust-free places at room temperature and 50–80% humidity. If stored too dry, it becomes brittle. When stored in a damp area, it may develop mold and transfer spores to the butter. Even though parchment paper is a grease barrier, it is not sufficiently impermeable to prevent oxygen penetration, and its transparency allows some light to penetrate which promotes oxidation. For superior product protection and longer term storage, laminates of aluminum foil are particularly useful.

Aluminum Foil Laminates.—Aluminum foil provides opacity and barrier properties. The material generally used to package butter consists of a thin gauge aluminum foil laminated to paper. The paper is added to effect better machinability and to strengthen the thin gauge foil. In Europe, a lamination of aluminum foil and parchment is used. The parchment provides wet strength and a sterile surface. Equally effective are laminations of foil and greaseproof paper and foil and tissue paper. The foil can be surface printed and overcoated with a scuff-resistant material. One advantage in printing aluminum foil as contrasted to parchment paper is ink transfer. Parchment paper must be printed with fat-insoluble inks. If the wrong ink formulation is used, ink might migrate through the occasional pinholes in parchment and cause discoloration.

The use of foil-plastic laminates is not recommended, because many plastics contribute off flavor or exhibit insufficient grease resistance. Economics for foil-plastic laminates are not favorable.

Recent developments in smooth-walled rigid aluminum containers offer butter producers a means for obtaining aesthetics as well as convenience. These packages are smooth in shape, colorful and hygienic. Half-pound or pound units with printed or unprinted plastic or heat sealable foil lids can be used at the table. Filling can be done during the peak season when milk is in glut supply and with proper storage and distribution, the packages can be sold throughout the year. Gourmet types of butter flavored with garlic or spices are also being marketed in formed foil containers.

Another novel foil package won the "Eurostar 1966" award with the following citation:

This package for margarine and butter can create a selling image for the product which remains on the table of the user until consumption. Although originally designed for margarine (or butter), the concept is equally applicable to other foods.

The citation package concept was "Cekacup" manufactured by Ceka-Christenssons Maskines and Patenter AB (Sweden). It too is a table ready convenience package. It consists of a carton and tray combination. The tray has a vacuum formed PVC inner portion and an aluminum foil laminate cover. The carton has a molded PVC lid. The entire package can be preformed by the carton producer. The butter packer needs only to fill and seal the package.

Portion Control

For further convenience in butter packaging the use of small portion control packages in formed foil is popular, particularly for air travel food service and for fine restaurants where a quality image is desired. Butter patties in iced water dishes leave much to be desired as to handling and sanitation. Paper trays with release-paper covers are also poor packages as they do not provide full sanitary protection. Machines are currently available for forming .001 in. to .002 in. thick aluminum foil into shallow (3/32 to 3/16 in.) trays. The trays can be colored or embossed if desired. Covering lids can be heat-sealable aluminum foil, PVDC coated cellophane, or other suitable barrier materials. Covers are heat sealed in place automatically after the butter patty is dropped into place, giving complete sanitary protection as well as convenience and quality protection. The full market for this concept has not been fully exploited. It undoubtedly will become available as a consumer item before long. Petersen and Fisker (1966) made an excellent summarization of the various materials used and proposed for individual butter packages.

Foreign Packages

The use of foil-wrapped butter has created increased sales on the Australian market. In an effort to improve the demand for butter and to fight the inroads made by margarine, two Australian butter producers recently introduced foil parchment wrapped butter. Both firms reported sharply increased sales as a direct result of the packaging. Marketed in the Australian state of Victoria by the Grippsland and Northern Cooperative Co. and Holdensen and Neilsen, the foil packages are priced competitively with conventional parchment wrapped butter. Even though the foil laminate cost three times as much as the standard parchment wrap to the producer, total sales increased by 20% in the first few weeks of introduction. The program was instituted specifically to boost the declining per capita sales of butter. Since foil wrappers are used extensively in Western Europe, the Australian producers felt the concept would offer advantages on the Australian market. The success has caused Grippsland and Northern to closely examine additional products including salted butter and new cheese concepts. Holdensen and Neilsen are using butter made by the Contimab machine. Grippsland and Northern utilize standard butter from country co-op factories. The con-

Courtesy of Chemische Werke Hüls Aktiengesellschaft

FIG. 60.   BUTTER GETS BETTER PACKAGING

Rectangular plastic tubs are filled with butter, hermetically sealed
with a foil laminated membrane, and closed with a protective clear
plastic lid, all automatically.

tinuous process used by H and N is claimed to give a better texture and
more easily spread product. The Victorian market is the keenest for
butter sales in Australia. There are six brands of butter as well as margarine. Both companies are still marketing their parchment wrapped
brands and the total sales of all wrapped butter is steadily increasing.

GHEE

In India and Pakistan butter is not used in the same way Western Countries enjoy it. Instead it is produced from cow or buffalo milk which has first been allowed to ripen (partially sour). The curd may be separated at this point and sold as *dahi,* or the curdled milk may be churned to produce butter. When a batch of butter has been accumulated—usually with some development of rancidity, it is converted to *ghee* by heating it at 212° to 234°F until all the water has boiled away and the curd turns a yellowish brown color. The fat is strained and then cooled slowly to 80° to 104°F when it is poured into containers and cooled slowly to room temperature. The final product is essentially 100% fat, yellow in color, salt free and rather strong in flavor. (Buffalo ghee is nearly white and has a higher melting point.) A more bland flavored ghee is accomplished if the butter is merely heated to 212°F and then allowed to stand until the water and curd particles settle out. The ghee is decanted and cooled.

Due to the high price and shortage of ghee these countries are now producing a substitute product called *Venaspati* made from hydrogenated fats (see Chapter 8).

Ghee is often made in the home and consumed from home containers of glass, earthenware, or tinplate. Commercial packages are usually tinplate.

CHEESE

No one knows who made the first cheese. The ancient Greeks revered cheese so highly that they believed it to be a gift of the gods. According to an ancient legend, it was discovered when an Arab merchant carried some milk in a pouch made from a sheep's stomach across the desert. The bouncing of the camel, the heat and the chemical action of the pouch caused the milk to separate into curd and whey. The whey satisfied the Arab's thirst and the curd was cheese.

Cheese is frequently mentioned in the Bible. It made its way from Asia and Africa to Rome and then carried by the Roman legions, it was introduced to Europe and England. The monks of Europe improved cheesemaking and in the 10th century, Italy became the cheese center of Europe. Gorgonzola was made as early as 879 in Milan. Roquefort was first mentioned in 1070 in the records of a monastery.

The Pilgrims included cheese in their supplies in 1620. In 1851, the first cheese factory in the United States was built in New York. Modern cheese technology started around 1870 when a commercial rennet preparation was offered for sale in Denmark. Dairy scientists perfected more sophisticated methods for cheesemaking and in recent years, significant advances have been made in mechanization and packaging.

Characteristics

Cheese is defined as the curd of milk (coagulated by rennet) separated from the whey and pressed into a solid mass. Most of it is made from cow's milk; however, smaller quantities are made from the milk of goats, ewes, camels, reindeer and buffaloes. Cheese is normally made by coagulation of the milk by rennet and lactic acid. The curd is then broken and most of the whey is removed. The partly dried curd is then ripened and sold as cheese.

Cheddar Cheese.—In making cheddar cheese, the curd is formed under controlled conditions of acidity, temperature and rennin concentration. Pasteurized whole milk is added to a vat together with a starter culture of *Streptococcus lactis*. After 30 min, rennin is added in a dilute acidic solution. The milk is allowed to set and stirring is stopped. Curd forms in about 30 min.

The curd is cut into small cubes (1/4 in.–1/2 in.) with special curd knives that are formed from wires strung along a frame. The cubes are then heated with continued stirring to separate the whey. The whey is drained off and the curds fuse together to form a cohesive rubbery mass. Next, the matted curd is cut into blocks. These blocks are turned at 15 min intervals and piled on one another 2 or 3 deep. During this time, additional whey is drained off and acid formation continues. The blocks are then cut into smaller pieces, salted, stirred, drained and pressed at about 20 psi pressure overnight.

After pressing, the cheese is removed and placed in a cool drying room at 60°F and 60% rh for 3 to 4 days. In order to prevent mold formation, the cheese wheel or block is dipped in hot paraffin then it is boxed and placed in the curing room for ripening.

Ripening takes place for at least 60 days and for a peak flavor it may be continued for 12 or more months.

Packaged chunk cheeses, such as cheddar, are usually marketed in gas or vacuum packaged pouches. A commonly used machine is the Hayssen RT which makes and fills flat pouches for gas packs. Rates of packaging are 60–100 per min and the machine can also handle sliced cheese. Another machine used for packaging is the Hudson-Sharp. This unit forms and fills flat pouches for vacuum packaging at speeds up to 20–40 per min. It will gas flush at 60–100 per min. Sliced and chunk cheese each require a different machine.

Types of Cheese

Natural cheese is made directly from milk, with or without ageing and ripening by bacterial action or molds. More than 400 different types of natural cheese are produced.

They can be categorized according to hardness and method of manufacture as follows:

(1) Low moisture (30%) very hard, ripened by bacteria,
    examples—Sapsago, Romano, Parmesan
(2) Moderate low moisture (38%) hard, ripened by bacteria,
    examples: (no holes) Cheddar, Gouda, Edam, Provolone
    examples: (with holes) Swiss
(3) Medium moisture (45%) semi-soft,
    examples: (ripened by bacteria and surface microorganisms) Hard, Trappist, Limburger
    examples: (ripened by bacteria) Brick
    examples: (ripened by interior blue mold) Roquefort, Blue Cheese
(4) High moisture (50–80%) soft,
    examples: (ripened) Camembert
    examples: (unripened) Cottage, Cream, Neufchatel, Ricotta, Mysost

Processed cheeses are produced from natural cheeses blended for uniformity of flavor, texture and cooking quality. They are ground together, melted, pasteurized and poured into molds. The first processed cheese was produced by Kraft in 1904. Examples of processed cheese include American Cheese, and processed Gruyere. Because of the addition of preservatives, processed cheese is much more stable than natural cheese. Process cheese spreads contain a higher moisture content to produce a more spreadable consistency at room temperature. Both spreads and solid processed cheeses may include other additives such as milk solids, wine flavorings and spices.

Packaging

In 1963, US cheese consumption amounted to over 1.5 billion lb of which 55% was packaged in laminates of paper, film and foil. Over 40% of these cheese packages involved the use of plastic films. The remaining packages consisted of cans, boxes and plastic containers.

Regional packers still handle a share of the cheese business; however, there is a trend toward centralized packing and long distance shipping of cheese products to points of sale. There is a need for durable materials and laminates which will withstand the rigors of rough handling incurred during shipment. Consumers also continue to demand more convenience features in cheese packaging.

Packaging Requirements of Natural Cheeses.—Any material to be used for packaging natural cheeses must (1) afford general protection, (2) prevent moisture loss, (3) improve appearance, (4) protect against microorganisms, and (5) prevent oxygen transmission.

Moisture vapor and oxygen barriers are critical in natural cheese packaging applications and as a result, materials used to package natural cheese all involve the use of polyvinylidene chloride (PVDC) to some degree. PVDC gives the final composite extremely low oxygen and moisture vapor transmission and, in combination with polyethylene (PE), the moisture transmission rate becomes even more insignificant.

Most natural cheeses require an oxygen barrier which will pass no more than 5 cc/100 sq in./24 hr at 73° F and 50% rh relative to oxygen.

An evaluation of a number of composite structures has produced the following data regarding their oxygen permeability:

Laminate	Oxygen Permeability
Cellophane—PVDC—PE	0.5 cc
Polyester—PVDC—PE	1.0 cc
Polyamide—PVDC—PE	0.5 cc
Oriented polypropylene—PE—PVDC—Cellophane—PE	0.5 cc
Oriented polypropylene—PVDC—PE	2.0 cc

It is obvious that all of these provide adequate oxygen and moisture barrier characteristics.

When fabricating pouches for natural cheeses on conventional bag-forming machines, a dust is normally applied to provide slip and easy handling. The dust is generally a derivative of corn starch, but since molds form on the surface of cheese, sorbic acid or its salts may also be applied as a dust up to 2.5–5.0 grams per thousand square inches of laminate.

Package Forms for Natural Cheese.—The traditional methods of natural cheese packaging were based on bandaging and dressing. Whole cheeses were treated at the farm or factory to protect them for retail distribution. In bandaging, a calico or cheesecloth bandage is applied subsequent to treatment of the cheese with greased muslin. The unit was either sewn or glued with a flour paste. Dressing involved the external treatment of the cheese with oil, fat or wax. In some cases, flour or bladders were used. Edam and Gouda cheese are famous for their wax dressing. They are washed in hot water, stained red and then waxed or immersed in a red wax.

Although bandaging and waxing are still used on certain cheeses, the trend is toward making most varieties in block form for use with flexible laminates. Traditional methods, such as described above, are expensive and offer limited protection to the cheese. They lead to the formation of a thick rind and are not suitable for modern mass-merchandising methods.

Rigid Containers.—Cartons, made of waxed paperboard, are used to package cheeses having no characteristic form. Cottage cheese is often packed in a rigid paperboard carton since it has a short life, readily loses moisture and must be consumed within a few days after distribution. Vacuum formed polystyrene tubs are now used almost exclusively for the consumer packaging of the cottage cheese. They offer superior protection from fat absorption and breakage. Ricotta cheese is also packaged in either waxed paperboard tubs or polystyrene tubs. Some grated cheeses such as Parmesan and Romano are sold in foil-paperboard composite containers and in glass jars. Since glass is fragile and heavy, it is rarely used for other cheeses. Dutch cheese ripened in glass jars was found to have developed an off-odor.

Plastic-lined, metal cans are used for special cheeses such as Camembert. Cans offer excellent protection but are expensive. In addition, the cheese frequently acquires a peculiar taste. Grated cheese has been sold successfully in seamless aluminum cans.

Flexible Packages.—Short-term storage (about three days) permits the use of a package made of either moisture-proof cellophane or Pliofilm. PVDC-coated cellophane and even PVDC film will hold the product for a longer period. These types of packages, however, are poor choices for long-distance shipping which requires more sophisticated composites with superior barrier properties.

Cellophane-PVDC-PE was probably the first major film composite used for long term storage as a packaging material by the natural cheese industry. The material is moisture-proof and it may be either vacuum-packed or nitrogen flushed with the PE film serving as a heat-sealing medium. The cellophane may also be reverse printed and adhesive laminated to the PE.

This combination of materials is fairly inexpensive and is very useful for packaging sliced Swiss, Muenster and other general types of cheese. However, cellophane composites have the disadvantage of poor abrasion and flexural characteristics which make them particularly unsuitable for chunk and brick natural cheeses.

A series of abrasion tests of cellophane composite packages, under simulated transportation conditions with both 8- and 16-oz portions of cheese, resulted in 21% failure in packages tested. Other wraps for 8–10 week storage periods include Cryovac film bags and cellophane-Pliofilm laminates.

Aluminum foil laminates are widely used for cheese packages. As early as 1930, several European soft cheeses were packed in aluminum foil. Natural cheese appears to ripen more rapidly in foil materials. Port du Salut has been wrapped in aluminum foil and Roquefort has been packaged in an aluminum foil laminate. Cream cheese is now packed in both 3-oz and 8-oz sizes in a laminate of foil-paper-foil. This material offers excellent folding characteristics coupled with protection.

Polyester-PVDC-PE materials offer more inherent strength and less

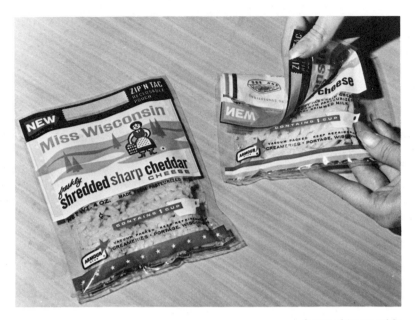

FIG. 61. FLEXIBLE FILM VACUUM PACKAGE FOR CHEESE

Laminated plastic film pouches provide good protection for vacuum packed
shredded cheese. This package features a "Zip'NTac" reclosure feature.

moisture sensitivity than most composite structures. There are various
grades of polyester film available, varying in their degree of orientation,
and these are designed for better abrasion resistance and transportation
durability than cellophane.

However, the most highly impact-resistant polyester will still fracture
when used in a lamination. Its cost is much greater than cellophane and
it still does not fill the cheese industry's need for a superior composite
material with increased abrasion resistance and flexibility.

The use of nylon as part of a lamination offers the cheese maker a film
with excellent flex and abrasion and both barrier and heat-sealing characteristics when used in combination with PVDC and PE. But it is
important to note that not all nylon films are alike and that there are
some which are much more suitable for cheese packaging than others.
Nylon 6 is superior in machinability and abrasion resistance and is less
expensive than other types.

Nylon-PVDC-PE is a cheese packaging composite material which is
about equal in cost to a polyester composite and yet superior in abrasion
and flex. The hydrophilic nature of nylon is altered when PVDC is

FIG. 62. CREAM CHEESE

This famous package illustrates the attractiveness of embossed aluminum foil-paper laminates. In the past 5 yr it had emigrated from the United States to the United Kingdom.

added to the final structure, and may result in difficulties when this material is handled by conventional packaging machinery.

A construction of oriented polypropylene-PE-PVDC-cellophane-PE lately has aroused quite a bit of interest as a packaging material for the cheese industry. It is a rather complex structure, but it offers excellent abrasion resistance and flexibility. The complexity of the structure has been found to retard its drawable character and packages made with this material do not conform tightly to product shape when vacuum packed. Since it is a rigid film, it has fairly good machinability but, on the other hand, it has a tendency toward delamination between film plies and it produces a loose package. Oriented PP gives the total structure excellent abrasion resistance and durability and the film may be used in the packaging of many chunk cheeses.

Elimination of cellophane will reduce the cost of the total structure to some degree. A material composed of oriented PP-PVDC-PE will provide good abrasion resistance, but the narrow heat-sealing range of the oriented PP can be a major difficulty. Recent technological advances by several film suppliers have resulted in improvements, and heat-set and stable films are now available that will yield a nonpuckered seal if machined under the proper conditions.

Rindless Natural Cheese.—When small portions of natural cheese with rind are prepackaged, considerable waste is obtained by discarding the rind. An additional problem is that the variable size and shape of natural cheese leads to additional waste. In order to reduce waste, rindless cheeses of uniform size and shape have been developed. Cheese is ripened in bags made from plastic films similiar to the types used for long term storage. Acetoglycerides and polyvinyl acetate emulsions have been used as coatings to produce rindless varieties of cheese.

Packaging Requirements and Package Forms for Processed Cheese.—As noted previously, processed cheese is a much more stable product than natural varieties. It is sterilized during processing, packaged in a molten state and is a homogeneous product with superior shelf-life and stability.

Due to the better keeping qualities of processed cheese, as well as its artificial conformation, packaging requirements are much less critical. Some of the composite films discussed above are used for processed cheeses. Foil composites with a heat-sealable layer are also used, with cellophane or paper laminated to the outer surface.

A great amount of processed cheese is packed in wax coated cellophane. Hot cheese is poured into preformed pouches and the sealed package is tight, mold resistant and inexpensive.

The wax-coated cellophane is treated with a mold-preventive agent to retard bacteria growth. Several converters are producing wax-coated cellophane materials for the processed cheese industry in substantial quantities.

Cheese spreads are packed in pry-open glass jars, in paperboard cartons with foil overwraps, and in rigid vacuum formed polyethylene tubs.

Packaging Machinery

Automatic packaging machinery has allowed cheese packers to eliminate all handwrapping operations and cheese packages may be formed from either rollstock or prefabricated pouches.

Use of vacuum packing is limited by the soft texture of some cheese slices which tend to cling together under vacuum. Paper inserts, which are placed between each slice of cheese, have done much to eliminate this.

Gas flush has been used fairly extensively and does not cause any deleterious effects on the product. Occasionally, however, a "balloon" pack may result, and if the product releases carbon dioxide, the seals will break and the package will rupture.

Many packers prefer a "cling" package—one under a slight degree of vacuum and not fully tight. There is no slice merger within the unit and a totally loose pack does not result.

Hayssen Manufacturing Co. offers a Model RT machine which is used extensively in the cheese packing industry. It may be used for both chunk and sliced cheese packaging and operates at speeds of 60–100 bags per min. The machine is equipped with a separate chamber for gas flushing which produces a "pillow" type package. Most of the composite films described above may be run on the RT machine and its rotary motion enables it to run at the high speeds necessary for profitable production.

The Hudson-Sharp Div. of Food Machinery Corp. offers an automatic high speed packaging machine capable of vacuum packing 20–40 packages per minute. The gas flush operation can run at a more rapid speed (60–100) and either a tight or "pillow" package can be produced depending on the method used in packing.

European producers of cheese packaging machines include Lerner Machine Co., Masson and Morton, Ltd., and the W. R. Grace machine used for the Cryovac process internationally.

OTHER PRODUCTS

Yoghurt

In recent years, yoghurt has become very popular in many nations. Originally a Balkan food, yoghurt is specially fermented milk. It was brought from the Balkans to France during the reign of King Francis I. Long associated with health-giving properties, yoghurt is made by reducing partially skimmed milk over slow heat. The milk is then divided into bowls and cooled. A culture is added and the bowls are stored for 28–48 hr before use. By the introduction of a wide variety of flavors, it is rapidly expanding sales in supermarkets.

In the United States, yoghurt is packed in coated paperboard containers with recloseable covers. European packaging trends are more imaginative. The product is automatically filled in polystyrene tubs and covered with 0.002 in. aluminum foil. The cover is clamped over the tub and is easily peeled off. The inherent brittleness of polystyrene may become a problem. It has also been reported that the terpene containing aroma of fruit yoghurt may have an effect on polystyrene. Fruit acids may also cause pitting of unlacquered aluminum lids. A polystyrene tub is nevertheless an excellent package for yoghurt. It is economical, practical and widely used. The unit is effectively displayed and promotes product distribution.

In Germany yoghurt tubs are now being sold in expanded polystyrene foam display trays. Each tray holds 20 tubs of yoghurt. Trays are stackable and serve not only as display holders and light weight shippers, but also holders during fermentation of the yoghurt. This saves extra handling and repacking.

FIG. 63.   YOGHURT PACKAGING

Flavored yoghurts are very popular. This package is made from a
molded plastic tub with printed foil lids.

Dried Milk

When raw liquid milk is dehydrated, milk powder is produced. The
product is highly susceptible to oxidation and moisture. Reconstitution
with water yields liquid milk. If the fat is removed as well as moisture,
dried skimmed milk (non-fat dried milk) results. This product is not so
sensitive to oxidation and does not require vacuum or gas packaging.

Most consumer packages consist of aluminum foil laminates. Paper-
polyethylene-foil-polyethylene comprise the vast majority of dried milk
portion packs sold. Larger retail units are packaged in cans, jars, or
lined paperboard cartons. Commercial bulk packaging is either in polyethylene bags inside corrugated cartons, or in 50- and 100-lb bags made
of 2-3 mil polyethylene liners and 4 to 6 ply kraft paper outers with
taped seam closures.

Courtesy of The Metal Box Co., Ltd.

FIG. 64.   DRIED MILK POWDERS

This dried nonfat skimmed milk package comprises an inner bag and
an attractively printed sealed paperboard carton called "Hermetet."

Fermented Milk

Fermented milk products other than yoghurt are not sold in large
quantities in supermarkets. They originated in other lands and have
exotic names from "Dahdi" of India to Egyptian "Laban Zebadi." Since
they are native and special ethnic drinks, they are packaged in a wide
variety of materials from sheepskins to glass bottles.

BIBLIOGRAPHY

ANON.   1966A.   Wrapping machine copes with irregular cuts of cheese. Pack-
aging News *13*, No. 12, 9.
ANON.   1966B.   Homogenized milk in light proof plastics sachets. Verpak-
king *19*, No. 2, 101–104. (Dutch)
ANON.   1966C.   European milk packaging. Package Eng. *11*, No. 12, 19–23.
ANON.   1966D.   Fills plastic bags, puts them in Kraft bags. Package Eng.
11, No. 11, 142.
ANON.   1967A.   Butterwraps. Food Technol. Australia *19*, No. 4, 179.
ANON.   1967B.   Plastics bottles are filled and closed at 3,000 per hour.
Packaging News *14*, No. 1, 14.
ANON.   1967C.   Cheese of all shapes wrapped and boxed at speed. Packaging
Rev. *87*, No. 3, 29.

Anon.   1967D.   Ice cream blocks wrapped at 100 a minute. Packaging Rev. *87*, No. 3, 31.

Anon.   1967E.   Plastics-lined milk pack replaces waxed cartons. Packaging News *14*, No. 4, 5–6.

Anon.   1967F.   With the dawn—the bottle is new. Canadian Food Ind. *38*, No. 4, 52.

Anon.   1967G.   The present state of the packaging of butter. Verpackungs Wirtschaft *15*, No. 6, 12–15. (German)

Anon.   1967H.   Production and wrapping of ice cream. Food Process. Marketing *36*, No. 429, 245–246.

Anon.   1967I.   Aseptic milk stays fresh for 3 months in film-foil pouch. Packaging News *14*, No. 4, 1–5.

Anon.   1967J.   Milk and soft drinks can be kept for six months in plastics containers. Packaging News *14*, No. 11, 12.

Anon.   1967K.   New techniques in packaging milk products. Emballages Dig. *10*, No. 94, 248–251. (French)

Anon.   1967L.   Plastics bags. Denmark's next largest milk packaging. Plastvarlden *17*, No. 10, 73. (Swedish)

Anon.   1967M.   Plastics mini-tube for cream for coffee. Emballages Dig. *10*, No. 95, 135–136. (French)

Anon.   1967N.   Now—heat sealed cup for milk. Packaging News *14*, No. 11, 1.

Anon.   1967O.   'Instant' milk is in jar with diaphragm seal. Packaging News *14*, No. 11, 6.

Anon.   1968A.   Dairy makes, fills, closes own plastics milk bottles. Packaging News *15*, No. 1, 1–8.

Anon.   1968B.   How Pergall milk paks boost milk consumption. Dairy Ind. *33*, No. 1, 23–26.

Anon.   1968C.   Packaging of butter in a PVC container. Emballages Dig. *38*, No. 251, 150–153. (French)

Anon.   1968D.   Form-fill-seal plastics pack for aseptic milk. European Packaging Dig. Mar., 5–6.

Anon.   1968E.   First UK form-fill-seal butter package doubles as a dish. Packaging News *15*, No. 3, 1.

Anon.   1968F.   New French cheese in a modern package. Tara No. 224, 233–234. (German)

Anon.   1968G.   Milk in plastic beakers, now a reality. Nord. Mejeritidsskr. *34*, No. 1, 15–16.

Anon.   1968H.   Bag-in-box packaging cuts indirect costs. Package Eng. *13*, No. 3, 79–80.

Anon.   1968I.   Pourer for milk powder pack scarcely cuts protection. Packaging News *15*, No. 6, 12.

Anon.   1968J.   Adams launch carton-packed butter. Packaging *39*, No. 459, 74–75.

Anon.   1968K.   How to make and use plastics containers for butter. Emballages Dig. *12*, No. 103, 212–228. (French)

Badings, H. T., and Radema, L.   1966.   The organoleptic quality of pasteurized milk packaged in polythene sachets. Misset's Zuivel *72*, No. 29, 674–681.

Botham, P. W.   1967.   Packaging and distribution of liquid milk. Australian Packaging *15*, No. 8, 33–72.

BRISTON, J. H. 1968. Progress in milk packaging depends on consumers. Inst. Packaging J. *14*, No. 98, 20–22.

DAVIS, J. G. 1965. Cheese. Vol. 1, Basic Technology. American Elsevier Publishing Co., New York.

DOLBY, R. M. 1966. Studies on film wrapped cheddar cheese. XVII Intern. Dairy Congress. D67–74.

DOWNEY, W. K., and MURPHY, M. F. 1968. Light-barrier properties of various butter wrapping materials. J. Soc. Dairy Technol. *21*, No. 2, 104–106.

GIBLIN, J. P. 1967. Cheese packaging in the USA Food Process. Marketing *36*, No. 429, 219–222.

GOFFIN, Y. 1966. High density polyethylene in disposable milk containers. Rev. Belge. Matieres Plast. *7*, No. 4, 307–310. (French)

GRINENE, E. 1966. Effect of some packaging materials on keeping quality of butter. Moloch. Prom. *27*, No. 5, 20–22.

GUSTAFSSON, T. H. 1967. Does bag milk have the last word in Finland? Plastvarlden *17*, No. 10, 74. (Swedish)

HANE, B. 1967. Demands on plastic milk cup and its closure. Plastvarlden *17*, No. 10, 75. (Swedish)

HANRAHAN, E. P. *et al.* 1967. Experimental equipment for cooling and packaging foam spray-dried milk in the absence of oxygen. J. Dairy Sci. *50*, No. 12, 1873–1877.

HOLZMANN, R. and PUTZ, T. 1966. Price of raw material and machine capacity are decisive. Verpackungs Rdsch. *17*, No. 12, 1656–1665. (German)

KALKSCHMIDT, J. 1967. Packaging in the dairy industry. Fette, Seifen, Anstrichmittel *69*, No. 4, 302–308. (German)

KILLINGSTAD, A. 1967. Milk package in Norway. Plastvarlden *17*, No. 10, 72. (Swedish)

KOCH, G. 1967A. Modern butter packaging in the 'Common Market.' Molk. -u Kas. -Ztg. *18*, No. 9, 275–278. (German)

KOCH, G. 1967B. Quality tests on fresh cheese in retail packs carried out in Kassel on 28th of February by the German Agricultural Society. Molk. -u. Kas. -Ztg. *18*, No. 40, 1513–1518. (German)

MACDONALD, L. H. 1967. Consumer packaging and its effects on the dairying industry. Australian Packaging *15*, No. 11, 25–28.

PARODI, P. W. 1966. The surface deterioration of cold stored butter. III. Absorbed taints. Australian J. Dairy Technol. *21*, 68–69.

PETERSEN, A. H., and FISKER, A. N. 1966. Investigations on the protective qualities of various packs for individual butter portions. Paper presented at XVII Intern. Dairy Congress, 113–120.

PFAB, W. 1967. Migration of phthalate plasticizers from lacquered aluminum foil into fatty foods. Dt. Lebensmitt. Rdsch. *63*, 72. (German)

RAMANAUSAKS, R. 1965. Packaging cheese in polymer films. Sov. Torg. No. 10, 26–27. (Russian)

RODER, H. E. 1968A. The one-trip plastics bottle for sterilized milk. I and II. Verpackungs Rundschau. *19*, No. 9, 1088–1098; No. 10, 1196–1208. (German)

RUTHERFORD, R. E. 1967. Returnable plastics bottles in an American dairy. Dairy Ind. *32*, No. 5, 371–375.

SACHAROW, S. 1966A. Flexible plastic films for cheese packaging. Svensk Emballagetidskrift *32*, No. 12, 8–10. (Swedish)

SACHAROW, S. 1966B. On the sealing methods of plastic films. J. Japan. Packaging Inst. *4*, No. 6, 39–44. (Japanese).

SACHAROW, S. 1967. Cheese packaging in the U.S.A. Neue Verpackung *20*, No. 3, 339–343. (German)

SHAUGHNESSEY, J. 1968. Research shows butter wrap may result in off-flavor. Food Can. *28*, No. 7, 23.

STORGARDS, T., and LEMBKE, A. 1966. Permeability to gas and light of packages used for fluid milk. Intern. Dairy Fed. Assoc. Bull. *4*, 35–75.

SWANN, L. W. 1967. Aluminum foil serves the dairy industry. Packaging *38*, No. 445, 114a–114d.

WILDBRETT, G. 1967A. Plastics in the milk processing industry—a critical examination. II. The significant properties of plastics in foodstuffs technology. Fette, Seifen, Anstrichmittel *68*, No. 8, 598–603. (German)

WILDBRETT, G. 1967B. Reactions of plastics on the quality of milk and its products. Dt. Lebensmitt. Rdsch. *63*, No. 1, 1–5. (German)

WOOD, S. 1968. PE bottles chalk up gains. Mod. Plastics. *45*, No. 10, 92–94.

CHAPTER 6

Fish and Shellfish

FISH

General Marketing Concepts

Although the sale of all seafood items in the United States in 1968 amounted to about $2-1/2 billion, the packaging of fish in the United States has been one of the most neglected areas of packaging technology. Most packaging advances have originated in Europe where fishing has remained a much more important part of the diet and where the fishing industry has received greater governmental support. Despite the rise of supermarkets in the United States with mass merchandising to the consumer, Americans eat only 10.8 lb of fish per person in contrast to 60 lb for Japan and 45 lb for Sweden. The nature of the product, the relatively small level of consumption, legal requirements, and the methods of processing and packaging of fish are all interrelated, and both cause and result of one another. US Law forbids the landing of fish at American ports in other than vessels built in the United States. In effect, US fishermen are barred from buying their boats at competitive prices in foreign countries. Many boats are at least 20 years old, and they are comparatively small and inefficient.

In 1966, US flag vessels harvested two million metric tons of fish, but this represented only ten per cent of the potential yield.

There are over 750 species of fish commonly consumed by humans. If combined with lesser known varieties the number would be well over 2000. To cover the subject adequately one must divide it into two major categories: fresh fish and processed fish. Each category requires special packaging materials and different handling procedures.

Fresh Fish

Fresh fish are one of the most perishable of all foods. The rate of spoilage doubles for every ten Fahrenheit degrees rise in temperature. They must be refrigerated or frozen immediately after harvest and kept refrigerated until eaten. Any departure from immediate and continued refrigeration will lead to flavor and texture losses. If the fish are of poor quality to start with, the quality loss will be even more pronounced. The best package cannot improve the quality of the contents. The fish must be of high quality prior to packaging.

Only eight per cent of the 14,000 fishing vessels in the United States are equipped with refrigeration machinery. The rest use ice. The result is that only the top of the catch is kept adequately fresh. After several

days of storage the bottom portion of the catch has deteriorated in quality.

The best means for preserving the quality of fresh fish until packaging
is on-vessel freezing. The freezer-trawler employs the same type of crew
as conventional vessels; however, it also includes a special group to
process and freeze the catch. The entire catch can be frozen whole and
thawed prior to packaging. In recent years, the use of freezer-trawlers
has increased. In 1965, about 11% of all Britain's catch was from
freezer-trawlers; this value rose to over 16% in 1966. The Russian and
Japanese fishing fleets are even more advanced. The quality of sea-
frozen fish undoubtedly accounts for the continuing popularity of seafood in these countries and the absence of such quality could be the
cause of its unpopularity here.

Packaging Requirements.—The rapid decline of retail seafood
stores has been coupled with a change in consumer buying habits. More
and more supermarkets now maintain fresh seafood counters as well as
offering pre-packaged "wet" fish. The maintenance of proper shelf-life
by packaging becomes an essential prerequisite for marketing success.
Close attention must be paid to the display cabinets used and to careful
and rapid handling of the product. What might be insignificant departure limits for other products could lead to a three-day shelf-life being
reduced to a few hours.

A suitable fresh fish package must: (1) reduce fat oxidation; (2)
reduce dehydration; (3) provide for less bacterial and chemical spoilage;
(4) eliminate drip; and (5) prevent odor permeation.

Fat Oxidation.—The highly unsaturated nature of fish oils and fats
makes them very susceptible to rancidification. Other spoilage occurs
also from chemical, enzymatic, or bacterial action. At room temperature
the lipid (fatty) components rancidify rapidly in the presence of oxygen.
The use of a good oxygen barrier package and cold storage drastically
reduce this action.

Dehydration.—Since fresh fish are "wet," excess moisture loss leads
to texture, flavor, and color changes. Poor storage conditions or a poor
barrier package will allow dehydration.

Bacterial and Chemical Spoilage.—Enzymes and bacteria present in fish cause rapid spoilage. Powerful proteases in the digestive tract
may, after death, perforate the intestinal wall and attack the surrounding flesh. This is why fish should be gutted and washed promptly or
immediately frozen. The characteristic "fishy" odor of trimethylamine
is not present in fresh fish. It is formed by enzyme action and by secondary enzymes produced by bacterial growth. The flesh of healthy fish
likewise is free of bacteria but the slime and digestive tracts contain
many bacterial flora. Immediately after death these bacteria begin to

grow and "spoil" the fish with development of off flavors and odors and pronounced textural changes. As with enzymes, immediate cleaning is the best means for preventing spoilage. Refrigeration alone is ineffective in preventing spoilage as many bacteria associated with fish (of the genera *Pseudomonas, Achromobacter,* Flavo-bacteria and others) are psychrophylic—that is they will grow at 32°F and below, some as low as 19°F. Freezing will retard spoilage provided the temperature is held low enough to prevent further bacterial growth.

Drip.—When a piece of fish is packaged, some juice may be lost and trapped in the package. The result is messy and unsightly. An absorbent pad may be placed in the package to soak up the drip. Another approach is to dip the fish in a polyphosphate solution long enough to swell the outer surface layer of cells. The rupturing of cell walls helps to prevent the escape of internal fluids.

Odor Permeation.—Odors from fish become stronger in permeable packages stored at 35°F within several hours. Although this odor may dissipate, a considerable amount of flavor volatiles will be lost. Foreign odors are easily picked up by fish. A good odor barrier is an essential part of a good package.

Bulk Shipment.—Formerly, iced fresh fish were shipped in wooden boxes by rail. In recent years, the decline in rail service coupled with a rise in consumer affluency has spurred the development of lighter and more economical shipping units. In order to ship fresh fish to markets with greater speed, air transport is required. Here a lightweight and protective container becomes a necessity. Air shipment is an emerging trend in fresh seafood distribution. In 1968, the volume of seafood shipped by air increased over 100% in Seattle alone.

The rigid-when-wet corrugated carton is acquiring a share of the wooden box market. Although several variations exist, corrugated boxes may be printed, are cheaper than wooden boxes, are better insulators, and are significantly lighter in weight. Most cartons used are wax coated and leak-proof. For further protection, inner insulation may be used. Slab polystyrene foam is effective in providing increased insulation. The choice between an insulated wax carton and one that is not is dependent on the season the shipment is made. In extremely cold weather, added foam insulation may not be necessary. The Bureau of Commercial Fisheries Technological Laboratory in Gloucester, Massachusetts recently developed a corrugated carton that will carry fresh fish successfully in all forms of transport with substantial savings. Insulation is used and very little ice is required.

Other types of cartons used for fresh seafood shipment include containers made of expanded polystyrene. They are very lightweight and offer excellent insulation. One French design consists of a container

Courtesy of St. Regis Paper Co.

FIG. 65.   FRESH FISH BULK SHIPPERS

Coated corrugated shippers for bulk shipments of fresh fish feature slabs of foamed polystyrene for insulation.

with a sliding lid. It offers a secure lid for the package and aids in stacking. Many other designs exist in European markets.

Retail Packaging.—The most popular package for fresh refrigerated fish consists of a shallow tray and transparent film overwrap. If the
product is large or irregular in shape, a direct film overwrap may be
used.

Tray packs are overwrapped with semi-moisture-proof cellophane,
polystyrene, oriented polypropylene or polyvinyl chloride. The tray
may be fabricated from molded pulp, foam polystyrene or clear polystyrene. Molded pulp is commonly used since it absorbs drip. A significant disadvantage is that it tends to deteriorate if too much moisture is
absorbed. Foam and clear polystyrene require the use of absorbent blotters. Finally, product identification may be achieved by means of a
pressure sensitive label or a printed overwrap. In order to obtain
improved rigidity for direct film overwrapped packages, a plastic or wax
coated paperboard insert is used. Occasionally, a vacuum pack is used

FIG. 66. RETAIL PACKAGING OF FRESH FISH

Film overwrapped trays and sealed film bags are used for prepackaged fresh fish fillets.

to improve shelf-life. Higher barrier films are required such as PVDC coated celiophanes, polyethylenes, polypropylenes, or polyesters.

Processed Fish—Frozen

A great deal of fish is sold in the frozen food section of the supermarkets in this country.

Between 1948 and 1966, the US frozen seafood pack increased from 164,855,000 lb to 435,952,000 lb. The bulk of the increase was due to the successful marketing of fish sticks, fish portions and shrimp. Fish fillets and steaks decreased in volume during the same period. The frozen seafood industry represents three per cent of the entire frozen food industry. In the institutional area alone the use of frozen seafoods has doubled between 1961–1966 to a present value of over $750,000,000.

One problem with frozen fish, but one that does not produce deterioration or spoilage, is "drip." This is liquid that escapes from the fish during thawing. Drip can be serious, however, since it represents a loss in the natural fish juices. Less drip occurs when frozen fish is thawed if the fish is fresh. In tests comparing rate of thawing, it was found that very rapid or very slow thawing gave the poorest product. Moderately slow or

Courtesy of Nixon-Baldwin Div., Tenneco Chemicals, Inc.

FIG. 67.   RIGID PLASTIC CONTAINERS

A formed rigid vinyl tray is used to package frozen breaded fish
sticks. A clear flexible film lid is heat sealed to the flange of the
container.

moderately fast thawing were recommended for highest quality frozen
fish.

The factors affecting the quality of frozen fish are: (1) moisture loss;
(2) oxidation; (3) rancidity; (4) change in odor and flavor; (5) loss of
volatile flavors; (6) enzymatic activity; and (7) loss of vitamins.

Glazes and Antioxidants.—For many years an ice glaze was used
on frozen fish. This consisted of dipping the frozen fish in cold water and
withdrawing them into air in a cold room. Thus, the adhering film of
water froze and glazed the seafood. Actually, glazing was not too satisfactory since the glaze was quite brittle and cracked or chipped easily.
In addition, it didn't last long, as it soon sublimed in dry atmospheres.
Glazing also increased the weight of the fish thereby increasing transportation costs. When a glazed fish thawed, the amount of drip increased and the fish flesh absorbed some of this moisture. This made
frying of the fillet more difficult because excess moisture caused the
grease to spatter. Ascorbic acid solution and ethyl gallate solution glazes
have proved to be more beneficial than water glazes. However, these
chemical glazes have not found wide acceptance commercially. At the

present time, most fisheries are not using glazes on fillets, as they believe that it is unnecessary. A glaze on properly wrapped fish to be held for short storage period is not necessary.

Packaging Requirements.—The optimum requirements of a package for frozen seafood, are: (1) flexible enough to fit the contour of the fish and leave little or no air space; (2) should not become brittle when cold; (3) should not deteriorate in cold storage; (4) puncture resistant; (5) moisture vapor proof; (6) impervious to oxygen; (7) easily filled.

Rancidity formation in frozen fish in storage is prevented through the use of gas packaging in a carbon dioxide or nitrogen atmosphere. Storage in carbon dioxide, however, produced a slight off-flavor to the fish that was not found in samples packaged in nitrogen gas. Vacuum packaging also serves as a means of shelf-life extension. It is also useful in combatting muscle toughening or protein devaluation. In addition, vacuum packaging even may aid in the manufacture of fish products. Herring fillets are packed in aluminum trays and covered with a mixture of breadcrumbs. When the tray is evacuated, the breadcrumb mixture spreads evenly around the product.

Packages for Frozen Fish.—Proper packaging for frozen fish is essential. Efficient packaging will help to offset the detrimental effects of oxygen and of dessication by barring the entry of air.

The package in most widespread use is a polyethylene, wax, or hot-melt coated carton with or without a waxed paper overwrap. There is a definite trend towards printed cartons throughout the frozen seafood industry. Although, the industry has lagged behind the times in both marketing and package design, changes are now appearing.

Wax coated cartons were the first packages used. Problems were encountered with wax flaking off and with dehydration through the score lines on the package. The arrival of polyethylene coated paperboard offered good flexibility and better moisture protection. Although this was an improvement over wax cartons, problems still remained in heat sealing, ink adhesion and delamination when in contact with fish juices.

At present, most cartons are coated with petroleum wax-resin blends (hot melts). If the hot melt is used on the inside of the carton, a high gloss is left on the product surface. This offers pleasing aesthetics. The hot melts are easily heat-sealed. Cartons may incorporate tear strip openings or other convenience features.

Overwrapped cartons for seafood have substantially decreased in use. One large remaining area lies with foreign co-packers. Over two-thirds of the US fish supply is imported. Co-packers produce many varieties of

FIG. 68. CARTONED FROZEN FISH

Cartons are used for frozen whiting fillets. These cartons feature
an easy opening end.

frozen fish in plain cartons. They then wrap the cartons in paper overwraps supplied by their customers. The overwrap may be coated with
wax, hot-melts or polyethylene.

Frozen fish may also be found in overwrapped trays displayed in the
frozen fish cabinet. Individual fillets are overwrapped with cellophane
or PVC with and without paperboard inserts.

Vacuum packaging is not generally done because of the added expen-
ses of operation and handling. In some cases, frozen fish is vacuum
packed at a low temperature and frozen. The most suitable material
must maintain an excellent oxygen barrier. Laminated structures such
as cellophane-aluminum foil-polyethylene or polyester-PVDC-polyethylene are desirable materials. As determined by peroxide values, rancidity formation decreases drastically in vacuum packs employing low
oxygen permeable laminates. Overall shelf-life is extended by almost
100%. Since vacuum packaging does not remove all the oxygen from the
fish, lipid oxidation still occurs—but at a much slower rate. Proper

Courtesy of Vita Food Products, Inc.

FIG. 69. CANNED FROZEN FISH

Frozen "Lox" smoked salmon in metal cans are available in both tinplate and aluminum cans. The latter feature pop top tear out panels in the lids.

freezing storage is essential for all packaged fish, whether it be vacuum packed or overwrapped with a plastic film.

A new frozen fish item is smoked salmon (lox) in tinplate or easy-open aluminum cans. Each layer of the sliced fish is separated by a parchment paper insert.

Processed Fish—Canned

Salmon, tuna and sardines form the main fish products currently canned. Tuna is canned in an oil or water pack. Salmon is usually canned in oil. Sardines are either smoked prior to canning or put up in brine and with tomato sauce.

The final quality of canned fish is dependent on the condition of the raw fish, the methods used during canning and changes which may occur during storage after canning. Fish to be canned must be as fresh as possible. In tuna processing, a proper refrigeration cycle is mandatory prior to canning. Fresh fish does not necessarily imply high quality. If salmon is caught late in the season, the fish may be soft and have a very low oil content.

Canned Tuna.—The choicest tuna species for canning is the albacore. Other species canned are the bluefin and yellowfin tuna. After the fish

TABLE 12

GENERAL TYPES OF CAN COATINGS

Coating	Typical Uses	Type
Fruit enamel	Dark colored berries, cherries and other fruits requiring protection from metallic salts.	Oleoresinous
C-enamel	Corn, peas, and other sulfur-bearing products, including some sea foods.	Oleoresinous with suspended zinc oxide pigment.
Citrus enamel	Citrus products and concentrates.	Modified oleoresinous
Seafood enamel	Fish products and meat spreads.	Phenolic
Meat enamels	Meat and various specialty products.	Modified epons with aluminum pigment.
Milk enamel	Milk, eggs, and other dairy products.	Epons
Beverage can enamel (non-carbonated beverages)	Vegetable juices; red fruit juices; highly corrosive fruits; noncarbonated beverages.	Two-coat system with oleoresinous type base coat and vinyl top coat.
Beer can enamel	Beer and carbonated beverages.	Two-coat system with oleoresinous or polybutadiene type base coat and vinyl top coat.

Source: Joslyn and Heid (1963).

are caught they are bled and gutted then washed with a water spray. Considerable care must be exercised to assure proper refrigeration.

In the factory, the fish are hung up and allowed to drain prior to cooking. The cooking process involves placing the tuna in pans which are in turn placed on portable racks. The racks are inserted into a steam chamber and exposed to 212°–216°F for 2-1/2 to 4 hr. Longer times are needed for larger fish. This allows all the natural oils to drain and permits thorough cooking of the flesh. After standing for one day, the skin, head and bones are removed and the dark meat is separated from the white meat.

The white meat is put on trays and cut to proper length for cans. The trays are then sent to the filling tables and the cans are filled. The main part of the fill consists of a thick piece of meat while smaller ends or bits are used to reach final weight. Olive oil, salt, soybean oil or water is added either before or after the meat is filled into the cans.

The canned tuna is given a 10–12 min exhaust and sealed. Since the cans may become greasy through handling, they are subjected to a weak alkali bath prior to being processed.

The process time and temperature employed for canned tuna is dependent on the size of the can. A typical No. 1 tuna can process when initial temperature is 70°F would be 95 min at 240°F and 80 min at 250° F. The type of can used may vary; however, enameled tinplate cans are usually employed. A commonly used coating is called "seafood enamel" which is a phenolic type material. Aluminum cans are particularly good for tuna as they maintain fresh color, odor, and flavor.

Canned tunafish is inspected by the voluntary inspection service of the industry as well as Food and Drug Administration officials. Some states also regulate both quality of fish and canning conditions. After processing, cans are cooled, labeled, and packed in shippers.

Errors in time or temperature are very rare in commercial canneries. However, constant vigilance and careful control are imperative. In 1963, three women in Detroit ate tunafish salad sandwiches. All became ill from botulism and two died. Investigation by health authorities at municipal, state and federal levels traced the source to one particular cannery and the use of cans having defective seams, which permitted *Cl. botulinum* organisms to enter the cans after sterilization, possibly from the cooling water. The cannery was quarantined and all product was located and withdrawn from the market. Despite the fact that the problem was identified and localized, the entire tuna canning industry was injured due to public loss of confidence. It is estimated the loss in sales was about $40 million.

Processed Fish—Smoked, Salted and Marinated

Smoking, or smoke-curing is a method of preservation effected by a combination of drying and the deposition of naturally produced chemicals resulting from the thermal breakdown of wood. Salting often accompanies smoking, or it may be used as a separate method of preservation. Marinated fish are treated with vinegar and salt. Spices and herbs are also often added. Most marinated varieties are sold in metal cans and glass jars. Examples include "Schmaltz herring," "Rollmops," "Lunch herring," and herring in cream sauces. Other types include herrings in dill, lemon and wine sauces. They are refrigerated during storage.

FIG. 70.   PICKLED FISH PRODUCTS IN GLASS

Marinated herring in various sauces is shown packed in glass jars
with metal screw on caps.

Curing is used to preserve the shelf-life of fish as well as to provide
improved flavor. A suitable packaging material must prevent moisture
loss or gain and be impermeable to organic volatile flavoring constitu-
tents. The odor and flavor of cured fish may be lost through package
permeability. In addition, there is a risk of the intermingling of flavors
between different products in a refrigerated cabinet.

Packaging.—Vacuum packaging can also be used for cured fish. It is
extremely important to maintain proper storage conditions during
refrigeration or freezing. In a fermented or lightly cured product, the
shelf-life is extended; however, the process has little effect on *Cl. botulinum* formation.

Most cured fish, i.e., kippers, smoked whitefish, smoked haddock,

Courtesy of British Cellophane Ltd.

FIG. 71.   FLEXIBLE FILM POUCHES FOR SMOKED FISH

Smoked salmon packaged in a "Cellophane" laminate can be
frozen or refrigerated.

are drier and more stable than fresh fish. They can be kept for several days completely unwrapped. For wholesale storage, they are packed in wooden boxes. Cured kippers are packed, for consumer sale, in overwraps of cellophane. Hygienic protection is obtained as well as proper brand identification. In addition, condensation within the package is reduced.

Smoked fish can also be frozen and cold stored. The shelf-life at a specific temperature is not as good as unsmoked fish. In smoked fatty fish, rancidity may develop in the frozen state and this must be carefully controlled. Since ice cannot be used, the temperatures during transportation of smoked fish are normally higher than for fresh fish. Proper storage conditions must be maintained at all times or there can be tragic consequences.

In 1963, a man and wife died from type E botulism caused by eating smoked whitefish bought in Michigan and a few weeks later four more people died in Knoxville, Tennessee also after eating Michigan packed smoked whitefish. Authorities traced the contaminated fish to inadequate smoking and subsequent mishandling. The fish were shipped in an unrefrigerated truck in summer and were routed first to Chicago where they sat for several days before going on to Knoxville. At no time was the shipment treated as perishable!

As a result several regulations have been adopted by the government agencies in the United States and Canada regarding the smoking process and inspection of plants. Canada prohibited the sale of smoked fish unless frozen. This incident all but destroyed the smoked fish industry, however it may recover in time under new regulations.

Recent developments in cured fish packaging include the use of polyamide-polyethylene laminated film and polyester-polyethylene laminated film. Kippers in butter sauce or smoked salmon may be inserted in a boilable pouch, evacuated, and frozen. Vacuum packaging prevents the pouch from floating to the surface during heating. High-density polyethylene film and polyester-polyethylene laminated film are being used. Smoked ciscoes have also been sold in polyester-polyethylene pouches.

Processed Fish—Sardines

The canning of sardines is an industry in itself which began in 1834 in Nantes, France and spread to other countries including Spain, Portugal, Norway, and Sweden. It was begun in this country at Eastport, Maine in 1875 and with varying success, due to supply of fish, in California. The sardine is by definition a small herring and originally received its name from the island of Sardinia. In the United States the eastern sardine is smaller and is packed in flat, oblong 1/4- 1/2- and 1-pound cans. The longer western (pilchard) sardine is packed in oval and round cans in 1/4- and 1-pound units.

Care is exerted in bulk handling of sardines in the fishing boats as damaged skins or broken fish are not acceptable for canning.

Preprocessing.—Fish are scaled, beheaded, cleaned, and de-tailed. (Some very small varieties packed in other countries are left with heads and tails.) All wastes and rejects are saved and processed into fish meal. After rinsing, the fish are brined to remove slime, blood, and water; to toughen the skin; to whiten the meat; and to salt the fish. Larger fish require a more concentrated brine and a longer period of immersion.

The fish are dried in warm air (100°F) for up to 90 min in order to dehydrate them and toughen the skins further.

Processing.—*Oil Frying.*—Fish are placed in wire baskets and immersed in hot (230°F) cottonseed or sardine oil for several minutes, depending on size. Frying is complete if the backbone can be removed easily and shows no redness. After steaming or hot-water washing to remove excess oil they are cooled, air dried, and packed. Quality depends on the quality and freshness of the frying oil.

Brine Frying.—Boiling saturated brine is used and the temperature is 220°–227° F. After 6 to 10 min fish are removed, sprayed with brine, cooled, and dried. This method is seldom used.

Precooking.—Fish are placed in cans and subjected to steam, super-heated steam, gas heat, or a combination in order to drive out liquid and oil, which are drained away.

Packing.—Fish are placed in cans according to sizes and weights and the desired sauces or oils are added. These may consist of tomato sauce, mustard sauce, olive oil, cottonseed oil or soybean oil. Oils are usually added hot (220° F), but tomato sauce must be added at 150° F and mustard sauce cold.

Vacuum Packing.—One other type of pack does not precook or fry the fish but instead after saucing, the can is closed in a vacuum chamber to an average can vacuum of 3 to 4 in. The result is cheap and simple but a rather watery product.

Sterilization.—After closure, cans are retorted for various times and at various temperatures depending upon can size, can shape, size of fish, sauce, temperature and initial product.

New Trends.—Two new developments have been reported for sardine packages. One is a vacuum packed flexible laminated plastic film-foil-film pouch and the other is an aluminum can in which the top has a scored panel which can be ripped out for easy access to the contents.

Processed Fish—Fish Meal

Fish meal is a product which has increased in importance annually. It is made by removing the liquid substances from fish and grinding the remainder. Present uses are feeds for poultry, pigs, and cattle. Current legislation is pending in several nations for its use as a potential source of protein in human diets. Under these circumstances the product is thoroughly sterilized and ground to a dry powder known as fish flour. Fish flour does not overheat (see below) because sterilization has destroyed bacteria.

Whole fish meal is a fatty and hygroscopic material and must be protected from oxygen and moisture. It is packed in air-tight containers such as multiwall paper sacks with an inner liner of polyethylene. Antioxidants such as BHT and Ethoxyquin may be added in amounts of 1 lb and 1/4 lb per ton respectively. The liner serves as a moisture barrier and retards oxidation. It also reduces the possibility of spontaneous heating in transit. The problem of spontaneous heating is a serious one and often necessitates the use of metal-lined, airtight containers. Laminates of aluminum foil as an inner liner would appear to be extremely useful for bulk packaging. In the UK, the British Ministry of Transport requires that fish meal containing less than 6% or more than 12% moisture be packed in containers devoid of paper or burlap.

Overheating of fish meal does not occur when its moisture content is less than six per cent, but when moisture rises to greater than twelve per cent, bacterial action occurs and this gives rise to overheating. In packing fish meal, the bags must be cooled and properly handled prior to shipment.

SHELLFISH

General Marketing Concepts

Shellfish are very similar to fish in respect to their flesh and how it is processed, packaged, and consumed. There are two groups of shellfish; the mollusks which include oysters, clams, scallops, mussels, abalone, and the crustaceans which include shrimps, prawns, lobsters, crayfish, and many species of crabs.

Most mollusks are taken alive and kept alive until just prior to their use. One exception is the scallop which has successfully been shucked and frozen.

A few oysters have been sold frozen but this is not prevalent practice. Crustaceans can be sold live or frozen and both methods are used extensively depending mostly on how close the market is to the fishing port. A few typical shellfish will be discussed in detail. Their characteristics are representative of what can be expected of similar seafoods.

Shrimp.—Some fresh shrimp are sold live at dockside. Most of the catch which is to be shipped to market is beheaded and iced at sea. A better quality product can be expected if the shrimp are frozen at sea.

When shrimp have been shucked, graded and deveined, they are usually placed in waxed cartons and frozen. Some of these cartons have an inner liner to aid in protecting the product against oxidation and loss of moisture. Other cartons have holes in the bottom and after the shrimp are frozen, the cartons are immersed in cold water to glaze the shrimp. Or the carton may be opened and the shrimp glazed by spraying with water.

Glazing presents a problem of extra handling and also adds about 1-1/2 lb extra weight to a 5-lb carton, thus increasing transportation costs. If a package with the advantage of easy filling, puncture proofness and high moisture-vapor proofness were used, it would eliminate the loss due to dehydration and would eliminate any need for the glazing process.

Experiments show that unglazed shrimp can be packaged successfully for 12 months storage at 0° F in a waxed carton with a satisfactory overwrap. Both glazed and unglazed shrimp from packages with no overwrap

Courtesy of Nixon-Baldwin Div., Tenneco Chemicals, Inc.

FIG. 72. COMBINATION PACK FOR COOKED SHRIMP

Cooked individually quick frozen shrimp packed in a thermoformed PVC plastic container featuring an auxiliary inner pack
age containing cocktail sauce. The inner pack is hermetically
sealed with a coated film membrane lid.

were slightly tough but had no off-flavor. Elimination of the glaze
reduced the defrosting time of the five-pound block of frozen shrimp
from 36 to 55% (under running water) and also reduced the shipping
weight of the box by 21%. Glazed shrimp packaged without an overwrap
showed high weight losses, disappearance of the ice glaze and desiccation of the surface shrimp. The principal adverse factor resulting from
dehydration appears to be texture change.

The development of individually quick frozen shrimp has expanded
sales at the retail level. IQF shrimp are packed in polyethylene bags of
various sizes. A more recent development has been thermoformed linear
polyethylene containers covered with a heat-sealable lid. The use of the
covering material allows for the printing and product identification. In
order to prevent crushing and other types of damage, an outer paperboard carton is used.

Canning of Shrimp.—The large shrimp caught off the East Coast
south of Savannah and in the Gulf of Mexico are not only sold fresh
and frozen but also are canned. The largest sizes are marketed as
prawns. The shrimp are graded for size and are sold by the barrel which
contains 200 lb. First they are iced to make peeling easier. Peeling consists of removal of the head, thorax, and shell leaving the muscular portion of the abdomen and tail. After a thorough washing they are placed
in wire baskets and parboiled in hot brine for about four minutes. This
changes the flesh to a firm white color and the surface to a bright pink.

After cooling, inspection, and removal of bits and fragments they are prepared for filling. Wet pack shrimp can be filled immediately. Dry pack shrimp are allowed to drain. A one percent brine is added to the wet pack.

Care must be taken not to overfill, as this can mat the shrimp and injure their appearance. Weights of fill for standard cans are specified by the Food and Drug Administration. After exhausting and closing cans are retorted. Retort processing varies from 12 to 30 min on wet packs depending on can size, initial temperature and process temperature (240° to 250°F). Dry packs vary from 50 to 80 minutes. C-enamel cans are recommended. This enamel is an oleoresinous type containing zinc oxide pigment which counteracts the action of the sulfur in the food and prevents blackening.

Scallops

Scallops are shucked at sea and the scallop meats are washed and packed in ice. It is usually 7–8 days before the meat is unloaded and rapid processing is extremely important. If excessive dehydration occurs, shelf-life decreases and flavor and nutrient loss occurs. A moisture proof, durable package is mandatory for scallop packaging.

Most commonly used packages include hot melt or polyethylene coated cartons in both small retail sizes as well as the larger five pound institutional carton. In spite of the increasing use of packaging, most fresh scallops are still sold iced and are individually weighed into containers by the retailer.

Crabs

Because of the seasonal nature of crab production, the fresh product does not appear on the market in uniform quantities throughout the year. During the months when the crab catch is large, some method of preservation must be utilized. Since the fresh characteristics of crab meat are lost in freezing, this method of preservation is used for short storage periods. However Blue crab meat does not retain its original quality for any length of time when in frozen storage. It loses its flaky texture, becomes rather tough and loses some of its flavor and color.

Crab meat wrapped in MSAT cellophane and stored at 0°F remains palatable for nine months before becoming undesirably tough. Cooked crab legs that have been frozen, ice glazed, packed in fiberboard cartons and stored at 0°F also remain palatable for nine months. Crab meat packaged in hermetically sealed tin cans and stored at 0°F is edible after 12 months. Covering the meat with one per cent or three per cent salt solutions delays the onset of toughening but does not extend storage life beyond 12 months due to off-flavors developing. The two most important factors which lower the quality of frozen crab meat are changes in color and texture. Crab meat that has been carelessly washed will show bluish discoloration after storage. Blue crab meat that has been packaged loosely with many air spaces in the package is likely to turn yellowish. In order to minimize toughening and discoloration, the meat should be packaged tightly in a moisture-vapor proof container. The freezing and storage of raw crab legs is not recommended. After only three months, the product becomes discolored and off flavored.

Cooked Alaskan King crab legs are marketed frozen in paperboard cartons sealed shut with hot melts and cellophane overwrapped.

Canning of Crabs.—The canning of crabmeat from Blue crabs (and lobstermeat) encountered difficulties in the early days, as plain tinplate cans reacted with the hydrogen sulfide released by the fish to produce a black discoloration. Parchment liners helped the problem and, later, enameled cans all but eliminated it. Acid dipping of the shellfish prior to canning also is helpful.

Very little crabmeat is canned in the US because the crabs are smaller than Japanese crabs and the meat is so torn up in removal as to be much less attractive. Crabs must be processed rapidly. Live crabs are steamed for 20 to 25 min then the meat is picked out and washed in a weak salt solution. C-enameled cans are used. The Japanese use double enameled cans and sometimes a parchment inner liner in addition. Dipping in one per cent citric acid acid solution and packing in acidified brine give best results in avoiding discoloration of the meat. A tight fill is necessary. After filling, cans are exhausted 10 to 15 min before closing. Retorting varies from 20 to 95 min depending upon can size and processing temperature ($230°-250°F$). Standard packs are 5-1/4 and 7-3/4 oz drained weight.

Oysters

Oysters are harvested by hand tonging or by dredging all along the East Coast, the Gulf Coast and the West Coast of the US. The West Coast fields are rather new having been started when it was discovered Japanese oysters could be cultivated in those waters.

Freezing of oysters should offer wide opportunities for expanding the market for these shellfish. Although oysters are highly perishable and are produced in quantity only during the colder months of the year, only a small amount is frozen. Bags made of cellophane, pliofilm, nylon, polyethylene, or other films and papers give comparable results as far as retention of moisture-vapor. A sealed moisture-vapor proof bag within a waxed carton that is then overwrapped, would be an ideal package for this product. An overwrap would provide added protection against leakage and normal handling.

Fresh oysters are shucked into paperboard or thermoformed plastic (polystyrene, or polyethylene) containers with friction or snap on lids and sold refrigerated. The product has a limited shelf-life; once shucked, the oysters should be used promptly.

Shucked oysters may show a pink or red discolored liquor when they are thawed. "Pink Yeast" organisms are responsible and they are capable of growing at temperatures of 0°F or lower. Thorough washing and sanitary handling during preparation will prevent this occurrence.

Canning of Oysters.—Oysters for canning are thoroughly washed, and dead oysters (open shelled) or damaged oysters are discarded. The oysters are then loaded into a wheeled "steamer-car" which will hold 12-1/2 bu, and thoroughly washed again. The cars are passed through a steamer where the oysters are steamed 5 to 10 min at 10 lb pressure. This kills the oyster and opens the shell enough to permit removal of the meat. Oysters are cooled, shucked, and inspected, discarding all that are discolored. The oysters are then conveyed to a washing tank where they again are washed to remove shell and grit. After draining for 15 to 30 min the meats are filled and weighed into cans. Slack weight must be avoided. Cans are topped with boiling brine, double seamed, and processed. Plain coke cans are used for oysters although C-enameled cans may be used if desired. Retorting times vary with can size and processing temperature.

Some oysters are smoked and then packed in a sauce. These come packaged in cans or jars and are usually sold as a gourmet item.

Lobsters

Most lobsters are sold live from chilled tanks of seawater, and connoisseurs demand it that way. Only a small quantity are frozen due to their cost and the toughening of the meat during storage. In order to facilitate removal of the meat from the shell, the lobster is immersed in boiling water for an interval of 15 sec to 5 min depending upon the thickness of the shell. A heating period of 1-1/2 min is sufficient for a one pound lobster. After this treatment, the lobster is cooled and frozen. Frozen, cooked lobster meat in sealed, waxed packages keeps less satisfactorily than whole precooked lobster. The storage life of the latter is reported to be 3-1/2 months at −20°F.

The spiny lobster from Africa is used for frozen lobster tails. The tails are removed from the lobster, usually wrapped in nitrocellulose coated cellophane and packed in a waxed board carton. The cartons are often frozen 2-1/2 hours in a plate freezer and stored at −50° F.

Canning of Lobster.—Small lobsters are boiled in salt water for 20–30 min then cooled. The meat is removed from the shell and packed in enameled cans using parchment paper liners. Three can sizes are standard 3–4 oz, 6–8 oz, and 12–16 oz.

Brine is added and the cans are exhausted to 140°F before closing. Retorting at 245°F varies from 35 to 60 min depending on can size.

An interesting package for lobster consists of an air transported unit sold directly by the lobster catcher to the consumer. A round metal container is used and seaweed forms the cushioning material for the lobsters. Steamer clams are added to the unit for use in a clam-bake. The lobsters are packed live by hand into the container with claws pegged so that they are not in a position to attack each other. Packaged ice is placed below and on top. The container serves as the refrigerated shipper and also as the cooking vessel. The outer package is double walled in order to insulate the contents. A shelf-life of 48 hr is obtained, which is enough for air transportation.

Lobster has also been vacuum packed in a polyamide-polyethylene laminate for ease in marketing.

Clams

Clams are treated essentially the same as oysters. In canning only C-enamel cans may be used as black discoloration is a common problem. Clam juice, which is collected during steaming of the clams, is also canned as a beverage, either straight or blended with tomato juice.

Other Fish Products

Fish Roe.—Shad or herring roe are by-products and can be sold fresh, frozen or canned. Fresh roe is packed in sealed polyethylene film bags and kept refrigerated. Frozen roe may be vacuum packed before freezing. Polyethylene coated polyester film or polyethylene coated nylon film may be used for this purpose.

Some fresh roe is canned. East Coast roe are usually minced and canned with brine. West Coast canners preserve the shape of the roe and can it with brine, oil, or sauce.

In the USSR, black caviar (beluga) is obtained while the fish is still alive. If the fish dies, the membrane of each roe sac breaks and all is lost. The fisherman slits the fish and carefully extracts the roe. As much as one and a half million eggs are wrapped in the thin membrane of the sac. It is also possible for one roe to weigh as much as 220 lb.

The roe is then given a cold shower and carefully put into a large wooden sieve. A caviar taster grades the lot for taste. Caviar from different fish are never mixed. Grades are based on color, taste, grain size and hardness. Each lot is packed in separate batches. In the sieve, the roe is mixed by hand. The membrane floats to the surface, while the eggs slip through into a wide container which holds exactly 12 kg (26.46 lb). The caviar is spread out evenly since the eggs may be crushed by their own

FIG. 73.   CAVIAR BULK PACKAGES FOR EXPORT

Iranian workers at the Iranian fisheries examining 2-kilo tins that will be
used to pack precious sturgeon roe.

weight. The final step involves sprinkling with very fine dry salt (three
per cent by weight).

Granular caviar is tightly packed in flat jars with pry-open tops. It is
also packed in wooden barrels. Barrelled caviar is given an additional
salt treatment (7 to 10% by weight). The exact amount of salt depends
on three factors—geographical position of the packing plant, season of
the year and quality of the roe. Oak barrels are used and they hold about
13 gal. They are given a thin coating of parafin wax on the inside and
brushed with drying oil on the outside. Granular caviar is also usually
pasteurized, which eliminates the need for antiseptics and increases
shelf-life.

Pressed caviar is prepared by immersing the roe in a hot saturated
brine for two minutes. It is then scooped out with a sieve into linen sacks
and squeezed under a press. The caviar turns into a dough-like mass.
After this operation, it is packed into oak barrels lined inside with white
fabric and wax paper. It is also packed in 1.6 kg (3.5 lb) cans.

Roe caviar is prepared from roe with immature or weak eggs unsuitable for granular or pressed caviar. It is inferior to the others in taste. Packaging is usually done in flat glass jars.

Whale Meat

Whales, dolphins, and porpoises are the only mammals that live entirely in the water. Most whales are captured for their oil, but whale meat is consumed abroad particularly in Japan, the USSR, and Scandinavia. Whale meat is very similar to that of other red meated animals except it contains tiny amounts of trimethylamine oxide and methylamines, which give it a slightly fishy flavor. It tends also to be somewhat oily. The fat is very high in unsaturates and hence readily subject to rancidification. The meat can be chilled, canned, or frozen, and packaging which would be suitable for fish will be suitable for whale. Whale meat is ground into a meal rather than being processed as steaks or roasts. Whale meat consists of approximately 70% moisture, 4% fat and the remainder carbohydrate and protein.

A comparison was made of the protection against dehydration and oxidation of products frozen in various sheetings. It was concluded that PVDC film was superior to polyethylene film, foil bonded to Kraft paper, and greaseproof paper. After six months, rancidity had developed even in the PVDC wrapped meat.

PREPARED FISH FOODS

A wide variety of fish and shellfish foods have been prepared cooked, and are usually sold frozen. Examples include breaded fish sticks, scallops, clams, or oysters; lobster, crab, and shrimp newburg; lobster thermidor; cod fish cakes; tunafish pot pies; the fish fillet TV dinner and others. Breaded items are packaged in waxed paper overwrapped car-

Courtesy of Vita Food Products, Inc.

FIG. 74. HERRING SALAD IN ALUMINUM CONTAINERS

Herring salad spreads are marketed in formed aluminum tubs with pull tab easy open feature.

tons with parchment liners or separations. Those which contain sauces are packaged in sealed plastic bags, which can be put in boiling water to reconstitute. The pouches are usually polyethylene coated polyester film, and are overpackaged in a printed paperboard carton. Pot pies and TV dinners are packaged in foil pans and trays, which are also enclosed in printed paperboard cartons.

A new package to appear recently has been "herring salad" which is prepared from herring, mayonnaise, celery, pickles, apples, onions and spices. It is marketed in formed all-aluminum tubs with an easy-open feature. The concept closely resembles the whipped cream cheese pack-

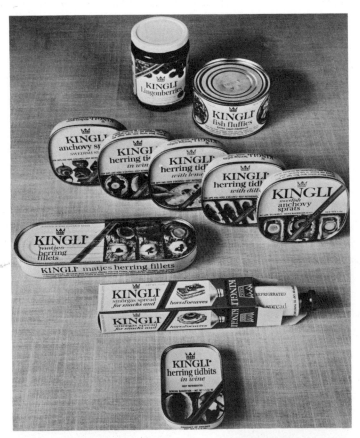

Courtesy of Vita Food Products, Inc.

FIG. 75.   FISH HORS D' OEUVRES

This picture shows several types of packages used for hors d' oeuvres fish products. Note particularly the collapsible metal tube used for the fish spread.

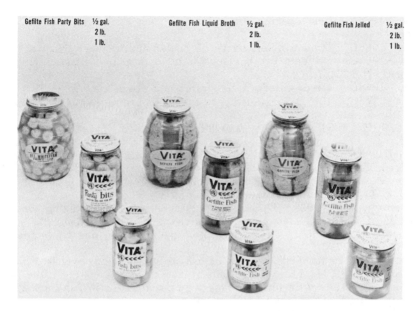

Gefilte Fish Party Bits	½ gal.	Gefilte Fish Liquid Broth	½ gal.	Gefilte Fish Jelled	½ gal.
	2 lb.		2 lb.		2 lb.
	1 lb.		1 lb.		1 lb.

Courtesy of Vita Food Products, Inc.

FIG. 76.   COOKED FISH PRODUCT IN JARS

Different shaped glass jars are used for packing several varieties of gefillte fish.

age and is one of the few new package innovations in prepared seafood dishes.

The collapsible metal tube has been used for anchovy spreads as well as salmon and other types of patés. It is an excellent package useful for hors'd'oeuvres preparation.

Gefillte fish is packed in glass jars and metal cans. It is a combination of two or three kinds of freshwater fish minced together with onions, carrots, matzo meal and spices. Combinations of fishes used include haddock, cod, whiting and halibut.

BIBLIOGRAPHY

ALMARKER, C. A.   1968.   Protection against light. Food Manuf. *43*, No. 1, 28.

AMBANELLI, G., and PIRATI, D.   1968.   The packaging in glass of products usually contained in cans. Ind. Aliment. *6*, No. 32, 77–80. (Italian)

ANON.   1966.   Fish cans of coated aluminium foil. Verpackungs Rdsch. *17*, No. 11, 1548. (German)

ANON.   1967A.   The prepackaging of fish. J. Refrig. *10*, No. 1, 5.

ANON.   1967B.   Plastics containers for pasteurized crab meat. World Fishing *16*, No. 2, 57.

ANON.   1967C.   Corrugated cases on the French fish market. Food Trade Rev. *37*, No. 7, 112.

ANON. 1968A. Easy opening fish cans. European Packaging Dig. No. *62,* Sep. 4.

ANON. 1968B. Oysters in flexi-packs. N. Zealand Packaging *5,* No. 6, 2.

ANON. 1968C. PS foam fish boxes—'Made to measure.' Plastics Rubber Weekly No. 234, 7.

ANON. 1968D. Keeps water out, flavor in. Food Eng. *40,* No. 2, 88–89.

CROSTON, W. G. 1967. Expanded polystyrene fish containers. Brit. Plastics *40,* No. 11, 112–115.

GIBLIN, J. P. 1968. Packaging fish in the U.K. Food Eng. *40,* No. 4, 120–125.

HEISS, R. 1968. The state of frozen fish packaging and the problems involved. Verpackungs Rdsch. *19,* No. 6, 41–44. (German)

JOSLYN, M. A., and HEID, J. L. (Editors). 1963. Food Processing Operations, Vol. 2. Avi Publishing Co., Westport, Conn.

MORTENSEN, J. P. 1968. Disposable polystyrene foam packaging for fish transport. Plastica *21,* No. 2, 56–58. (Dutch)

NACHENIUS, R. J. 1967. The packaging of fish meal. Fishing News Intern. *6,* No. 2, 20–27.

SACHAROW, S. 1966. Prepackaging of fish in plastics film. Svensk Emballagetidskrift *32,* No. 11, 14–16. (Swedish)

SACHAROW, S. 1967. Bacteria spoilage is a critical factor in fish packaging. Paper, Film, Foil Converter *41,* No. 2, 56–62.

SACHAROW, S. 1968. Prepackaged fish market is considered undeveloped; converters could help. Mod. Converter *12,* No. 21, 20–22.

VYNCKE, W. 1967. Tests on the temperature of fresh fish during transport. Rev. Gen. Froid *58,* No. 3, 275–280. (French)

WUNSCHE, G. 1968. Can problems in the long term storage of tinned fish. Verpackungs Rdsch *19,* No. 9, 1074–1081. (German)

CHAPTER 7

Fruits and Vegetables

INTRODUCTION

Fruits and vegetables originated in many different parts of the globe and for thousands of years accompanied man's travels. The earliest record of the grape and olive show them to have been cultivated in Transcaucasia about 6000 BC. Lentils, cucumbers and figs were Stone Age foods from the Himalayas and Mesopotamia. Cabbage, broad beans, and spinach were raised in the Bronze Age in the Mediterranean area, Asia, and Afghanistan.

The Myceneans, Greeks, and Romans spoke highly of the quality of Persian melons and peaches, Russian plums and cherries, Arabian asparagus, and Syrian eggplant.

FRESH FRUITS AND VEGETABLES

The market for fresh produce is very large and varied. Even the most popular items number well over 100. It is therefore impossible to discuss all types in this book. Rather, it will be necessary to discuss general categories with a few examples, seeking, where possible, a common denominator.

In 1966, $9,751,450,000 worth of fresh produce was sold for direct consumption in the United States. This does not include exported goods, which were processed into other foods; or those which were preserved by freezing, canning, or the like. Of this total about $6.45 billion were vegetables and $2.3 billion were fruits.

During the 16 year period from 1950 to 1966 the population of the United States increased 30%. During the same period the total pounds of fresh fruits consumed declined 3.6% (16.5 billion pounds to 15.9 billion pounds) and the total pounds of fresh vegetables consumed increased only 10% (17.4 billion pounds to 19.2 billion pounds). From the same figures it is obvious that the per capita consumption has shown a slow but steady downward trend.

This is attributed to the industry's failure to make fresh produce competitive with canned and frozen products in terms of convenience and to find ways of preserving fresh quality long enough to satisfy the consumer.

Some experts say the decline in consumption of fresh fruits and vegetables is due to the increase in working women who do not have as much time for food preparation. Hence, also, the increase in demand for convenience foods. During the same period advertising for canned and frozen foods was tremendously increased, and the money spent on research and development by food processors more than doubled. Fresh produce companies could not compete either with advertising or research. It became apparent however that something would have to be done to improve the convenience of fresh produce.

Product Characteristics

Because fruits and vegetables are living organisms even after harvesting, they can remain fresh only as long as normal metabolism continues. Metabolism involves the absorption of oxygen which breaks down the carbohydrates in the product to water and carbon dioxide. If the availability of oxygen is restricted, the chemical reaction changes and small quantities of alcohol are produced. This results in off-odors and -flavors and a breakdown of plant cells. This series of events is called anaerobic decay and can spoil fruits or vegetables within a few hours.

Fruits and vegetables are very high in moisture, ranging from 75–95%. Their equilibrium humidities are as high as 98%. Under any normal atmospheric condition they will dry rapidly. This causes wilting and shriveling due to loss of rigidity and shrinkage of the cells. Proper packaging is able to prolong the storage life of fresh fruits and vegetables by preventing wilting. Platenius (1946) showed that serious wilting occurred if the loss of moisture exceeded 10% of the product weight. The rate of moisture loss varies with the product and water vapor permeability of the packaging film. The use of small perforations for oxygen permeation has an insignificant effect upon moisture loss.

A very prevalent type of spoilage of fruits and vegetables is that caused by microorganisms such as yeasts, molds, and bacteria. These organisms can cause destruction by growing on the exterior of the product or they may invade the interior through a surface bruise or cut and cause internal decay. This is why careful handling and packaging is so important in the preservation of freshness and quality.

Normal ripening of fruits and vegetables causes alterations in color, texture, odor, and flavor. At some point for each type ideal ripeness is achieved. Beyond that point the product becomes overripe and quality deteriorates. The primary goal of fresh produce merchandising is to deliver the product to the consumer at such a point in the ripening scale, that it will achieve perfect ripeness at time of eating. This of course is extremely difficult to do. In practice the produce is delivered somewhat underripe at time of purchase and the consumer delays consumption until ripening has completed.

Since all of these processes are very highly sensitive to temperature, they can be slowed by storing the produce under refrigeration. Each fruit or vegetable has an ideal temperature for storage. If this temperature is not used, deleterious results occur, for example—tomatoes will not ripen if chilled below 40°F, bananas will turn black below 50°F and potatoes develop a sweet flavor below 41°F.

An additional reason for the necessity of refrigeration for fresh produce is the heat generated due to metabolism or respiration. For example, green beans, sweet corn, broccoli, green peas, spinach, and strawberries generate from 15,000 to 50,000 Btu per ton per 24 hours of storage at 60°F. Even when chilled to 32° 40°F they still evolve from 2500-17,000 Btu per ton per 24 hours. This heat must be taken into consideration in the design of refrigeration equipment.

Some types of fresh produce give off volatile compounds during ripening which will impart unacceptable odor and flavor if not allowed to escape, or they may prematurely ripen the fruit.

Packaging Requirements

For the most part, packaging cannot delay or prevent fresh fruits and vegetables from spoiling. Incorrect packaging can accelerate spoilage. However packaging can serve to protect against contamination, against damage, and, of most importance, against excess moisture loss. Too much of a moisture barrier will cause an excessively high relative humidity in the package and result in accelerated spoilage due to microorganisms or in skin splitting on some fruits.

PREPACKAGING OF PRODUCE

The packaging of produce is not new but there is little recorded about it. In very early times most communities raised and consumed their own, and had no need to package it. Later in times of plenty, surpluses were taken to markets where they were vended. Transportation was by boat or horse-drawn cart or similar means. Little or no protection was given—perhaps covering of sorts—a goatskin in early times, a mat or cloth or tarpaulin in later periods. Any convenient box, barrel, basket, or sack served as a container for transporting smaller loads.

Not until 1856 was any attempt made to give additional protection to a fresh fruit. In that year paper was used to wrap and package fruit. In 1879, oranges that were shipped from Australia to England were wrapped in paper, and by 1880 the practice was commonplace. Yet, wrappers were not used for vegetables until about 1900.

In 1895, US apple growers began to ship boxes of paper-wrapped apples. Over the years they built a reputation for quality fruit, free of scald and bruises. In 1907, Samuel Frazer sold apples in New York City in paperboard boxes containing 6 or 12 units. In 1910, a New York cooperative tried prepackaging potatoes in ten pound bags and half-peck containers but the public was not ready. Not until the late 1920's and

FIG. 77. A SCENE FROM LES HALLES PRODUCE MARKET IN PARIS

Fresh produce traditionally was shipped to market in wooden crates and baskets.

early 1930's did some prepackaging of potatoes, onions and oranges by shippers and retailers meet with some success. In the late 1930's and early 1940's special terminal prepackaging plants began to prepackage tomatoes, kale, salad mixes, spinach, and soup mixes; and growers began to package mushrooms and cranberries. During the 1950's, the trend continued with apples, celery, and other citrus fruits joining the group. But general acceptance by produce vendors was slow. They could see no good reason for investing in equipment in the face of a slow consumer acceptance. They wondered about the real cost of the operation.

With considerable foresight, in 1944 a group of companies pooled their interests and sponsored a project to evaluate prepackaging of all items in a produce station. This became known as the Columbus Experiment. Merchandise was brought to the warehouse in conventional containers and there it was washed, trimmed, and repackaged in consumer units. Moisture proof cellophane was used as wrapping material and each unit was labeled, coded, and price marked. All produce was kept refrigerated from start to finish. Results were conclusive. Prepackaging together with refrigeration was proven to reduce distribution losses by as much as half, to save labor costs, to lengthen shelf-life, and to achieve good consumer acceptance.

A typical market was found to lose as much as 30% in conventional instore handling, trimming and reconditioning of produce, whereas prepackaged produce had negligible losses. Very little markdown or discarding was needed.

Shelf-life of some products was dramatically improved. For example, prepackaged spinach was 90% saleable when bulk spinach had totally spoiled. Prepackaged cauliflower was 100% saleable after five days, whereas ordinary cauliflower without refrigeration was 50% unsaleable in the same period.

Carrots, radishes, parsnips, beets, turnips, and all root crops were extensively prepackaged in the 1950's, some at shipping points and some at specialized packaging plants (terminals). In the early 1960's bananas, pears, asparagus, grapes, cauliflower, more apples, more citrus, snap beans, celery and early potatoes were prepackaged at the shipping point. Cut-up squash and parsley were packaged at terminals, and soft fruits in the retail stores.

From 1960 to 1965 great strides were made in developing new and improved packaging materials and machinery for their application. Film manufacturers began to produce different kinds of films aimed at providing necessary combinations of strength, heat sealability, heat shrinkability, moisture retention, gas permeability, clarity, breathability, stiffness, durability, fogging resistance, water resistance, and printability. The goal was to extend the shelf-life of fruits and vegetables and also to offer the consumer much more effective and attractive packages. By selection of materials and by redesign, containers have been developed which reduce injury and bruising during shipment and display.

Much research has been conducted on the use of controlled atmospheres in storage and transportation of produce to delay deterioration on long shipments in specially equipped trucks, railroad cars, or even ships. One method used was storage of apples and pears in boxes sealed within polyethylene liners. When held at low temperature the respiration of the fruit reduced the oxygen and increased the carbon dioxide in the container thereby slowing ripening of the fruit by as much as several weeks. On removal from storage, bags must be perforated to avoid anaerobic decay and off-flavor development at ambient temperatures.

Advantages of Prepackaging [1]
Permits brand identification, advertising, price and product information, and recipes.
Reduces waste throughout the marketing system.
Reduces transportation and handling charges.
Reduces retail labor costs.
Increases quantity sold per customer.
Speeds up sales in stores.
Increases maintenance quality and shelf-life of produce.
Fits in with trend toward unitization and self-service retailing.
Makes a cleaner, neater and more attractive display.
More economical and convenient handling.
Provides a service and becomes more of a convenience item.
Decreases possibility of accidents in produce area—thus lower insurance rates.
Decreases spoilage losses and markdowns.
Impulse buying is stimulated because packaged produce is attractive.
Product sanitation is improved because contact with dirt, insects, and consumer handling is prevented by the package.
Pilfering of small produce is minimized.
Fresh appearance is maintained longer.
Reduces possibilities of errors in weighing and pricing at retail level.
Permits customers, regardless of the time of day, to select produce which has not been "picked over" and handled.
Permits an increase in variety of produce displayed in smaller areas being alloted to the produce department.
More convenient storage of consumer packages in the home.
Provides for easier inventory control and proper stock rotation.

Disadvantages of Prepackaging [1]
Cost of prepackaging in relation to cost of product may be high.
A general lack of knowledge as to what, how and where to prepackage.
Resistance to change—wholesalers—retailers—growers and shippers —consumers.
Requires high standards and uniformity.
Quality control difficult on perishables.
Not all produce is adaptable to prepackaging.
Requires considerable investment.
Consumers like to pick out own merchandise.
Difficult to package in sufficient sizes for all families.

---

[1] From an unpublished work by J. S. Raybourn, Produce Packaging and Marketing Association, Inc., Newark, Del.

Deterioration of product causes loss of merchandise as well as packaging costs.

Increased refrigeration requirements throughout market system.

Packaging of produce is currently being done at three primary levels: shipping point or source level in the production region, terminal packagers or central warehouses at the distribution level, and in-store or retail level. Each level at which packaging is done has some advantages and disadvantages and it is the general consensus that packaging will continue to be done at all levels for the foreseeable future

Source Level Packaging
Advantages [1]
Saves transportation charges on waste discarded before shipping
Allows brand identification and national advertising.
Cheaper labor and property costs.
Can standardize and grade for specific standards.
Better control over product and how it reaches consumer.
Large scale permits mechanization and improved skills in product packaging.
Producer gets larger share of consumer's dollar.
Producer-shipper becomes more familiar with consumer demands and can better adjust to the market.
Less bruising of product in transit.
Number of times the produce is handled is reduced.
Fewer commodities handled should permit greater specialization leading to more efficient operations at source level.
Facilitates the adjustment of supplies to demand because of greater choice of alternate markets.
Permits packer to grade and sort to various standards and to take advantage of area preferences for particular grades of merchandise, and size and type of packages.
Disadvantages [1]
Seasonal marketing may preclude use of machinery and facilities or seriously limit the efficiency.
Some deterioration will occur in the marketing channels due to the perishability of the produce.
Additional investment and costs which prepacker must pass on to purchaser.

Distribution Level Packaging
Advantages [1]
Greater efficiency thus lower costs due to larger volume, mechanization, specialization, skilled personnel and year around operation.

Can maintain a more complete line and achieve a satisfactory selection of quality and price by buying from different parts of the country for various commodities.

Closer to retail market than grower-shipper. Can adjust to changes.

Reduction in investment for equipment and materials at retail and source level.

Quality control improved over source packaging since merchandise is packaged hours rather than days before reaching retail outlets.

Permits more careful grading for quality and ripening than source packaging, and package sizes can be varied to meet local consumer demands.

Can justify considerable investment in machinery and labor saving devices due to large volumes handled, and relatively long packaging seasons.

Disadvantages[1]

Higher rent than at source and space usually at a premium.

Union labor costs are higher, also fringe benefits, than at source level.

Transportation costs are higher since non-usable parts of produce are shipped from source.

Cost of waste disposal is relatively high.

Retail Level

Advantages[1]

Can package better for store's clientele since preferences can be determined easier.

Operations can be changed to meet demands.

Can price commodities when packaged. This is difficult to accomplish consistently at other points.

Make more efficient use of labor by smoothing out labor peaks during the week.

Quality control can be better established and maintained since inspection and trimming is done just prior to sale.

Less chance of spoiled merchandise being put on shelves.

Disadvantages[1]

Small scale operation cannot justify expenditure for full mechanization.

Labor costs are very high—additional personnel required.

Space is at a premium.

Personnel training and knowledge insufficient for wide variety of products needed.

Additional investment for equipment, materials, and space.

Time required for packaging to permit adequate stockage may not be available.

Packaging operations are basically a hand operation, on a small scale, so packaging costs are relatively high.

Trends in Produce Packaging

As mentioned previously, produce has been sold in package form throughout the United States for a number of years, however, it was not until 1964 that the sale of packaged produce approached the quantity of the sale of produce in bulk throughout the country. In 1964 packaged produce accounted for 46% of the sales, while bulk produce accounted for 54% of the sales. In 1965 prepackaged produce became the predominant method of sale accounting for 55% of the sales, while bulk sales dropped to 45%. In May of 1967, it was estimated that prepackaged produce accounted for 56-1/2% of the total sales, and, in November of 1967, it was estimated that packaged produce accounted for 57.5% of the total sale of produce throughout the country.

The following is a tabulation on the trends in variety and amounts of produce prepackaged by all sources prior to receipt at retail stores, and compares the periods from 1955 through 1964.

	1955 %	1958 %	1964 %
Apples	25	35	60
Asparagus	5	7	15
Bananas	1	5	10
Beans, lima		3	5
Beans, snap	5	5	5
Beets	20	20	50
Blackberries	100	100	100
Blueberries	100	100	100
Broccoli	5	5	10
Brussels sprouts	75	90	100
Cabbage	5	5	5
Carrots	80	90	95
Cauliflower	5	20	70
Celery	10	20	30
Cherries	5	3	3
Chinese cabbage		10	5
Collards	5	10	20
Corn	5	10	7
Cranberries	100	100	100
Dandelions		3	10
Endive	5	10	10
Escarole	5	10	10

	1955	1958	1964
	%	%	%
Garlic	90	90	95
Grapefruit	10	20	40
Grapes	3	5	2
Kale	75	75	85
Leeks		5	5
Lemons	15	20	30
Lettuce	5	5	15
Limes	5	15	25
Mushrooms	50	60	25
Mustard greens	25	20	30
Okra		5	5
Onions, dry	20	35	75
Onions, green	5	5	15
Oranges	20	20	40
Parsley	1	5	15
Parsnips	25	35	80
Peaches	5	7	8
Pears	5	5	15
Peas, green	5	5	5
Peppers, sweet	5	5	5
Plums-prunes	5	5	5
Potatoes	35	55	75
Radishes	50	75	80
Raspberries	100	100	100
Rhubarb	5	15	35
Shallots		5	10
Spinach	75	80	90
Squash	1	5	10
Strawberries	100	100	100
Sweet potatoes	1	5	15
Tangerines	5	5	5
Tomatoes	60	65	60
Turnips-rutabagas	25	50	65
Watercress		10	95

The supply of some commodities has increased considerably over 1955, while in others there has been a reduction, so actual tonnage packaged may be greater or less than indicated by the percentages. It is the trend which is important, and there is every indication packaging prior to the retail level is gradually increasing.

Packages and Packaging Materials

Bulk Shippers.—In the early days of shipping produce barrels were used and the fruit at "the bottom of the barrel" left much to be desired. Also due to the close packing and heavy shaking occasioned during transport, unless the fruit was resorted on delivery, rapid spoilage could be expected, for everyone knew it took only "one rotten apple to spoil the barrel." Careful studies over the years have proven the virtue of specially designed bulk shippers. For example, potatoes are shipped in corrugated boxes, and apples and similar fruit are shipped in molded pulp trays. New deep pocket trays are proving to be even more effective than the former shallow cups. Easily bruised fruits are protected by using foamed polystyrene trays, shrink films, or paper wrappers to prevent their jostling against one another. Round baskets were found to be wasteful of shipping space and have given way to rectangular boxes. Jumble packing has been studied and found to reduce bruising by as much as 50% over place-packing or careful facing of fruit. Jumble packing has also proven to save money due to reduced labor costs, product losses, or more efficient usage of shipping space.

Jumble packed fresh peaches in 1967 saved $3.3 million dollars over conventional methods. Jumble packed Washington cherries in 20-lb boxes instead of faced 15-lb boxes saved $200,000 a year. Similar packs for apricots and prunes are expected to save $450,000 and $900,000 annually.

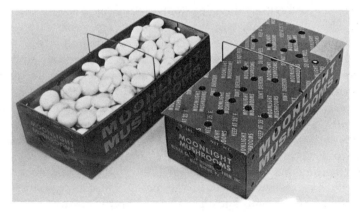

Courtesy of St. Regis Paper Co.

FIG. 78. SPECIALLY DESIGNED MUSHROOM SHIPPER

This corrugated shipper eliminates splintering and puncturing of the product—a frequent occurrence with previously used wooden shippers. Product brand name can be attractively displayed, and the handle pushes down to facilitate stacking.

Courtesy of E. I. du Pont de Nemours and Co.

FIG. 79.   CORRUGATED SHIPPERS COATED WITH PLASTIC
FOR MOISTURE RESISTANCE

Pretrimming of produce in the field also saves on shipping costs. Trimming carrot tops and cauliflower jackets in the fields saved almost half the space and required far less icing to deliver these products to the market. The savings amounts to $4.5 million per year. There is great potential for similar savings in other products such as celery, and lettuce. Over $6.5 million could be saved on these two products alone.

Air shipment of produce is becoming more and more frequent. In 1967, 40 million pounds of California produce was air shipped, an increase of 65% over 1966. During the first quarter of 1968, the total was nearly double that of the same period in 1967, mostly due to strawberries. Over 150,000 lb of asparagus were air shipped to Europe in 1968. It is not unreasonable to expect air shipped produce will amount to 10% of the total shipments by 1971. Special light weight shippers are being developed which utilize fiberboard and foamed polystyrene for strength, weight-saving, protection, and insulation, thereby saving on amount of ice required.

With the increased use of mechanical harvesting equipment more and more bulk packing will be jumble packed in large capacity bins which can be palletized and shipped to terminals for prepackaging. Whether these bins will be throwaway shippers, or reusable metal or plastic containers remains to be established. Standardization of sizes may ultimately be expected, but lack of standardization on dimensions of transport equipment is delaying this desirable goal. There is no question that it will result in substantial savings in warehousing, inventories, and handling costs.

Containerization may also lead to the elimination of damage due to extra handling. Ideally produce will be loaded into shipping containers and these into van containers and then not unloaded until arrival at the receiver's warehouse. One eastern apple shipper made three van container shipments to Great Britain and one to Panama with reports that the apples arrived in excellent condition. Again, more standardization will be required before this can be adopted on a wide scale.

Retail Packages.—The materials used for retail packaging of produce are extremely varied, however they can be categorized as films, backings, boxes, bags, bands, closures, and labels.

Films.—The first study involving the use of plastic films for fresh produce was conducted by Platenius (1943). He based his ideas on the observation that the changeover from normal respiration to anaerobic fermentation occurs when the oxygen concentration in the atmosphere has fallen below three per cent. He believed five per cent to be the minimum safe permissable concentration for oxygen in the atmosphere and then determined the rates at which oxygen was utilized by other commodities at this level of concentration. A wide variation existed and he found that no existing film possessed enough oxygen permeability in a completely sealed package. Thus, he formulated the concept of "vents." He demonstrated that two small holes, each 1/8 in. diameter, in a package were sufficient to prevent anaerobic fermentation under all conditions.

Films are purchased in the form of sheets, rolls, bags, and sleeves and may be nonshrink or shrinkable types. In the nonshrink category coated cellophanes are still probably the most widely used but recently have given way to plasticized PVC at the in-store level. They come in a wide assortment of sizes and may have a coating on one or both sides of nitrocellulose or polyvinylidene chloride to provide moisture-proofness, heat sealability, and water resistance. Because cellophane is only semipermeable to oxygen, it must be perforated in order to allow normal respiration. A few small holes the size of a pencil point are usually sufficient. Some stores have a roller studded with sharp needles which can be pushed across a stack of film sheets to perforate. Of course if the package uses the film only as a cover or a sleeve, leaving openings elsewhere for respiration, perforating is not necessary. There are three major US suppliers of cellophanes each offering many types. It is also possible to purchase imported cellophanes and to purchase special laminations from converters.

Polyethylene is available as a nonshrink film from many suppliers. It is usually low density and easily heat-sealed. It is not as transparent as cellophane, however this property has been improved and research continues. Polyethylene is the most widely used film for bagging applications and excellent for firm products like oranges, potatoes, onions, and carrots. It must be perforated for those products requiring moderate to high respiration. Polyethylene sleeves are used for celery, and small bags are used for lettuce. It is readily heat sealable and quite strong and tear resistant.

Cellulose acetate film is highly transparent and "sparkly" making an attractive package. It is relatively high in permeability to oxygen and carbon dioxide. Consequently perforations are not needed on produce having low respiratory rates. It has been used as an overwrap for tomatoes in cartons and may also be used as the window in window cartons.

Pliofilm (rubber hydrochloride) has occasionally been used for wrapping fresh fruits and vegetables. It is usually confined to heavy items where its strength plays an important role. It is a good barrier and highly moistureproof. It must be perforated to permit respiration, adequate moisture loss, and prevent fogging or early bacterial spoilage. It can be heat sealed. It is usually used in a bag form where the heated film is stretched to give a tight wrap.

Polystyrene film is similar to cellulose acetate in that it is clear and crisp and has high permeability. It is also used extensively for lettuce and tomatoes, and is also available as a heat shrinkable film.

Perforated polypropylene is being used by a number of growers as contour wrap on produce such as lettuce, cabbage and cauliflower.

Other heat shrinkable films are used which have been mentioned before in other chapters. These include irradiated polyethylene, polyvinyl chloride, polyvinylidene chloride, and polypropylene. All find considerable general use in shrink wrapping of fresh produce and preferably are used as sleeves or complete overwraps. In the latter instance, perforations are needed.

Backings.—This term is used to embrace a wide variety of tray type containers used to hold produce in conjunction with film overwraps or sleeves. They are made from molded pulp, formed paperboard, formed plastic, molded plastic foam, laminations, and aluminum foil. Each is designed for a particular item and quantity. One may hold four round fruits like tomatoes or oranges. Another may hold 2 ears of corn, still another 6 ears of corn or 5 or 6 large baking potatoes. Some provide maximum view of the contents, others provide maximum protection against handling damage.

Courtesy of American Viscose Div., FMC Corp.

FIG. 80. SOFT PVC FILM STRETCH WRAP

Another type of PVC film packaging for produce is illustrated by American Viscose's PVC P-21, a soft vinyl film which is stretched tight around the product.

Boxes.—In this category are included wooden, pulp, and plastic "tills" for small fruits like berries and cherry tomatoes; window cartons; lidded boxes; and cups. Window cartons offer more opportunity for brand identification and decoration, but the traditional "till" still hangs on. Pulpboard is taking over from wood, but some injection molded plastic tills are becoming popular.

Bags.—This category includes all bags (other than plastic film) and includes regular kraft paper sacks, double-walled paper bags, window bags, fiber net bags, plastic net bags, paper mesh and plastic mesh bags.

GCMI ITEM NO.	CAP. O/FLOW FL.OZS.	WT. MAX. OZS.	A	B MAX.	D	E	F	SPECIMEN FINISH
15-46	14 $\frac{3}{4}$	6 $\frac{5}{8}$	4 $\frac{1}{2}$	3 $\frac{3}{64}$	2 $\frac{1}{4}$	2 $\frac{23}{64}$	$\frac{3}{4}$	63-1710
15-47	15	6 $\frac{5}{8}$	4 $\frac{1}{2}$	3 $\frac{1}{8}$	2 $\frac{1}{4}$	2 $\frac{15}{32}$	$\frac{3}{4}$	66-1710
15-53	17	7 $\frac{1}{4}$	4 $\frac{11}{16}$	3 $\frac{1}{4}$	2 $\frac{3}{8}$	2 $\frac{15}{32}$	$\frac{13}{16}$	66-1710
15-71	28 $\frac{3}{8}$	11 $\frac{1}{4}$	4 $\frac{7}{8}$	4 $\frac{3}{32}$	2 $\frac{1}{4}$	3 $\frac{9}{64}$	1 $\frac{3}{64}$	83-1710

NOTES :—

1. WHEN OTHER FINISHES ARE USED, CAPACITY, WEIGHT AND HEIGHT SPECIFICATIONS ARE ADJUSTABLE WITHIN THE REQUIREMENTS OF THE FINISH USED. SEE NOTE 2.

2. HEIGHT (DIMENSION 'A') IS BASED UPON USE OF SPECIMEN FINISH SHOWN.

3. THE SPECIFICATIONS SHOWN MAY VARY MODERATELY ACCORDING TO COMMERCIAL TOLERANCES AND INDIVIDUAL MANUFACTURER'S PRACTICE.

4. 'B' DIMENSION IS VARIED TO MAINTAIN CAPACITY.

From Joslyn and Heid (1963)

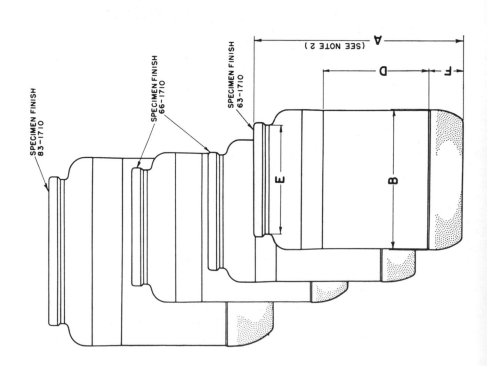

Courtesy of Reynolds Metals Co.

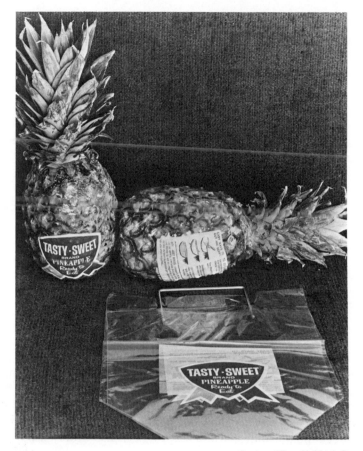

FIG. 82. SHRINKABLE REYNOLON PVC FILM FOR PINEAPPLES

A printed heat shrink PVC film provides brand identification as well as helping to preserve freshness of the product. Tear resistant Reynolon films can survive the sharp contours of this spiny fruit.

The mesh and net bags have long been used in packaging potatoes, onions, and some citrus fruits. A new plastic mesh netting made from high density polyethylene has found an excellent application in field-bagging of grapes and citrus fruits. Prepackaged grapes confined by the netting showed reduced losses of 1-1/2 to 2-1/2 lb per standard bag. Grapes are easily visible and can be washed or stored right in the bag.

FIG. 81. STANDARD DIMENSIONS OF SHOULDER-TYPE FRUIT AND VEGETABLE JARS

FIG. 83. ITALIAN INGENUITY CREATES AN APPLE BASKET
FROM A CARTON

This container is die cut and folded from a two side hot melt
coated white kraft paperboard. The carton is easily assembled,
compact, stackable and provides its own carrying handle. This
was a winner in the "Oscar for Packaging 1968" contest sponsored by the Italian Institute of Packaging.

Bands.—A number of different materials are used for banding produce items, including simple elastic rubber bands, printed paper bands
that can be heat sealed or otherwise locked together.

Closures.—Twist ties are the most popular method of closing bagged
goods; but cellophane adhesive and stapling are also widely used.

Specific Applications

The particular type of package used depends upon the shape and
perishability of the product. There are five main classifications—soft
fruit; hard fruit; stem products; root vegetables; and green vegetables.

Soft fruits are highly perishable and easily subject to anaerobic spoilage. They bruise and squash easily which leads to rotting. They are best
packaged in semi-rigid containers with a cover of cellophane, cellulose
acetate, polystyrene or other suitable film cover. Adequate ventilation is
a must to avoid fogging. Handling must be careful and avoided as much
as possible. Shelf-life is limited due to individual damage and decay.
Some berries under ideal conditions only remain top quality for 2 or 3
days. Typical soft fruits are currants, cherries, grapes, blueberries,
strawberries, raspberries, plums, etc.

Hard fruits are better able to resist damage from handling. They are
also less perishable and have lower respiratory rates. Shelf-life is weeks

Courtesy of E. I. du Pont de Nemours and Co.

FIG. 84. "VEXAR" HIGH DENSITY POLYETHYLENE NETTING

This netting helps reduce damage and losses in soft products
such as grapes.

rather than days. The most common package is an open tray or boat
with a plastic film overwrap or sleeve. Hard fruits may also be bagged in
perforated polyethylene film or in nets. Examples of hard fruits are
apples, bananas, citrus fruits, peaches, pears, and tomatoes.

Stem products are highly perishable as they rapidly lose moisture.
They should be bagged or wrapped in moistureproof cellophane or polyethylene with ventilation, or they should be banded or sleeved with
shrink film. Typical stem products are celery, rhubarb, and asparagus.

Root vegetables are not highly perishable. They can be stored for long
periods; however, it is desirable to protect them against moisture losses.
They are washed, graded and sized prior to packaging, which is usually
in durable polyethylene bags. Typical root vegetables include carrots,
parsnips, turnips, radishes, rutabagas, onions, beets, yams and potatoes.

Courtesy of American Viscose Div., FMC Corp.

Fig. 85. Cellophane Bagging Helps Sell
Prepackaged Mixes

Prepackaged combinations of vegetables are already marketed.
Future trends will probably include paring, dicing and other types
of preparation now performed by the housewife.

Potatoes are susceptible to light which causes greening. This can be
avoided by printing the film, or tinting it an amber color.

Green vegetables tend to dry out readily and wilt. Moistureproof
packaging helps to keep them crisp. However, they have high respiratory rates and are very sensitive to anaerobic spoilage. Ventilation is
extremely important. This group includes brussels sprouts, cabbage,
lettuce, broccoli, and cauliflower.

Equipment

In order to package fresh produce either in-store or at warehouses
special equipment has been developed which wrap packages, shrink
film, heat seal film, fill bags, close bags, apply bands, weigh, compute
price, price mark, and print labels. Other machines make ice, insert
staples, tie bags, convey product, chop, slice, sort and grade, and dispose
of waste. Each represents an investment but overall the savings to the
stores more than justifies the expense.

Future Trends

There are several areas where improvements can be expected and these have already been mentioned. They include improved standardization of bulk shippers, more air shipment, controlled atmosphere storage, and of course improved design packages and new packaging materials. The most likely area for the future development of fresh produce sales lies in convenience packaging. Imaginative efforts can help fresh produce compete with canned, frozen, and dried products.

Some examples are as follows: prepared vegetables for soups or stews; tossed salad; washed and bagged spinach, collards, kale, and other greens; washed and bagged beet greens; chopped endive; precut green beans; peeled winter squash; peeled parsnips (sliced); peeled and diced turnips (or sliced); sliced melons, and melon balls; sliced tomatoes; citrus salad; chop suey mix; washed and wrapped baking potatoes; prepared cole slaw; and prepared relish tray.

It is also quite likely that more effort will be placed on attractive merchandise displays, advertising, suggested recipes, and instructions on how to prepare them. With such an effort perhaps the per capita consumption of fresh produce can be put on the increase again.

There will also be a greater demand for produce to be packaged at the grower level or distribution level. This concept is based on the promise that high volume packaging increases production and thereby reduces unit cost. It also enables better supervision of the packaging and a greater degree of quality control. The industry foresees more mechanization of the produce packaging operation at the grower and distribution levels. In-store packaging will reduce proportionally, since this level will no longer be able to compete economically with the packagers at the grower and distribution levels.

A recent study made by the Agriculture Department stated that food stores could save more than $13.6 million by adopting assembly line packaging of produce at the warehouse level. The lowest costs for packaging produce after it is shipped in bulk to stores averages 7.2 cents per package. If warehouse packaging is used, this cost decreases to 4.8 cents. It is expected that by 1970, 60 to 75% of all fresh vegetables will be packaged prior to arriving at retail food stores.

FROZEN FRUITS AND VEGETABLES

In 1967-1968, over $862 million worth of fruit and $659 million worth of vegetables were frozen. The freezing of fruits in this country for later use in making preserves, jellies, pies, etc. began in 1905 on the East Coast. Pacific Northwest berries were frozen in 1910. The commercial freezing of vegetables started in Oregon in 1929. Although it was known in the Great Lakes Region that fish left on the ice by winter fishermen froze rapidly and commanded a premium price due to higher quality, few people related this fact to the freezing of fruits and vegetables. Slow and unreliable methods of freezing resulted in such poor qualities that many states and municipalities passed laws requiring merchants to inform their customers when they were selling thawed frozen foods. "Cold storage" was not a connotation of quality. Today frozen food products have widespread acceptance and are one of the fastest growing segments of the food industry.

Product Characteristics
Frozen fruits and vegetables have a few properties that differ from the fresh products.

First is the fact that most frozen foods have been prepared in order to eliminate the cost of freezing unnecessary components. Stems, skins, tops, hulls, pits, etc. are removed and discarded and the remaining product may be treated whole or sliced, diced, shredded, chopped, or even puréed. During this preprocessing some alterations can be encountered unless proper precautions are taken. Vegetables lose flavor if there is a delay between harvesting and blanching. Corn loses up to 1/2 of its sugar if held 24 hr at 86°F and peas up to 1/3 if held 6 hr at 77°F after picking. Spinach, lima beans, asparagus and mushrooms lose sugar rapidly. Corn, peas, asparagus, mushrooms and potatoes lose appreciable amounts of glutamates on standing. These again represent losses in flavor. Rapid handling and cooling is the only way to minimize those losses.

Bananas, peaches, apples, and pears turn brown when exposed to air after peeling. This seems to be related to the presence of catechol-tannins, which are readily oxidized by oxygen in the presence of polyphenoloxidase enzymes. Freezing with nitrogen avoids this discoloration as it displaces oxygen and slows enzymatic action. Ascorbic acid also helps prevent browning. Hulled strawberries deteriorate rapidly on exposure to air and to drying and acquire a stale taste.

Blanching of vegetables in steam or hot water is used to inactivate certain enzymes, such as catalase, which cause off-flavors. Blanching also preserves the color of the vegetable so that in later cooking it will not wash out. Blanching should be kept to a minimum however as it leaches out valuable nutrients and vitamins. It is not desirable to blanch to the point the vegetable is cooked—as frozen cooked vegetables don't keep as well.

The freezing process, unless very rapid, forms large ice crystals within tissue cells. These cause structural breakdown of the product and limpness or mushiness. Fluctuations in temperature of storage can cause further growth of internal ice crystals. Even after freezing some fruits are still sensitive to oxygen and will darken or discolor, or develop substandard flavors.

Freezing and subsequent thawing have a combined effect of shortening the cooking time needed for fruits and vegetables. The chemical actions and enzymatic actions that occur rapidly at room temperature in the fresh products also take place at freezing temperatures, although much slower. At 0° F and below the changes are not serious for reasonable periods of time (several months to a year). Upon thawing, however, the deterioration is much faster than before freezing. Freezing and thawing can cause some fruits such as plums to become very sour—hence sugar is usually added. Other fruits may "weep" or leak fluid.

Because of their acid pH fruits and berries will not support the growth of dangerous pathogenic organisms such as *Clostridium botulinum,* but they are very likely to ferment or mold.

Packaging Requirements, Materials and Packages

Frozen produce packaging is principally aimed at protection against moisture loss and convenience in handling. Some protection against light is desirable but not always bothered with. Protection against oxygen is not an essential, but may be employed on particularly sensitive fruits.

Earliest frozen products were packaged in waxed cartons with moisture-proof cellophane liners. Clarence Birdseye indicated that E. I. duPont developed coated cellophanes to meet his requirements.

Frozen berries were shipped from the West Coast in paraffin coated wooden barrels of 5-, 10-, 30-, and 50-gal. capacity. Lacquered tins with slip covered lids were used for cold pack fruits, and bulk quantities (30 lb units) of IQF (individually quick frozen) vegetables such as lima beans and peas were packaged in tins with friction lids. In the early days of frozen foods processors were afraid to use tin cans, even though they would have made excellent packages, because they were afraid the housewife would get confused and leave frozen tins on the shelf where the product would spoil. By the time the public was familiar with proper handling of frozen foods the composite can made from wax impregnated paperboard and tinplate ends have proven to be as good in performance and less costly than the conventional can. Glass containers have been used for frozen products but they have not been popular.

By far the most commonly used package for frozen foods has been paperboard cartons and containers. Paper cups, tubs, and cylinders formed from waxed solid sulfite or sulfite backed manila papers have provided rigid, water-tight sanitary containers that stacked easily, when full, and nested conveniently, when empty. No setup equipment is required. The container is merely filled and closed. Paper spiral wound containers have been made up to sizes large enough to hold five gallons of frozen food.

For a limited time a few duplex paper and plastic film bags were used for frozen foods.

The folded paperboard carton or box has been the work horse of frozen produce packages until very recent years. Several styles have been provided; end opening, top opening, and two piece; tuck flaps, glued flaps, and locked flaps; in sizes ranging from 8-oz capacities up to 10 lb. The boxes were made of good grade bleached sulfite on white patent coated manila waxed for moisture-proofness. Board thicknesses ranged from 14 to 16 points in small sizes to 17 and 24 points for institutional sizes. The boxes were used with liners, or outerwrappers, or even both. End-filled cartons were convenient for loose frozen smaller fruits and vegetables, while the top opening cartons were best for bulky items like spinach and broccoli. One carton was made in which the liner material was laminated to the inside of the carton. Liner materials were moisture-proof cellophanes, rubber hydrochloride, laminated coated wet strength glassines, and parchment paper. Outer wrappers were cellophane or waxed paper, and later foil-wax-paper.

Courtesy of Reynolds Metals Co.

FIG. 86. ALUMINUM FOIL LAMINATED OVERWRAP

Foil overwraps provide attractive packages and added protection against freezer burn and cavity ice for traditional frozen food cartons.

Newer Packages

Several materials had very strong influences on the packaging of frozen produce. By far the most important was the introduction of polyethylene films, coatings, and laminations. Polyethylene has much better moisture barrier properties and low temperature properties than wax and wax coated papers or any of the previously used materials, except aluminum foil laminates. Its transparency, low cost, ease of opening, and ease of reclosure quickly brought polyethylene to the forefront in packaging of IQF vegetables. In fact, one third of the frozen vegetable packages sold at retail are now packaged in polyethylene film bags.

The polyethylene coated polyester film bag became the work horse of the "boil-in-bag" prepared vegetable package. Vegetable preparations with cream sauces, and butter sauces, in different combinations such as peas and mushrooms, or buttered corn with red peppers were put into "boil-in-bags," and two bags were placed in a folded outer carton. The consumer needs only to place the unopened bag in a pot of boiling water for a few minutes in order to bring the food to serving temperature. Whatever extra cooking may be needed is accomplished during the reheating cycle. The latest use for the "boil-in-bag" is the "thaw pack." This is the same package but the contents are frozen fruits usually in a sugar syrup. Immersion in boiling water is used only to thaw the ice crystals. The package is opened and the fruit served when it is still thoroughly chilled. Rapid thaw is accomplished because the frozen slab of fruit is relatively thin.

One of the objections to the oblong composite can was the difficulty in opening it. New opening devices such as the aluminum pull tab and the plastic film tear out seaming tapes have made this package a competitor again.

The folding carton is getting a new lease on life as it now comes with a hot melt (polyvinylidene chloride) coating. This makes the package a much better barrier and also permits the ends to be sealed shut. Opening is done by tearing out a strip along perforations. For heat sensitive products cold-sealing adhesives can be spotted on the end flaps. The improved barrier characteristics eliminate the need for liners or for overwraps, which more than offsets the cost of the coating. Production line speeds in filling and closing are also increased. Similar coatings on the interiors of corrugated shippers give added product protection in several ways. The coatings prevent the shipper from weakening by absorption of internal moisture. They are smoother and therefore do not abrade the packages packed inside. They also give added moisture and gas barrier protection to the food.

Another innovation has been the ability of converters using new printing presses with better web tension controls to produce high quality

Courtesy of The Metal Box Co., Ltd.

FIG. 87.   MODERN FROZEN FOOD PACKAGES FEATURE PRINTED
CARTONS WITHOUT OVERWRAPS

The use of hot melt coatings or other moisture barrier coatings provide
sufficient protection to allow elimination of carton overwraps.

process printing on extensible films both by flexography and by gravure.
The better body and stiffness of polypropylene and its excellent clarity
may find it competing with low density polyethylene for the bagged
frozen vegetable market. The future will also certainly bring new frozen
foods into the market place.

Insulation through the use of foams may be the innovation of the
future. Excessive temperature fluctuation is the chief cause of the formation of cavity ice in frozen foods. The use of foamed polystyrene
sheeting or molded polystyrene foam as an outer container for a multipack of pouches of frozen vegetables or fruits would result in both an
attractive and a functional package.

Boil-In-Bag Corn

Whole-grain corn used for freezing is harvested from 1-2 days less
mature than the variety to be canned. The corn is harvested by mechanical harvesters and large sloping bins with slatted bottoms hold the raw
corn.

The storage bins are connected to a conveyor which transfers the corn
to the husking operation. Between 30–50 tons of raw unhusked corn is
held in these storage bins.

A double husker is used to husk the corn and it is then inspected,
washed, and the ears conveyed to cutting machines where ends are
trimmed and kernels are then cut from the ear. The kernels are rinsed

and then conveyed to steam blanchers where blanching is done in about 2–4 min.

After the blanchers the corn is screened, flumed, and washed to remove bits of husk, silk etc. It is then conveyed to picking tables where women pick out all foreign matter. Boil-in-bag corn is IQF frozen in trays which are placed on racks.

A commonly used packaging machine for boil-in-bags is the Bartelt form-fill-seal type. The packaging operation consists of:

(1) A woman controls the level of incoming corn on an overhead conveyor. She inspects it for discoloration and damage. A control handle is used to reduce the product flow.

(2) On the machine, a roll of polyester/PE passes over guiderails and is heat-sealed at bottom. The film then passes through two side sealers and cooling bars. A preprinted eyemark controls the position of the side seal.

(3) The pouches are notched or slit for easy tear opening and severed from the web. Individual pouches are grasped by clamps mounted on an endless chain and thus brought to the filling head.

(4) The Bartelt hopper receives the frozen corn and feeds 4 oz into 6 volumetric cups around outer edge of a turret. The turret rotates 60° for each cycle.

(5) The pouch is opened and each pouch receives a charge about half full. It moves along the line where it is then completely filled by a second filler turret.

(6) A timed interval valve operates a nozzle which injects butter sauce into each pouch. Then a vacuum draw snorkel removes the air.

(7) Just after the snorkel is withdrawn, a flat bar seals the top seal and then a round seal bar applies a second seal on the top.

(8) Each filled pouch then drops down a slide and into individual cartoner conveyor pockets.

(9) A cam-operated arm pushes the pouch into a carton, the end is closed and the filled carton moves to tray assembly line and thence into freezer storage.

DRIED FRUITS AND VEGETABLES

Because dried fruits and vegetables do not spoil and because tremendous savings in weight and volume are realized, drying has a very strong economic reason for being a preferred method of preservation. In war time when shipping space is vitally important dehydration of foods is imperative. In peace time, some foods are not saleable when dried because they have undesirable color, flavor, odor, and appearance, or because they are difficult to prepare. Nevertheless, over $1.3 billion of dried fruits and vegetables were sold in the US in 1967.

The most popular of dried produce items are fruits which have a high level of sugar. These include apricots, cherries, dates, figs, grapes, loganberries, peaches, pears, plums, and some raspberries and strawberries. Most fruits when dried to a moisture content of 23% or less will not mold. This is especially true if they have been treated with sulfur dioxide. Packaging is therefore aimed at keeping moisture out and preventing volatilization of flavor and odor constituents. There are two basic procedures used in drying by heat, sun drying and oven drying.

Sun Drying

Apricots, figs, peaches, prunes (plums), raisins (grapes), pears, and dates are commonly sun-dried in California. Sun-dried fruits are also imported from countries surrounding the Mediterranean Sea.

Pears, peaches, and apricots are exposed to sulfur dioxide fumes for 3 to 6 hr before being spread out on trays to dry. Some prunes and grapes are dipped in lye to crack the skins slightly. This makes them dry faster. Care is taken not to dry too much as this would be an economic loss (low weight), while underdried fruit will spoil. Usually the last stages of drying are done in the shade away from the full sun.

Cherries, berries and walnuts are also sun-dried in small amounts.

Artificial Dehydration

Dehydration through proper use of heated air ovens produces a dried product which, when cooked, more closely resembles the cooked fresh fruit in color and flavor. There is less chance for spoilage from lack of moisture control and fewer losses to insects, rodents, birds, or the weather.

Unlike dehydrated fruits, many dehydrated vegetables have not been a popular product. This is probably because earlier processors didn't bother to blanch. This is essential. If peroxidase and catalase enzymes are not destroyed by blanching; flavor, color, odor, and texture will deteriorate during drying and during storage. Best results with vegetables are attained by drying to 5 percent moisture content or less. The principal vegetables that have been heat dried are potatoes, white and sweet, onions, carrots, pumpkin (marrow squash), tomatoes, peppers, pimientos, sweet corn, string beans, peas, spinach, cabbage, cauliflower, brussels sprouts, okra, celery, rhubarb, garlic, horseradish, and various beans.

Some vegetables are sulfite treated in addition to blanching to further preserve color. In addition to the fruits that are also sun-dried other fruits are dehydrated by artificial heat. These include apples, bananas, and on a limited scale pineapple and citrus fruits. Some fruits have been advantageously soaked in fruit syrup before drying.

Some dried fruits and vegetables are pulverized and sold as small flakes or fine powders. Powders can be reconstituted as juices or used as flavorings. Flakes are exclusively used as the latter. Dried onion, parsley, garlic, etc. are used in soups and sauces. Dried potato granules or flakes have been used extensively for instant mashed potato recipes.

Dehydrated Potatoes.—Potato granules consist of precooked dehydrated potatoes reduced to about 70-mesh. When reconstituted with hot water or milk, they resemble mashed potatoes.

The processing method generally used consists of the "add-back" method. Potatoes of high solids content are washed, lye-peeled, trimmed, sliced and steamed until well done. They are then gently mashed and mixed with dried potato granules in the ratio of 15:85 (by weight) This is done to obtain an overall moisture content of 35–40%. After standing for a tempering period, further agitation in another mixer induces granulation. The mixture is then dried first by flotation in a stream of heated air to 12% moisture and then in a secondary drier to 7%.

After this the dried product is put through a coarse 70-mesh screen to remove any lumps or dried skin. Of the finished dried product 85% is recycled and 15% goes to packaging.

Care must be exerted not to rupture individual cells during agitation and mixing lest the product become sticky.

Packaging is done in unsupported opaque polyethylene film bags made on form-fill-seal machinery. Some packages are made from paper-polyethylene-foil-polyethylene laminates. During filling moisture control is quite critical. The flakes are filled from a hopper, and through an inner tube inserted into the bag. Bags are filled and sealed automatically. High moisture barrier laminates are used as the product must be kept dry during normal storage. Although many packers merchandise the product in bags, some are also placed in sealed liners or pouches and merchandised in folded cartons.

Freeze dried fruits and vegetables are being used to a limited extent as whole or chunk additives to cereals and soup mixes; and as flakes and powders to a variety of prepared dishes, sauces, gravies, etc. Because they more rapidly resorb water than oven dried products they are used extensively as condiments, flavorings, and "instant" beverage powders.

Package Requirements

In the packing of dried fruits or vegetables one of the greatest problems is insect infestation. Products must be thoroughly fumigated to destroy eggs, larvae, pupae, or adult insects, or if practical they may be heat treated to deinfest. Packages should be treated to prevent infestation also, and should be made from material that insects will not penetrate. It is also necessary that the package protect the dried product from moist environments otherwise it may cake, or develop mold.

Package Forms and Materials

Whole dried fruits used to be packed in large 10-, 25-, and 50-lb wooden boxes for shipment to bakers who used them in pies, cookies, and the like, or they were used in candies and confections. Other bulk containers have been used including corrugated cartons, spiral-wound paper canisters, large tins. For retail consumption, packages for dried fruits include paperboard folded cartons with foil laminated liners and/or overwraps. Coated cellophane bags and polyethylene or polypropylene bags are now also being used. Dried fruit flakes or powders are marketed in friction lidded tins, plastic screw-capped glass bottles, or foil laminated flexible pouches. Desiccants can be included in hermetically sealed containers or the packages can be evacuated prior to sealing. Cellophane-polyethylene-foil-polyethylene laminates have been the most commonly used flexible materials and paper-polyethylene-foil-polyethylene has also been used. For small packs of dried fruits used as confections, Saran coated polypropylenes are very effective. Dried legumes such as beans, peas, lentils, etc., require very little protection. Simple plastic bags are now the most popular package. Here again cello-

Courtesy of British Cellophane Ltd.

FIG. 88. DRIED PRUNE PACKAGE

"Cellophane"-polyethylene lamination is used to provide a convenient package for dried fruit which requires good moisture barrier protection.

Courtesy of Reynolds Metals Co.

FIG. 89. FOIL PROVIDES PROTECTION FOR POTATO GRANULES

Instant potato granules require the high level of protection provided by vacuum or nitrogen gas packaging. The use of aluminum foil laminates in inexpensive flexible pouches provides the requisite barrier properties.

phanes served well for many years but were supplanted by low density polyethylene film.

CANNED FRUITS AND VEGETABLES

Canned fruits and vegetables account for the second largest proportion of the total crop—second only to fresh produce. Over $5 billion of canned foods are produced annually in the US. While individual crops vary from year to year, it is fairly easy to establish an over-all listing of the individual products which are sold in the largest volumes. Table 13 lists the most popular canned fruits and Table 14 lists the most popular canned vegetables in the US.

Canned Peaches

For canning peaches should be picked 4 to 6 days from canning maturity. Upon arrival at the cannery, they are graded according to size either mechanically or manually. When optimum ripeness is reached (firm flesh, orange yellow color) they should be canned promptly. Most canned peaches are packed as halves or slices; however spiced peaches are canned whole. Halving is conducted by automatic machines, which also pit the fruit.

Table 13

MOST POPULAR CANNED FRUITS

Fruit	Number of Cases[1] (000 omitted)	Cumulative %
Peaches	32,764	32.8
Apples and applesauce	16,737	49.5
Pineapples	14,982	64.5
Mixed and fruit cocktail	13,726	78.2
Pears	5,633	83.8
Apricots	4,051	87.8
Cranberry sauce	3,307	91.1
Olives	2,451	93.5
Citrus fruit	2,103	95.6
Plums	1,314	96.9
Cherries, sour	946	97.8
Cherries, sweet	503	98.3
Spiced peaches	436	98.7
Figs	395	99.1
Blueberries	296	99.4
Blackberries	239	99.6
Raspberries	107	99.7
Other berries	129	99.8
All others	—	—
	100,119	

[1]Case of 24 size 2½ cans (holds 26 oz approx.)

Table 14

MOST POPULAR CANNED VEGETABLES

Vegetable	Number of Cases[1] (000 omitted)	Cumulative %
Tomato products	81,413	30.0
Corn	44,152	46.3
Green and wax beans	37,667	60.2
Green peas	33,588	72.6
Beets	12,665	77.3
Sauerkraut	9,400	80.8
Asparagus	9,263	84.2
Sweet potato	8,756	87.4
Spinach	8,031	90.4
Carrots	5,100	92.3
Pumpkin and squash	4,015	93.8
White potato	3,508	95.1
Lima beans	3,089	96.2
Leafy greens	2,757	97.2
Mixed vegetables	2,706	98.2
Field peas	2,083	99.0
Carrots and peas	1,893	99.7
Others	803	100.0
	271,407	

[1]Case of 24-303 cans (holds 15 oz approx.).

For Clingstone peaches, a lye peeler is used. A hot lye solution is poured over the skin side of the halves at 215°–220°F for 45–60 sec. They are then washed with cold water to remove lye and peels.

Freestone peaches are peeled by steaming. The halves are placed pit cavity down on a belt which passes through a steam scalder. They are then cooled with water sprays and the skins are removed. Other methods used for Freestone peaches are scalding in hot water followed by hand peeling. Some varieties are lye peeled.

After being peeled, the peaches are inspected. Good halves go directly to filling lines, damaged fruit is diverted to cutting for pie fillings, good halves that are to be sliced are diverted to slicing machines. The cans are filled as rapidly as possible to avoid discoloration caused by exposure to air. Peach halves are filled in-line automatically while slices are hand-filled in order to remove defective pieces.

The cans used are tinplate with enameled electrotin plate ends. If colored spiced peaches are packed, inside enameled electrotin plate cans are used.

Syrup is then added by rotary and straight line syrupers or prevacuumizing syrupers. If hot syrup or decorative syrup is used cans may be closed without exhausting. Otherwise the cans are usually thermally exhausted by a steam exhaust for approximately six minutes at 190° –195°F. They are then closed by either the "Steam-Vac" closure or mechanically. A positive headspace of 5/16 in. is necessary.

Processing is conducted immediately after exhausting so that the heat of the exhaust is not lost before processing begins. The conditions employed vary according to use of water or syrup pack, type of peach, size of can and initial temperature prior to processing. For Freestone peaches with continuous reel type cookers, the following is recommended:

Can Size	Exhaust	Time and Temp in Cooker
301 × 411	6 min at 200°–210° F	14–16 min at 210° F

After processing, the cans are water cooled until the temperature of the contents reaches 95°–105°F. Cans are then sent to labellers and then to packers and warehousing.

Canned Green and Wax Beans

Beans used for canning are harvested when average maturity and size of the crop appears to be optimum. If they are to be held for any length of time prior to canning, the beans should be spread out to permit ventilation.

Mechanically harvested beans are cleaned by shakers, declumpers and air cleaners. This removes dirt, leaves and vines. They are then washed and snipped. Mechanical snippers are used. A "pregrader" separates the beans according to thickness and length to minimize snipper loss. After snipping, the beans are inspected and defective beans are removed manually.

The beans are then put through size graders for separation based on size. Small sizes are usually packed whole. Larger sizes are cross cut into various lengths or Frenched (longitudinally sliced). They are then washed and blanched (170°–180°F) for 2–3 min. Cooling follows by sprays of cold water.

The filling operation varies according to the type of bean. Whole beans are filled with a semi-automatic "hand-pack" filler. The cans are placed in a tray which is covered with a metal plate pierced with specified apertures. The tray is filled with beans and agitated until most of the beans fall into the cans. Hand operators complete the fill.

For cut beans, circular disc filling machines are commonly used. The "vertical pack" involves packing whole beans parallel for insertion into the can. After the cans are filled, brine is added at 200°F. An exhaust is normally employed for larger size cans and is also recommended for smaller sizes. It is important that the cans be well filled with brine in order to prevent discoloration of the top layer of beans. The fill-in weights are then added and the cans are sealed.

Beans must be processed in order to cook them and sterilize them. They are cooled quickly after sterilizing and processing. Recommended process time and temperature depends on the type of bean canned and can size. For whole or cut green or wax beans, the following is used:

Initial Temp of Food		Cooker Temp[1]	
	240° F	245° F	250° F
70° F	21 min	16 min	12 min
120° F	20 min	15 min	11 min

[1]For No. 2, 307 × 409, and smaller cans.

Product Characteristics

Though much has been done to improve on techniques of processing, and to develop better varieties of fruits and vegetables the very nature of the canning process sacrifices some of the desirable characteristics of the fresh product in order to insure prolonged preservation without fear of spoilage and with complete safety from food poisoning. Since much trouble has been undergone to destroy spoilage bacteria, it is of course extremely essential that the package prevent later reentry of other spoilage organisms. Consequently, packages for canned fruits and vegetables have been the sealed tin can and the sealed glass jar or bottle.

So much has been written on the can and the glass jar that it would be redundant to repeat it here. Sizes, shapes, and closures have been improved upon as have the basic materials. Today we have better glass, better tinplate, better enamels and seaming compounds, we even have better aluminum cans, but the basic package is the same today as it was 100 years ago.

New Trends

Much work has been conducted for a number of years to develop a flexible aluminum foil laminated package which can be used to replace the can or jar, the so-called autoclavable pouch. Such a package will have to undergo heat processing to render its contents sterile; however, due to the rectangular rather than cylindrical geometry of this package, it can be brought to sterilization temperature more rapidly. A shorter "cook" time in the sterilizer means less degradation of the food. Results are foods having odor, flavor, color, and texture more like that of fresh or of frozen produce rather than like the typical canned product. The convenience of the flexible package, its economics, and the superior quality of the food should make it a success in the market place. Only a few products have been marketed to date, but more can be expected as more and more research programs are brought to completion. Most autoclavable laminations contain a polyester film as the outer component, aluminum foil, and an inner plastic component. Differences lie in the thicknesses selected, in the choice of laminating adhesives, and in the inner food contacting resin. Polyamide films, polycarbonates, polyvinyl chlorides, polypropylenes, and modified high density polyethylenes have all been evaluated for this purpose and each new plastic resin introduced is given immediate attention.

BIBLIOGRAPHY

ANON. 1967A. Packaging system for potatoes. Packer 4, No. 1, 36–37. (Swedish)

ANON. 1967B. A 'standard' way to tight fill pears. Produce Marketing 10, No. 2, 20.

ANON. 1967C. Container switch pays off in pears. Produce Marketing 10, No. 2, 28–29.

ANON. 1967D. Shrink wrapped cucumbers keep fresh longer. Verpackungs Rdsch. 18, No. 2, 178–179. (German)

ANON. 1967E. Berries by air, latest results. Produce Marketing 10, No. 3, 15–16.

ANON. 1967F. Swedish cartoning systems cover full range of Italian frozen vegetables. Packaging News 14, No. 3, 17–18.

ANON 1967G. Fibreboard streamlines fruit production. Australian Packaging 15, No. 4, 22–24.

ANON. 1967H. Experiment in apple packing. Australian Packaging News 8, No. 4, 2.

ANON. 1967I. Better bananas, faster with new packaging methods. Packaging News 14, No. 5, 65–67.

ANON. 1967J. Moisture-repelling fibre tray ousts wood for lettuces. Packaging News 14, No. 5, 67.

ANON. 1967K. Fruit packaging breakthrough. Paperboard Packaging 52, No. 9, 50–57.

ANON. 1967L. Canned fruit and vegetables. Canning Packing 37, No. 440, 17.

ANON. 1967M. Keeping moisture where it belongs. Mod. Packaging 40, No. 13, 133.

ANON. 1967N. Uses PVC tubs, foil lids for pre-cooked potatoes. Food Drug Packaging 17, No. 9, 1–6.

ANON. 1967O. A sleeve-wrap for Church's celery. Produce Marketing 10, No. 11, 26.

ANON. 1968A. Orange aid . . . cartoning "decisions" are automatic. Mod. Packaging 41, No. 2, 96–97.

ANON. 1968B. New tomato packs a great success. New Zealand Packaging 5, No. 2, 11.

ANON. 1968C. Golden dates in new style pack. Food Trade Rev. 38, No. 3, 111.

ANON. 1968D. Still more scope for plastics in produce packaging. Packaging News 15, No. 5, 28–40.

ANON. 1968E. Strength of moulded tray brings cut in cost of apple cases. Packaging News 15, No. 5, 41–42.

ANON. 1968F. The right pallet box for soft fruits. Mech. Handling 55, No. 6, 936–937.

ANON. 1968G. Certificate of long life for dessert grapes. Emballages 38, No. 252, 105–106. (French)

BRISTON, J. 1968. Fiberboard box growth stunted for horticultural produce? Converter 5, No. 6, 44–45.

DEDINI, V. 1967. Polyethylene containers for the harvesting and transport of citrus fruits. Mater. Plast. Elast. 32, No. 12, 1251–1256. (Italian)

DEULLIN, R. 1967. Trends in the use of refrigerated containers for the transport of tropical fruit by sea. Fruits 22, No. 8, 368–370.

FORSYTH, F. R. et al. 1967. Controlling ethylene levels in the atmosphere of small containers of apples. Can. J. Plant Sci. 47, No. 6, 717–718.

JOSLYN, M. A., and HEID, J. L. (Editors). 1963. Food Processing Operations, Vol. 2. Avi Publishing Co., Westport, Conn.

MONZINI, A., and GONNI, F. L. 1966. Principles and use of plastics in retail packages for fruit. Imballagio. 17, No. 129, 3–21. (Italian)

PLATENIUS, H. 1943. Effect of oxygen concentration on the respiration of some vegetables. Plant Physiol. 18, 671.

PLATENIUS, H. 1946. Films for produce . . . physical characteristics and requirements. Mod. Packaging 20, No. 2, 139.

PRESTON, L. N. 1967. Flexible packaging. VII. Fresh fruit and vegetables. Food Packaging Design Technol. May, 36–41.

SACHAROW, S. 1967. Fresh produce . . . growing market for flex packs. Paper, Film, Foil Converter 41, No. 7, 41–42.

VEERAJU, PI, and KAREL, M. 1966. Controlling atmosphere in a fresh fruit package. Mod. Packaging 40, No. 2, 168–254.

CHAPTER 8

Fats and Oils

INTRODUCTION

Olive oil pressed from the fruit of the olive tree was a primary cooking oil for many civilizations bordering the Mediterranean Sea. The Minoans had a prosperous trade with Phoenecia and Egypt through their export of olive oil. The ancient peoples used olive oil for salads, cooking, lamp light, and cosmetic purposes. Olive plantations existed in Spain and Portugal in 1700 BC. Since it took 30 years to reach full production, the destruction of an olive grove by war or nature was a great tragedy.

Sesame oil pressed from seeds was of oriental origin. It was used by Chinese and Indian cooks for many centuries, being "discovered" by Portuguese explorers in the 1500's. The Chinese also discovered the oil of the soy bean. The peanut which yields peanut oil was grown first in Brazil. It was transported to Africa by slave traders and from thence to Virginia. Palm oils were found in various parts of the globe. The coconut palm being the most important, perhaps, of all.

Modern methods for extraction of oils, and for refining or hydrogenating them have led to many new and valuable products used both in foods and in other industries as well. Cottonseed oil, corn oil, and more recently, safflower seed oil have found great acceptance in table oils and margarines. Useful oils have also been pressed from peach, prune, apricot, cherry, and almond pits as well as tomato and grape seeds.

The ability to convert unsaturated oils to solid fats (margarines or shortenings) stems from the discovery of a German worker by the name of Normann in 1901 that these products could be reacted with hydrogen in the presence of a nickel catalyst.

Hydrogenated vegetable shortenings almost replaced lard in American kitchens, but they themselves came to be challenged by refined cooking oils.

World production of animal fats and oils probably exceeds 60 billion pounds. Of this amount about 20–25% are nonedible and used as lubricants or drying oils. Some of the edible varieties are used for other purposes, for example peanut oil is used as a lubricant and coconut oil is used in soaps. Many of the edible fats and oils are further processed before reaching the consumer. Great quantities are consumed by the baking industry and the candy and confectionery industry. Nevertheless the retail sale of cooking fats and oils in grocery markets in the United States amounts to over $1.3 billion.

237

OLEOMARGARINE

Oleomargarine or margarine is a fatty food which is made to look like butter and to be used as a butter substitute. Different countries have different regulations but principally they all agree margarine must not derive its fat content except to a minor extent from milk fat. The product is based on the invention of a French chemist in 1869, who converted fresh beef tallow into two fractions. The higher melting fraction was a hard white fat and was called oleo-stearine. The lower melting fraction was a soft yellow fat which looked, smelled, and tasted like butter and was called oleo-margarine. After separating the two fractions he mixed the oleomargarine with skim milk, water, some bicarbonate of soda and some macerated cows' udder. He then agitated the mixture until he had an emulsion looking like cream. Upon churning the "cream" he obtained a product which he named "Butterine" and later "Margarine."

As prices and availibility of beef tallow fluctuated, other readily available fats and oils were substituted. These included peanut oil, olive oil, sesame oil, coconut oil and palm kernel oil. With the introduction of hydrogenation processes other cheaper vegetable and animal oils became usable, such as cottonseed oil, soybean oil, palm oil, and sunflower oil. In more recent times corn oil and safflower seed oil have come into general use in the United States.

Modern production plants for margarine are now operating with continuous processes including emulsification, cooling, plasticizing (kneading), and final packing—all out of contact with air to reduce oxidative degradation. The production of margarine is carried on in many countries.

In most countries the production of margarine is rivaling that of butter and in many of them it exceeds butter production substantially.

Product Characteristics and Packaging Requirements

The different countries making margarine use whatever natural or hydrogenated oil is most economically available to them. Russia and Hungary, for example, use sunflower seed oil. Poppyseed oil and tea seed oil have been used in the Orient and pumpkin seed oil and tobacco seed oil have been used in the Balkan countries. Lard, tallow and whale oils are used extensively in European countries such as Norway and Russia.

In the production of margarine much care is devoted to the selection of blends of fats and oils which will give the most desired properties in the final product: (a) stability; (b) consistency; (c) plasticity (or spreadability); (d) minimum variation in softness with temperature; (e) flavor; (f) odor; and (g) color.

Care must be taken during manufacture to avoid excessive exposure of the fats to air at temperatures of 104°–122° F as oxidation will occur in a matter of a few days.

Manufacture

Margarine is made largely from vegetable oils that have been hydrogenated or crystallized to achieve proper spreading texture.

Two mixtures are usually made. All of the oil and fat soluble ingredients are mixed together while the other mixture consists of water and all water soluble ingredients. The melted oil mixture and the water mixture are then pumped into a high speed refrigerated cylindrical mixing chamber. This causes a uniform distribution of the water phase throughout the oil phase. The oils crystallize and solidify with the water droplets trapped in the mass. In order to develop optimum sized fat crystals, proper temperature control is essential. Additional agitation then further subdivides the water droplets and promotes crystallization.

The resultant semisolid margarine is continuously extruded and packaged in-line. Margarine is packed in many different shapes and containers. Most margarine sold in the United States is in the form of 1/4 pound sticks. These are commonly wrapped on a Benhil machine (Benz and Hilgers). It wraps 50–60 packages per minute and has separate molding and wrapping units. The margarine is dropped into a stainless steel hopper where it is compacted and pressed towards the extrusion orifice. Molding chambers are mounted in a revolving drum. The block of margarine is expelled and wrapped. Since there are several molding chambers on the machine, the cycle of operations is continuously repeated. The machine works well with parchment or aluminum foil laminates. It is equipped with coding devices. In most cases, automatic cartoning and boxing lines are integrated in-line. These boxes then go by chutes to a storage room for distribution.

Milk solids are still used in margarine in most countries to impart desirable flavor and aroma, to eliminate a fatty taste, to stabilize the emulsion, and to retard oxidative flavor development. They help the formation of brown sediment during frying which is a characteristic of butter. Milk fractions are ripened through use of bacterial or chemical additives, and adjusted to the optimum pH (about 5.4–6.0). If pH drops as low as 4.3–4.5 a fishy taste can develop.

Where salt is employed it is added principally for taste and also to help preserve the product. Tastes vary among and within countries. Salt content may be as low as 0.1% and as high as 5%.

Some countries permit chemical preservatives to be added to prevent hydrolytic rancidity formation. Many different chemicals have been tried but most now use sorbic acid in concentrations of about 0.1% or less. Its use provides good shelf-life at 60°F and 60.5% rh for up to eight weeks.

Coloring of margarine used to be a controversial subject. Some countries insist on the use of the natural coloring materials such as beta-carotene. Others have prohibited the use of colors used in butter. Some dairy states in the United States passed legislation prohibiting any coloring matter to be added. Now it is allowed almost everywhere but a few states levy high taxes on the colored product.

Legal regulations on margarine packaging have been passed to regulate not only the purity, safety, and composition of the product but also to guarantee means for its ready identification. Some countries require indicator additives, such as sesame oil or starch to be added. The sesame oil develops a strong red color when heated with hydrochloric acid and mixed with a few drops of alcoholic solution of furfural. Starch gives a characteristic blue color when it reacts with iodine.

Since margarine is now processed in vacuum or in nitrogen atmospheres to minimize oxidation, it is very desirable that the package protect against oxygen reentry. Control of evaporation of moisture is very important as the guaranteed minimum net weight of the package must be clearly indicated. Since the margarine is a pasteurized product and is susceptible to bacterial or mold attack, packaging materials should be free of contamination and should protect against contamination. Light, particularly ultra-violet light, is not good for margarine. It can cause a change in color, it can destroy vitamins and it catalyzes oxidative rancidification. Like butter, margarine is very susceptible to absorption of foreign odors and must be protected against them and against loss of its own aroma.

Package Forms and Packaging Materials

Margarine has been packed in many differently shaped and sized containers. Bulk packages have included barrels, kegs, tubs, and circular or oval containers. Present bulk packaging is primarily in large cubical or rectangular boxes. These have been parchment-lined wooden boxes, but are mostly corrugated paperboard containers today. Parchment liners are now being replaced by polyethylene film bags and the containers themselves may be hot melt coated with polyvinylidene chloride.

Retail packaging also has varied in the past with round rolls, pats, and prints as well as the now more popular cubes and bricks.

Some margarine packages are made strictly for economy. They appear to be attempts to duplicate the cheapest butter packages. Thus one finds a 1/4 or 1/2 lb chunk wrapped in cellophane or a 1/2 or 1 lb block wrapped in butter parchment.

Some export packages are tinplate cans. The newer soft spreading margarines are being marketed in paperboard, plastic, or formed aluminum tubs and trays.

Those margarines packed in tin containers may be filled hot (pourable semiliquid) or cold (extruded solid). Containers must be of good quality and free of pinholes in order to avoid rust or corrosion. Cans may be enameled and often also include parchment liners. Most canned packages are vacuum packed or nitrogen packed.

When the intimate wrap for margarine prints is a parchment paper, it is important that the paper be free of pinholes and free of materials that might be extracted by the margarine and cause adverse changes. Metals such as copper are particularly undesirable as they catalyze degradation reactions. Good grades of 25 lb per ream or higher vegetable parchments, imitation parchments, with a suitable level of moisture—not too dry but not more than nine per cent—are used. They should be free of mold spores and placed in close contact with the surface of the margarine to eliminate air pockets.

Paper when manufactured comes off the calendar stacks and driers mold spore free. Most paper manufacturers will guarantee mold-free stock out of their door, but will not guarantee the paper will be mold-free on receipt as they have no further control over possible recontamination. Converters of packaging materials who coat or laminate papers follow the same rule. Careful handling and storage by the margarine packager and good plant sanitation will minimize chances for mold spore contamination. Sorbic acid or sorbate dusts may be put on wrappers or incorporated into coatings or adhesives for more positive control.

The better grades of margarine have been packaged with foil laminates as intimate wraps. These can be of various combinations depending on length of storage, desired degree of protection, and cost. Foil-parchment and foil-wet-strength-tissue have become the most common materials; however much experimentation has been done with film-foil-parchment combinations. Films have been polyvinyl chloride, polyvinyl acetate, polyvinyl propionate, polyethylene, polyamide, polypropylene, and polyester.

The most popular over all packages have consisted of foil-wet-strength-tissue inner wraps and foil-paperboard laminated outer cartons. These result in a very high quality appearance of the package. In recent years rigid thermoformed polyvinyl chloride, polypropylene, high density polyethylene, and acrylonitrile butadiene styrene plastic containers pigmented to make them light-proof have been introduced. The

FIG. 90.   FOIL-PARCHMENT LAMINATE FOR ENGLISH MARGARINE

Illustration of a 1 lb brick of margarine containing 10% butter
packaged in a wax lamination of embossed aluminum foil and
a 17 lb vegetable parchment.

margarine is filled in a very soft state, a snap on cover (high density
polyethylene) is pressed on or a peelable lid is heat sealed and the product is chilled in storage rooms.

Special varieties of margarine are produced for special uses. For
example regular table margarines are not the same as margarines made
for pastry or for cake. Different blends of fats are used and different
processing procedures. Cake margarines usually have a high percentage
of hydrogenated fats, while pastry margarines have some added oils.
Other margarines include cream margarine used for ice cream manufacture and whipped margarines, meat flavored margarines, and cheese
flavored margarines. The margarines made from peanut oil, cottonseed
oil, sunflower seed oil, soybean oil, corn oil, or safflower seed oil have
greater percentages of polyunsaturated fatty acids in their molecules
than butter, lard, or palm oil margarines. These are believed by some

FIG. 91. MARGARINE PRINT WRAP AND CARTON

Aluminum foil is used both in the intimate wrap for a single print
1/4 lb stick and in the registered embossed protective carton.

medical authorities to be less likely to cause atherosclerosis (hardening
of the arteries). This medical theory received a great deal of publicity
and the consumer demand for the soft polyunsaturated margarines
skyrocketed. To distinguish them from regular margarines new air
formed foil containers were designed. The containers have attractive
colors and shapes so that they can be placed directly on the table. Again,
snap on lids are employed, although some work is being done to develop
hermetically sealed containers with peelable membrane heat sealed lids.

In India, a margarine called *Vanaspati* is produced which is similar in
consistency to *ghee*. Vanaspati is made from hydrogenated vegetable oils
with a minimum of five per cent sesame oil, added vitamin A and special flavorings. The product is packed in small 1/5 lb tinplate containers, or in 40-lb bulk tin containers.

Courtesy of Reynolds Metals Co.

Fig. 92. Aluminum Tub Container for Soft Margarine

This formed aluminum container is attractive enough for table
use and can later be reused as a candy cup or dish.

COOKING FATS AND OILS

Solid Fats

Butter, lard, chicken fat, beef tallow, and to some extent the solid fats
of other animals such as sheep, goats, buffalo and the like have long
been used as fats in cooking. Butter and lard have been used for shallow
and deep frying while the other fats have been used chiefly in basting
roasts and in making gravies. Chicken fat, lard, and butter have also
been used in baking as solid shortenings.

When the hydrogenation process was commercialized in 1911, it
became possible to change vegetable oils into solid fats. A number of
such products came on the market in competition with lard. The products looked like lard but had a blander flavor. Advertising stressed better digestibility of the new vegetable fats. Most of the vegetable shortenings were packaged in tins with a key wind tear strip opening device and
a lid so shaped that it could be replaced. The 1-lb tin was a popular size;
however, 1/2-lb and 2- to 3-lb larger capacity containers were used later.

One of the latest packages developed for solid shortenings is a spiral-
wound composite can. The body is made from a two-ply, skived, spiral
wound kraft paperboard with an outer coating of white pigmented polyethylene. There is also an inner liner made from foil, and polyolefin

laminated to the kraft body. The top is a coated aluminum lid with a ring tab which allows the whole top lid to be removed. The bottom is 80-lb tinplate. With this package, which can be hermetically sealed, shortening can be shelf-stored as long as two years.

Liquid Oils

In the section devoted to margarine it was pointed out that there are many types of edible oils available from a wide variety of sources. Most of these oils are also suitable for use in cooking both as frying oils and as salad oils. They can be modified if desired by blending, partial hydrogenation, interesterification or transesterification, or polymerization.

Blending.—Blending of several oils may be done for economic reasons, to improve color, flavor, odor, stability, digestibility, viscosity, melt point, smoke point, chill point, or any other property. The only requirement for such blends is that the package be clearly labeled as to the compositions of its contents.

Packages and Packaging Requirements.—Most all oils have some degree of unsaturation and are therefore susceptible to oxidative rancidity which is catalyzed by light—particularly the blue and ultraviolet wavelengths. For this reason they have traditionally been packaged in tins and colored glass bottles (amber to brown). Both types of containers are still used. Olive oil and corn oil are not as sensitive as some of the others and with today's improvements in refining and use of antioxidants some packers feel they can safely use clear glass. Very recently, with the development of good heat stabilizers that meet FDA approval and prevent thermal degradation during extrusion, clear and colored polyvinyl chloride resins have been used to make blow molded bottles for cooking and salad oils. Other plastics have also been used such as ABS modified styrenes, and polycarbonates. Injection molded polycarbonate bottles have been used in Japan for salad oils.

In the market place one finds the old style lacquered tin cans now equipped with molded plastic caps and pour spouts. A one-half-liter can of walnut oil packaged in France contains a tinplate top with a fitted plastic collar opening. Inside the collar is a high density polyethylene spout with a ring tab. By pulling on the ring tab the end of the spout can be torn open. A rigid plastic screw cap protects the spout and serves as a reclosure feature.

In another corner on a gourmet shelf one finds a series of clear glass bottles with conventional screw-on caps with a pigmented shrinkable film overlay neck band. They come in 8 oz and 16 oz sizes and contain refined salad oils: almond oil, rice bran oil, peanut oil, cottonseed oil, avocado oil, safflower seed oil, soy bean oil, sesame seed oil, walnut oil and corn oil are listed.

In Europe and Canada, molded PVC bottles are being marketed in 8, 16, and 24 oz sizes and even in 128 oz sizes for institutional use.

PVC Packages.—The first types of rigid PVC sheet and film were produced by solvent casting and by calendering. Solvent casting involved high solvent costs and calendering equipment was very expensive (one to four million dollars). Extrusion methods involved much less costly equipment and processing but unfortunately at extrusion temperatures the polymer degraded. Heat stabilizers were developed, and by 1964 rigid clear bottles of PVC were available for use in nonfood products. In Europe, PVC plastic containers of all shapes were adopted for

Courtesy of Air Reduction Co., Inc.

FIG. 93. BLOW MOLDING OF PVC BOTTLES

Special new blow molding system brings parison to the mold rather than the conventional mold-to-parison system.

food use for a wide variety of products from beer and wine to salad oils. In September 1967, an English soft drink bottler switched to PVC.

In the United States, however, FDA regulations prohibited the use of the tin stabilizers. A great deal of research was conducted and a new family of modified PVC resins was developed which incorporated polypropylene molecules into the polymer chain. These new resins permit extrusion at temperatures 20° to 50°F lower than conventional PVC polymers and hence can be used very effectively with FDA approved calcium-magnesium-zinc stabilizers.

By the latter part of 1908, the FDA had removed its objections to tin stabilizers and given them limited approval. Packages made from tin stabilized PVC's will undoubtedly move into the market place soon. However there are still some hurdles to clear. FDA requires no more

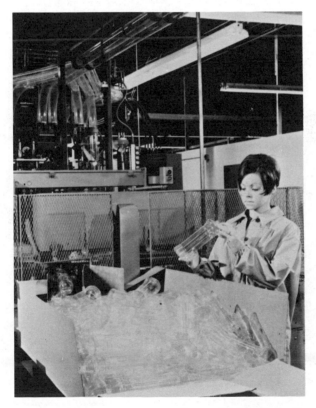

Courtesy of Air Reduction Co., Inc.

FIG. 94.   PVC BOTTLE INSPECTION

Blow molded PVC bottles made by the Air Reduction Co.

than one part per million of the dioctyl tin to be absorbed by the food product. FDA also does not permit use of tin stabilized PVC bottles for dairy products, carbonated beverages, or malt products such as beer.

Another competitive producer is about to introduce a PVC bottle made from previously sanctioned materials which does not require tin stabilizers to prevent yellowing.

It is expected that several major producers of polystyrenes and acrylic bottles will switch to tin stabilized PVC in the near future.

Blow molded plastics can be formed in a variety of attractive and interesting as well as functional shapes. They have many advantages including their light weight and lack of fragility. The latter property assures no fragments chipping off into the product and there are less breakage and clean up losses in shipping and in supermarkets. For light sensitive products the PVC can be colored, usually an amber to brown shade; other colors including opaque whites are possible. Cap closures are usually screw thread and molded from a plastic such as polypropylene. Air is removed at time of bottling and a polyvinylidene chloride liner is used in the cap to insure against oxygen reentry. Thickness of

Courtesy of Air Reduction Co., Inc.

FIG. 95.   PVC BOTTLE FOR OLIVE OIL

Gloria Packing Corp., a division of C. Pappas, Inc., have adopted
blow molded PVC bottles for their olive oil and vinegar.

Courtesy of E. I. du Pont de Nemours and Co.

FIG. 96. MEXICO SUPERMARKETS FEATURE EDIBLE OIL IN
TETRAHEDRAL SHAPED CONTAINER

Laminations of EVA-wax, paper, polyethylene, and "Surlyn"
ionomer resin coatings are used to make a tetrahedral shaped package in half liter size for edible oil.

the plastic wall is controlled as thin as possible for economic reasons but
strength and permeability requirements put a minimum limitation on
how thin.

Several US companies have already announced they will use polypropylene modified PVC packages. The Gloria Packing Corporation of
Boston, is packing olive oil and wine vinegars in 4-, 8-, 12-, and 15-oz
containers, the Uddo and Taormina Corporation (Progresso Foods) of
New York are packaging a grated cheese product, and Morton Salt
Company of Chicago have just announced a flavoring product in an
opaque white PVC jar with lag screw cap lid.

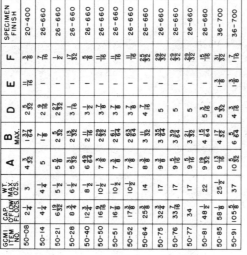

GGMI ITEM NO.	CAP. OFLOW FL. OZS.	WT. MAX. OZS.	A	B MAX.	D	E	F	SPECIMEN FINISH
50-08	2¼	3	4 3/32	1 37/64	2 5/32	11/16	3/8	20-400
50-14	4¼	4¼	5	1 7/8	2 9/16	—	7/16	26-660
50-21	6 19/32	5½	5 5/8	2 5/32	2 27/32	—	½	26-660
50-28	8¾	6½	5 31/32	2 13/32	3 1/16	—	17/32	26-660
50-40	12¾	8½	6 49/64	2 11/16	3½	—	5/8	26-660
50-50	16 9/16	10½	7 3/8	2 29/32	3 7/8	—	11/16	26-660
50-51	16 7/8	10½	7 3/8	2 61/64	3 7/8	—	11/16	26-660
50-52	17 3/8	10½	7 3/8	2 63/64	3 7/8	—	11/16	26-660
50-64	25 3/8	14	8 3/8	3 11/32	4 7/16	—	25/32	26-660
50-75	32¾	17	9 3/8	3 35/64	5	—	29/32	26-660
50-76	33 7/16	17	9 5/16	3 39/64	5	—	29/32	26-660
50-77	34	17	9 16	3 21/32	5	—	29/32	26-660
50-81	48½	22	9 9/32	4 15/64	5 3/16	—	15/16	26-660
50-85	58 5/8	25½	9 13/16	4 7/32	5 9/32	1 3/8	31/32	36-700
50-91	105 5/8	37	10 5/32	6 5/64	4 3/8	1 3/8	7/16	36-700

NOTES :—

1. WHEN OTHER FINISHES ARE USED, CAPACITY, WEIGHT AND HEIGHT SPECIFICATIONS ARE ADJUSTABLE WITHIN THE REQUIREMENTS OF THE FINISH USED. SEE NOTE 2.

2. HEIGHT (DIMENSION 'A') IS BASED UPON USE OF SPECIMEN FINISH SHOWN.

3. THE SPECIFICATIONS SHOWN MAY VARY MODERATELY ACCORDING TO COMMERCIAL TOLERANCES AND INDIVIDUAL MANUFACTURERS PRACTICE.

4. 'B' DIMENSION IS VARIED TO MAINTAIN CAPACITY.

From Joslyn and Heid (1963)

FIG. 97. STANDARD DIMENSIONS FOR SHORT LINE, ROUND FOOD LINE GLASS CONTAINERS

With the door now open for PVC bottles and jars in the food packaging field the Society of the Plastics Industry now estimate the market for such containers could be as high as 1 billion units by 1970 or over $100 million dollars. At present FDA limits usage to temperatures below 150° F.

Other New Trends.—Another approach has been the use of flexible cartons, such as "Tetra-Paks," for edible oils. These are made from ethylene vinyl acetate wax coated paper-polyethylene-paper-ionomer resin coating. The interior ionomer coating is heat sealable and oil resistant. A strip of ionomer film is used to cover the raw edge and thus avoid oil wicking along the longitudinal seam of the package.

Flexible laminations of film-foil-films may be expected to be utilized as carton inner liners for edible oils. Interior plastic components will probably be ionomers, polypropylenes, or polyamides. If made heavy bodied and properly folded, a fin sealed bag can stand by itself. In a suitably designed carton with tear away upper portion it could be used as a pourable dispenser as is done in Europe for milk packages.

BIBLIOGRAPHY

ANDERSON, A. J. C., and WILLIAMS, P. N.   1965.   Margarine, 2nd Edition. Pergamon Press, London.

ANON.   1967A.   Plastics packs for margarine. Packaging *38*, No. 451, 80–82.

ANON.   1967B.   Ionomer resin coatings tailors Tetra Pak for edible oils—will motor oil, food follow? Packaging News *14*, No. 1, 1.

ANON.   1967C.   Margarine goes into PVC tubs. Packaging News *14*, No. 12, 1–50.

ANON.   1967D.   Fiber can for shortening. Packaging Innovator *2*, No. 5, 5.

ANON.   1968A.   Clear PVC bottles for dressings, oils. Food. Eng. *40*, No. 5, 51.

ANON.   1968B.   Clean-pouring PVC bottle for edible oil. European Packaging Dig., No. 58, 3.

ANON.   1968C.   PVC for food; it begins. Mod. Packaging *41*, No. 8, 101.

CRUESS, W. V.   1958.   Commercial Fruit and Vegetable Products, 4th Edition. McGraw-Hill, New York.

ECKEY, E. W.   1954.   Vegetable Fats and Oils. Reinhold Publishing Co., New York.

KIRSCHENAUR, H. G.   1966.   Fats and Oils, 2nd Edition. Reinhold Publishing Co., New York.

CHAPTER 9

Food Flavorings and Condiments

INTRODUCTION

Salt

In the Middle Ages salt was so highly regarded it was served at noble dinner tables in magnificent gold cellars. A man's social position was measured by whether he sat above or below the salt at banquet tables.

Wars have been fought over salt and battles lost for lack of it. Salt making was undertaken at Syracuse, New York because it became so scarce and expensive during the war of 1812. This production helped save the Union in the Civil War, whereas the South lost nearly all its salt production. By 1863 salt was literally worth its weight in gold in the South. Today, salt is so low priced that cost of transportation is a major factor in its production.

Fruit, Vegetable, Herbal, and Spice Flavorings

Spices come mostly from the seeds or leaves of plants, but a few are exceptions. Ginger and turmeric come from roots and the clove is a flower bud from a tree. Cinnamon is the inner liner of a bark. The nutmeg is a fruit pit. Allspice and peppercorns are dried berries. The fruit pod is used for paprika, cayenne, and chilli peppers. The outer covering of the nutmeg gives us mace. The rarest of all spices, saffron, comes from three hair-like filaments (stigma) from each blossom of a crocus. It takes 225,000 stigma to make a pound of saffron. Other spices are also very tiny in size such as the celery seed (750,000 to make a pound) or the poppy seed (900,000 to make a pound).

Spices were literally worth their weight in gold in early times. Rents, dowries, taxes, and tributes could be paid with spices. When Rome was ransomed by Alaric the Goth, he demanded 3000 lb of peppercorns as part of the treasure. Arab traders monopolized the trade in spices from the East and the price was exorbitant. King Edward I of England spent over $100,000 in one year for spices for his table. Europeans sought cheaper ways to obtain spices by establishing their own trade routes. Marco Polo in 1270 went overland to China, and established an overland trade route, but when Constantinople fell to the Turks in 1453, Europe was completely cut off from the spice trade. The quest for new routes by sea led the Dutch, Portuguese, Spanish, and English explorers to the rediscovery of the ancient sea lanes and led Columbus to the Americas.

253

Production

Salt mining is conducted at many different sites to reduce transportation costs. There are three basic methods for obtaining salt and all are used commercially.

Solar Evaporation of Sea Water.—Solar evaporation of the oceans and the Great Salt Lake is done by pumping water into ponds in which it is evaporated by solar heat. When the concentration has reached the saturation point by evaporation, the brine is moved by gravity to crystallizing ponds where the salt crystallizes out. The remaining liquid is returned to the sea (or lake). An entire year's supply is gathered during the summer and refined year 'round. Final processing consists of drying, coarse screening, milling, and fine screening. Final products are graded and then packaged. Additives are added to enhance free flowing, or to improve nutritional values. For example special salts with mineral additives are made for cattle feeding.

Mechanical Mining.—Underground mining of rock salt is the second oldest form of salt production. Underground deposits are the remnants of prehistoric inland seas which deposited their salts in vast beds when they were dried up by the sun and winds. Such salt beds formed during several geological periods and were covered with earth and rock to great depths—sometimes several thousand feet below the surface. Concrete-lined shafts are sunk to the salt bed and tunnels are cut into the bed forming great rooms. Solid pillars of salt are left standing in these rooms to support the roof. By this procedure up to 65% of the salt can be excavated. The coarse lumps are sent to crushers and screeners below ground.

Some underground deposits are salt domes. Unlike the strata of the previously described deposits which may be from 8 to 18 ft thick, salt domes may be a mile in diameter and up to 16,000 ft thick. Here the mining technique is similar but is called quarrying. Rooms are cut down vertically as well as horizontally and may reach depths from floor to ceiling as much as 90 ft. This makes percentage removal higher. Final processing is the same.

Evaporated Brine.—In some mines it has been found more convenient to sink two holes to the salt bed and to pump water down one hole and brine up through the other. The brine goes to huge settling tanks where impurities are removed chemically. Then the brine is pumped to vacuum pans where three successive pans at successively lower pressures are steam heated to boil the brine. The "pans" are three stories high, and as the agitated brine boils, cube shaped crystals form and settle to the bottom where they are removed as a slurry. The slurry is pumped to a filter where it may be completely or partially dried. After drying it is milled, screened, and graded.

Marketing Information

About 33 million tons of dry salt or salt brines are produced in the United States annually. Of this amount 22 million tons are consumed by the chemical industry. About 4 million tons are used to control ice and snow during the winter. The food processing industry uses about 2 million tons and 1 million tons reach home kitchens. Other quantities are consumed in animal feed mixtures. Salt is used by restaurants, hospitals, and hotels, in swimming pools and in water softeners. There are an estimated 14,000 different uses, and uses are growing.

Product Characteristics and Packaging Requirements

Salt is very soluble in water. In moist atmospheres, the surface of a dry salt crystal will absorb water from the air. The basic aim in salt packaging is to keep it dry, free flowing, and clean. Ease of opening and dispensing together with reclosure features are very desirable in consumer packages.

Packaging

Bulk packaging of salt is usually in multiwalled bags in various sizes up to 100 lb. The modern package contains a polyethylene liner.

The consumer package for fine grained table salts for many years has been a 26-ounce capacity spiral wound paper canister with paper ends. A pivoting metal pour spout was attached to the top lid. The opening was sealed against moisture and accidental spillage by a wafer of glassine paper. A printed paper label completed the package. A modernized version of this package has an injection molded plastic opening in addition to the plastic pour spout. The opening has perforations making the can a large shaker dispenser.

A smaller 4-oz capacity spiral wound canister with a decorative printed foil label (free of brand identification) is sold as a table ready shaker dispenser. The top is molded plastic and is twisted to move a shield away from the perforated openings. Three of these salters are sold as a unit in a paperboard tray which carries brand identification. The whole unit is overwrapped with cellophane or various shrink films.

Other dispensers of blow molded plastic have been designed as table ready dispensers. A miniature version is marketed for picnic and patio use. Small individual servings have been available for several years in polyethylene coated paper packets. One variety has fluted paper or fluted foil-paper laminate as one side of the packet. This adds stiffness and supposedly easier opening.

Courtesy of Morton Salt Co.

FIG. 98.   UPDATED VERSION OF AN OLD PACKAGE

Morton Salt Co. has moved with the times in adapting plastics to the pouring features of the package top and in modernizing their famous brand identification—the girl with the umbrella.

Courtesy of Morton Salt Co.

FIG. 99.   TABLE DISPENSERS FOR SALT AND PEPPER

The use of imaginative package design has brought these point of sale packages to the table.

Future Trends

We already have seen a trend toward more individualized portion packaging and greater use of plastics. This undoubtedly will continue. More decorative packages for table use will be introduced. More varieties of salt can be expected including finer or coarser grains and different flavors. Other constituents may be added such as tenderizers, and flavor intensifiers for use in patio cooking.

FRUIT AND CITRUS FLAVORINGS

The use of concentrated juices and extracts of fruit for flavoring of other foods is principally limited to beverages, candies, and baked goods (Chapters 10, 11, and 12). A few fruits and fruit flavorings are used in the preparation of condiments; these are treated later in this chapter. Some fruit extracts are used also for coloring principles. A few, such as lemon and orange extract, are packaged for home use, most however are packaged in bulk for industrial consumption.

Product Characteristics

The skin and peels of fruits carry most of the flavor, and the flesh, most of the color. Some added flavoring may be extracted from seeds. Ideal flavor and aroma is achieved when the fruit reaches complete ripeness and the sugar content reaches its maximum. The total flavor is a compound mixture and is derived from several parts of the whole fruit; some from expressed juices, some from extraction of the fruit flesh, and some from aromatic constituents extracted from peel, seed, or stems.

The flavoring essences are for the most part quite volatile and are chiefly alcohols, ketones, ethers, acids, and esters. As such they are generally only slightly soluble in water but readily soluble in alcohol. For this reason alcohol is usually the extractive medium. Straight distillation or vacuum distillation of fruit juices and fruit extracts in order to concentrate the juice will volatilize nearly all of the flavor constituents. Some means must be used to capture them in order that they may be added back or used separately as flavorings. Different fruits have different sensitivities to heat. Only those temperatures that will not deteriorate juice quality can be used. On the other hand it is difficult to capture highly volatile components in a vacuum system. Especially when the volatiles are mixtures of some 50–100 different chemical compounds each of which may be present in as little as one part per billion and rarely more than ten parts per million. Frequently the major flavor producing component is present in the smallest amount. For example, total esters in Concord grape juice are about 50 ppm concentration but methylanthranilate, the major flavor contributor, is present only in about 1–1.5 ppm concentration. Present practice is to heat juices or pulp mixtures at atmospheric pressure at maximum permissable temperature for a brief period to volatilize essences and capture them, then the feed stock is quickly chilled and further concentration is done under vacuum at lower temperatures. When extracts are the goal, alcohol may be added to the original feed stock to create lower boiling azeotropic mixtures and to help capture the essences.

Essence recovery is achieved through condensation of fractionated vapors and by scrubbing of noncondensable gases. Alcohol is recovered and reused.

Alcoholic extracts containing eight per cent alcohol can be further concentrated by freezing. The ice crystals will be almost totally free of essence and alcohol. Added sugar however makes this procedure more difficult.

Packaging Requirements

Most fruit essences are fairly stable at room temperature and can be stored for several months. Color principles may be light sensitive requiring protection. The chief package requirement however is prevention of losses of flavor and aroma through evaporation or permeation. The package material also must be resistant to deterioration by water containing some fruit acids and some ethyl alcohol.

Packages

Bulk packaging of fruit essences is in large glass containers or in paperboard or fiberboard drums with polyethylene liners. For long term storage the product is usually frozen in which case either the latter mentioned package or tin cans are used.

Retail packaging of essences is almost exclusively in glass bottles. Most of these are now also enclosed in printed paperboard cartons.

Future Trends

While there has been little if anything done in this direction for direct retail packaging of essences, it might be possible to absorb an essence onto an inert water soluble porous carrier. Freeze drying or some other means would be used to remove water and the result might be a highly concentrated free flowing granular solid or powder.

VINEGAR

Vinegar is well documented in ancient writings so again we must assume its origin is lost in the prehistoric era.

Vinegar is a fermentation product. Its name comes from the French *vinaigre* meaning sour wine. This is probably the classical source of vinegar. A wine that is allowed to ferment too long will convert some or all of its alcohol content to acetic acid.

Any sugar or starch containing material can be fermented to an alcohol and then to a vinegar, provided the sugar content exceeds nine per cent.

Product Definitions

Cider Vinegar.—Cider vinegar or apple vinegar is the product made from fermentation of apples. About 7.5% of the total US apple crop is diverted to vinegar production. In the United States it is the only product that may be labeled "Vinegar" without qualification. In addition to acetic acid it must contain (when undiluted) at least 1.6 gm of apple solids per 100 cc and of the latter not more than 50% may be reducing sugars.

Grape or Wine Vinegar.—Grape or wine vinegar is the product made by fermentation of grapes. In the United States it must contain more than one gram of grape solids and 0.3 gm of grape ash per 100 cc.

Spirit, Distilled, or Grain Vinegar.—Spirit vinegars are made from fermentation of distilled alcohols.

Malt Vinegar.—Starchy cereal grains, usually barley, may be converted to sugars using an enzyme, diastase, which is contained in malt.

Other vinegars are made from various fruit culls including oranges, prunes, peaches, pears, pineapples, bananas. Sweet potatoes make a pleasant vinegar, whereas white potatoes produce a disagreeable flavored vinegar which must be distilled. Low grade honey will make a vinegar if yeast is added to convert it to alcohol. Some red colored vinegars are made from grapes with the skins included during fermentation, or they may be made from bad batches of red wine which have turned sour.

There are also a number of flavored vinegars produced. These are specialty products, such as "spiced vinegar," tarragon vinegar, chili vinegar, rosemary vinegar, and fine herb vinegar.

Manufacture

Vinegar manufacture from fruits consists of crushing, grating, or chopping the fruits, or gathering the already pressed pomace from juice manufacture, and fermenting the mass. Cereals or starchy vegetables are crushed, and heated with water to gelatinize the starch and make a mash. The mash is then cooled and mixed with malt. The malt converts starch to maltose (sugar). Yeasts are added which convert the sugars to alcohols. Care must be taken to use pure strains of yeasts as wild strains can cause undesirable side effects. Sulfur dioxide is added to the fermentation process to control wild yeasts and increase yield. Since fermentation creates heat, some cooling may be required. When fermentation to alcohol is complete (not needed when wine is used), the liquor is separated, diluted, and sent to different reactor tanks where acetic acid producing bacteria *(Mycoderma aceti)* are added. These convert the ethyl alcohol to acetic acid. Again temperature control is desirable (85° F being optimum); and since heat is generated, cooling is required.

After fermentation to vinegar and clarification, the product is aged in barrels to eliminate harsh flavors and to develop full bouquet. Final filtration leaves the product ready for packaging.

Package Requirements

Vinegar is corrosive to metals and is adversely affected by dissolved metallic salts such as iron, copper, or tin, hence metal containers are to be avoided. It also is susceptible to contamination by further growth of vinegar bacteria, other wild strains of bacteria or certain types of insects. Pasteurization, micropore filtration, or addition of sulfur dioxide help prevent growth of microorganisms.

Packaging

Bulk packaging of vinegar has always been in wooden barrels or large glass bottles. Oak formerly was used but it has largely been replaced by spruce. Interiors of the barrels are heavily coated with paraffin. Polyethylene lined drums should serve just as well since polyethylene cap liners have been used in closures for bottles with no difficulty.

Retail packaging has been in clear glass bottles. Screw cap closures have largely replaced the crown cap. Metallic caps must have liners that will not allow penetration of volatile acid with subsequent corrosion. Some bottles have shrinkable plastic covers applied over their caps and necks.

The latest trends in vinegar packaging are the use of plastic bottles. Polystyrene has been used in England, and blow molded PVC bottles are now being evaluated.

VEGETABLE FLAVORINGS

There are several hundred different flavoring materials which come from plants, trees, and shrubs other than the fruits and berries previously described. Some are used in cooking as direct additives, as in the case of the spices and herbs. Others are extracted or distilled and used as flavorings or coloring materials for prepared foods and beverages. In the latter type the botanical product is extracted with or without heat with mixtures of alcohol and water. In commercial dealings the various seasonings are classified as spices, aromatic seeds, and herbaceous plants (herbs). The most commonly used of each of these classifications are shown below.

Spices	Aromatic Seeds	Herbs
allspice	anise	basil
cassia	caraway	garlic
cinnamon	cardamon	laurel (bay)
cloves	celery	marjoram
ginger	coriander	mint
mace	cumin	onion
nutmeg	dill	oregano
pepper (black)	fennel	parsley
pepper (white)	fenugreek	rosemary
pepper (capsicums)[1]	mustard	sage
turmeric	poppy	saffron
	sesame	savory
	star anise	tarragon
		thyme

[1]Cayenne, chili, red, paprika peppers.

Product Characteristics

Two basic characteristics are common to all herbs and spices: they will spoil if not dried, as they are subject to molds and enzyme actions; and they will lose flavor if not kept cool and in closed air tight containers, for their principal flavor constituents are volatile essential oils. On the other hand, freezing of spices is not recommended as they are said to lose flavor. Many spices are chopped fine or made into grains or powders. If allowed to absorb moisture they will cake and not flow freely from the package. This is a great disadvantage as usually very small quantities are used. Insect or rodent infestation is also a problem that must be dealt with in shipping and storage of spices.

Product Preparation

Spices, seeds, and herbs are dried at or near the point of harvest else they might spoil in shipment. On arrival at a destined port they are inspected, graded, partially cleaned, and sold to spice-makers. The spice-makers grind, blend, and package the product for distribution and sale to consumer and industrial markets. Nearly 40% of the spices produced are consumed as manufactured products such as cured meats, baked goods, pickles, canned foods, beverages, and condiments. Oleoresinous extracts may also be made as described earlier.

Package Forms and Materials

In early days one could buy dried herbs in loose bundles tied with string or in loosely woven cloth bags. The finer imported spices were sold in glass or earthenware jars, in tinplate boxes or in metal lined wooden

Courtesy of McCormick and Co., Inc.

Fig. 100. Paprika in Glass Retains Flavor and Aroma

Brand identification is maintained by McCormick and Co.
through standardized bottle shapes, distinctive plastic tops and
foil label design.

chests. Small retail packages consisting of paperboard cartons with
paperliners were used for many years in grocery stores. Small glass jars
with screw-on plastic tops replaced the cartons and probably are responsible for the major upswing in spice consumption. Spices in glass do not
lose flavor so rapidly and are easy to dispense through molded plastic,
perforated inserts in the mouth of the jars. Some dried vegetable products (some of which were freeze-dried) are now marketed from the spice
shelf in identical packaging. These include powdered, granulated, or
minced garlic or onion, chopped chives, chopped green (sweet) peppers,
celery flakes, grated lemon or orange peel, and mixed vegetable flakes.
Other mixtures sold include flavored salts, flavored sugars, apple pie
spice, barbecue spice, chili powder, crab or shrimp spice, curry powder,
herb seasoning, Italian seasoning, mixed pickling spice, poultry seasoning, and pumpkin pie spice. One also finds cream of tartar, meat tenderizers, monosodium glutamate, and charcoal flavoring.

New Trends

There are only a few novelty packages on the market but perhaps they
are indicating new trends in spice packaging.

A polystyrene vial with a rigid plastic friction lid cover is available
containing one gram (1/28th of an ounce) of dried whole saffron. This is

Courtesy of McCormick and Co., Inc.

FIG. 101.   SPICE MIXES

Preformulated spice mixes are packed in foil pouches for seasoning special foods such as Mexican style tacos.

packaged in Spain. A similar product from France contains one-half
gram of powdered saffron in a glass vial with a cork closure.

A meat tenderizer is available in solution in a polyethylene squeeze
bottle. A lemon peel flavorant (for cocktails) is packaged in a squeeze
bottle with atomizing top. Two ounces of the product containing 85%
alcohol is packed in a four-ounce bottle. The air left in the bottle atom-
izes the product when the bottle is squeezed.

TABLE 15

TIME REQUIRED FOR ODORS TO PENETRATE 1-MIL MYLAR POLYESTER AND 0.8-MIL
CELLOPHANE FILMS[1]

| | Time in Hours for First Olfactory Perception | |
Source of Odor	Mylar	300 MSAD Cellophane
Vanillin	141	20
Pinene	140	20
Methyl salicylate	116	24
Ethyl butyrate	20	0.5
Propylene diamine	92	1

[1] Nagel and Wilkins Reprinted with permission of *Food Technology* Vol. 11, 180–182. (1957).

An injection molded polypropylene jar is being used to package a dry mustard powder. Four ounces of the mustard powder is now packaged in a jar which weighs only 20 gm. The old glass jar that was replaced weighed 100 grams. The unbreakable plastic jar doesn't require dividers in shipper cartons. Thus the user saves three ways—on shipping weight and volume, size of shipper, and breakage. The jar has a recessed bottom to accomodate the top of another jar when stacking on the shelf, and a protruding ring just below the lid to protect paper labels against damage.

Many spices are also being sold in single portion aluminum foil laminates for easy use in cooking. They are usually contained in four-sided sealed pouches. Dried onions are packed into such pouches for consumer use.

Vanilla Flavorings

Cortez' expedition to Mexico City in the early 1500's not only resulted in the discovery of chocolate (Chapter 11) but also in the discovery of the vanilla bean. Actually the chocolate drink enjoyed by Montezuma was heavily flavored with vanilla.

The vanilla is a type of orchid and has a beautiful lime-yellow bloom that lasts less than a day. The raw bean is picked when ripe, but has very little flavor. Beans are dried in the sun, then "sweated" to initiate enzymatic reactions and a slight fermentation. This process is repeated for several months until full cure has been attained. Beans are then totally dried, inspected, graded, and packed in waxed paper lined tin boxes containing about 7–8 kg.

Vanilla flavor is extracted from the dried beans by soaking in alcohol at room temperature. Three repeated soakings each for five days are made. The residue is washed with water and pressed to get the last traces of flavor. The diluted wash waters are blended with the concentrated extracts. The extract is aged in glass or stainless steel containers and then bottled. Instant powdered vanilla is made by mixing the extract with sugar.

SYNTHETIC FLAVORINGS

Chemists have been studying natural food flavorants for many years and have successfully purified and identified many of their component ingredients. In many instances these have proved to be relatively simple compounds, but even some of the most complicated ones have been synthesized successfully. The natural flavorants are blends, for example over 124 different compounds have been identified in orange aroma and over 137 in strawberry flavor. Synthetic anise flavor is now used in place of natural anise in the United States by a factor of ten to one. Synthetic vanillin consumption is three times that of natural vanilla. Synthetic strawberry and grape flavors are used in several-fold ratio over their natural counterparts. Most synthetic flavorants are consumed by the food industry and are therefore packaged in bulk. A few are formulated into alcoholic extracts, bottled and sold for home cooking.

PICKLED FOODS

Pickles are undoubtedly as old as vinegar and spices. We know that cucumbers have been cultivated for more than 3000 years, and sauerkraut dates back to ancient China. The use of vinegar in Egyptian cookery and their knowledge of fermentation processes leads to the conclusion that pickling also originated before written history. Even in Roman days pickling factories existed in Spain.

A wide variety of fruits and vegetables can be preserved by pickling. They may be preserved as a single pack, that is all of one kind, or they may be a mixed pack of several types. When coarsely chopped they are used in relishes or chutneys and when finely comminuted they are added to sauces, spreads, and dressings. Within each category there is a wide variety of possible combinations of sizes, cuts, pickling process, and final composition. For example cucumber pickles range in size from tiny cocktail gherkins to large whole dill or "kosher-style" pickles. They may be packed as long quarter strips, as cross-cut chips, in odd chunks, or chopped as a relish. They may be finally processed as "sweet," "sour," "dill," "bread and butter," or "mustard style" pickles. Similar variations can be applied to each of the other fruits and vegetables listed, although fruits are usually pickled sweet and spicy.

Pickles in the form of relishes, condiments, salad dressings, sandwich spreads, and other condiment type sauces accounted for $1,629,970,000 worth of business in grocery stores in the United States in 1966. This figure does not include sauerkraut which is treated as a canned vegetable.

Cucumber Pickles and Relishes

Manufacture and Product Characteristics.—Pickling varieties should be firm in flesh and picked when underripe. Bruising is to be avoided. A short length of stem is left attached when harvesting. After sorting and preliminary grading for size the pickles are cured in a ten per cent brine during which the cucumbers ferment and cure. This process takes from 4 to 6 weeks. Salt concentration must be maintained and replenished as needed during the brining operation. Sometimes about one per cent of dextrose sugar is added to control the character of the fermentation. Air is excluded as much as possible to allow the lactic acid bacteria to multiply. In the early stages carbon dioxide gas is generated and must be allowed to escape. When cured, pickles are translucent and have a yellow or olive green color. They are then stored in a higher concentrated brine until ready for final processing.

Final processing consists of a careful grading for size, a water soak to remove salt, and then the addition of vinegar. Water soaking takes several days and several separate baths. Alum or calcium chloride additives in the final wash bath are used to harden the texture and turmeric to improve the color. "Vinegaring" consists of soaking the pickles in one or more baths of vinegar of the desired strength to give a final equilibrium concentration of 2.5–3.5% acetic acid. If packed at this point, the pickles are "sour pickles."

Sweet pickles differ only in the final vinegar soak, where a spiced sweet vinegar is used.

Dill Pickles.—Dill pickles may be categorized as fresh pasteurized, genuine and imitation. Genuine dills consist of larger size pickles which are fermented in 40–50 gal. barrels with dill weed, dill vinegar, salt and dill spice.

The dill weed is placed on the bottom of clean barrels and these barrels are then filled half full with cucumbers. Mixed dill spice and salt is then added. After shaking, more dill spice, salt and dill vinegar is added on top. The barrels are then filled with water or a 29° salinometer brine. They are then placed on the side with bung up. Fermentation takes place best at 80°–85° F for 6–8 weeks.

In order to insure adequate shelf-life, the pickles are usually pasteurized. The pickles are placed in jars and covered with brine from the original fermentation process. Cloudiness in the brine is removed by filtration. The brine is heated to 160°–170° F and poured over the pickles. After capping and sealing, they are pasteurized at 165°F for 15 min followed by prompt cooling.

Most pickles are packed in glass jars. If cans are to be used heavily lacquered cans are essential.

Packaging.—Pickles can be canned in heavily lacquered cans if heated to 200° F for about ten minutes prior to closing. An additional process at 200° F for about ten minutes is an extra precaution.

Most pickles are packed in glass jars and sealed, preferably using a vacuum. Vacuum is attained by steam injection into the head space just prior to capping. Brines are usually filtered to remove cloudiness, or they may be replaced with fresh brines. Jars come in 6-, 10-, 16-, and 20-oz sizes. The larger half-gallon and gallon sizes are mostly used by restaurants and institutions. Lacquered metal caps are used with cork or coated paperboard liners with a hot melt sealing compound around the rim.

Plastic bottles as well as pouches have been used for pickle products. Pliofilm pouches have been used to contain dill pickles. Since dill pickles are perishable unless pasteurized, this unit is intended for a short shelf-life. An additional problem in using plastic pouches for pickles is the possibility of acetic acid diffusion through the walls of the package. Unless pickles are pasteurized, they rely on acetic acid for bacteriological stability. Loss of acetic acid through the package walls can reduce shelf-life and cause contamination with adjacent products. Other materials used for pickle pouches include polyamide-polyethylene pouches.

Other Pickles and Relishes.—There are only minor differences in the handling, processing, and packaging of pickles and relishes made from other fruits and vegetables.

New Trends

An 8-oz jar of miniature ears of corn in a pickled brine is now being replaced in the New York market with a polyethylene coated polyester bag in a printed paperboard carton.

The development of suitable flexible materials for sterilizing will probably cause significant development in flexible pickle packages either in transparent film pouches or packages fabricated from alumi-

Courtesy of E. I. de Pont de Nemours and Co.

FIG. 102.   "PURE-PAK" MILK CARTON NOW A PICKLE PACKAGE

This version of a familiar carton is adapted to pickles through the addition of a foil lining.

num foil laminates. Rilsan (Nylon-11) is under development for a sterilizable pickle package. There is also activity in Europe in using aluminum foil laminates.

A recent trend in pickle packaging has been the adaptation of the "Pure-Pak" milk carton to package dill pickle slices and sweet pickle relish. An inner aluminum foil lining is used to insure flavor retention and adequate shelf-life.

OLIVES

Olives are not only one of the oldest fruits used by man but they are one of the first to have been processed to improve their usefulness. The fruit grows only in climates that are similar to the Mediterranean, consequently most of the World crop comes from Spain, Italy, Greece, and Portugal. There is a substantial crop now produced in California—over 2-1/2 million cases of canned ripe olives are processed there. Another 1-1/2 million cases of green olives are imported annually.

Product Characteristics

Fresh olives are extremely bitter and unpalatable. The bitterness can be removed by dilute alkali at room temperature. The classical method was to treat them with wood ashes (more likely a solution of potash obtained from leached wood ashes). Such fruit is palatable but will spoil.

Processing

Green or ripe olives can be pickled or canned. The chief differences are at what stage they are picked and how they are lye treated. Both types are stored in brine until ready for processing. Sometimes a partial fermentation occurs. The fruits are lye-processed in repeated baths of dilute lye to set color and texture and to remove bitterness. Fermentation is done in large vats of brine (sometimes sugar is added). This is a lactic acid fermentation similar to that for cucumber pickles.

Final Packing.—After fermentation the fruits may take several alternate paths. They may be canned; after being pitted and stuffed they may be pickled in fresh brine and packed; or they may be pickled and packed whole or chopped.

Canned olives are packed in brine, exhausted, sealed and retorted at 240°F for about 60 min. Pickled olives are packed in brine and lactic acid and bottled with care to exclude air.

Packages.—Enameled tinplate cans are used for canning. Glass bottles and jars are used for pickled olives. It is reasonable to expect that plastic bottles or jars may eventually be used for this purpose.

An interesting concept for olive packaging is the "Doypack" concept

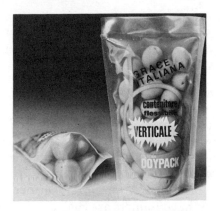

Courtesy of Istituto Italiano Imballaggio

FIG. 103. "DOYPACK" STAND-UP OLIVE PACKAGE

An Oscar winner for 1968 at the Italian Packaging Institute was this flexible film gusseted pouch, which because of its construction will stand without toppling.

marketed extensively in Europe. Polyester-polyethylene laminates are used to form a stand-up pouch containing a gusset. A novel feature of this package is that the gusset can be made from a different laminate than the body of the pack. The entire package can now be formed on a newly developed form-fill machine.

SAUERKRAUT

Sauerkraut is the fifth largest volume vegetable canned in the United States. It is also very popular in northern and eastern Europe. It is one of the few pickled foods that is normally cooked before eating and served as a main course vegetable rather than as a condiment.

Product Characteristics and Manufacture

Sound firm heads of cabbage are selected (green, leafy open headed cabbage is not suitable). After a slight wilting the outer leaves are stripped and the core is reamed and shredded. Then the entire cabbage is cut into thin shreds. The shreds are mixed with salt (about 2-1/2% by weight) and placed in a fermentation vat. Heavy pressure is applied, which together with the salt extracts the cabbage juice, and soon the liquid covers all of the shredded cabbage. First stage fermentation converts the sugars to lactic and acetic acid, ethyl alcohol and mannitol and evolves carbon dioxide gas. Second stage fermentation sets in when acidity reaches one per cent with no evolution of gas. Then when acidity gets still higher, a third stage fermentation begins with evolution of carbon dioxide and continues until fermentation is complete. Total acidity reaches more than two per cent and contains lactic and acetic acids in about 4 to 1 ratio. After fermentation the sauerkraut may be stored in the fermentation vats almost indefinitely so long as air is excluded and temperature is maintained at a cool level.

Packing and Package Forms

Most sauerkraut is canned, some is put up in jars, and an increasing segment of the market is put in plastic bags and sold refrigerated. Canned or jarred kraut is heated to 160° F to purge carbon dioxide and to pasteurize. The hot kraut is placed in the container, filled with brine, juice, or plain hot water to adjust salinity or acidity, and then the container is exhausted and closed. After a brief further processing to bring the contents to 180° F the containers are cooled and labeled.

Bagged sauerkraut is sold in printed low density polyethylene film bags containing 1, 1-1/2, or 2 lb. Since the kraut has not been pasteurized chemical stabilizers and bleaches such as sodium bisulfite and sodium benzoate may be added.

New Trends

Considerable effort has been made in Germany and the United States to develop a polyester-aluminum foil-polyolefin laminate flexible package. The foil makes an attractive package and shuts out harmful light. Since the kraut can be pasteurized in the package, no preservatives are needed. This package has been sold commercially in Germany for several years and is finding wide acceptance.

PICKLED WALNUTS

Green (unripe) walnuts are picked before they have formed a hard shell. They are softened by steam, pricked with a needle and steeped in a pickling liquor containing vinegar, sugar, water, mushroom ketchup, caramel and spices. They are then drained, packed in jars and covered with malt vinegar. Shelf-life is up to two years or more.

CAPERS

Capers are the flower buds of a certain plant, which have a unique flavor. They are grown in France and brined there. "Pickling" merely consists of draining the brine and adding malt vinegar. The product is usually sold in small screw-top glass jars, however a new package has just been introduced. It consists of green colored plastic bottles. These are lighter weight and use only about half as much shelf space. Furthermore breakage is eliminated.

CONDIMENT SAUCES (VINEGAR BASED)

There are several basic types of sauces used as condiments. They can be thick or thin, vinegar based or oil based, and sweet, sour, salty, spicy, or peppery. Ketchup would be an example of a thick sauce, while Worcestershire sauce would be a thin one. The aforementioned sauces and various fruit chutney's are principally vinegar based with little or no oil; whereas most all salad dressings are oil based.

Thick Fruit Sauces

Thick fruit sauces are made from finely minced fruits, vegetables and spices suspended in sugar, salt, and vinegar. Vegetable gums or flour starches are added to give "body."

After finely comminuting the ingredients all are brought to a boil. Thickeners and final flavorings are added, and the product is simmered about half an hour. Some sauces are matured in oak casks for several months. Finally the sauces are bottled and capped, usually using a hot fill.

Thin Sauces

In order to keep them thin and watery in consistency thin sauces are made from vinegar extracts of the various ingredients. The ingredients are ground, macerated, or crushed, boiled in vinegar and then filtered or strained to remove solids. The different extracts are then blended, matured and bottled.

Packaging

As with thick sauces the thin sauces are packed hot in bottles and capped so as to create a partial vacuum. In former years a cork covered glass stopper was used. The modern bottle has rolled metal or molded plastic screw cap with a hot melt coated paperboard liner. Sometimes the bottle neck is made very narrow as with hot pepper sauce and sometimes it is restricted by means of a plastic insert to prevent measuring out too large a dose.

New Trends

Thin sauces have been packaged in blown polyethylene bottles and the newer PVC bottles. Some bottles that formerly were overwrapped with paper are now being overwrapped in clear plastic film.

Thick sauces have been packaged in aluminum collapsible tubes in Europe but this has not caught on in the United States probably because of the "tooth paste" image.

Polyethylene squeeze bottles have also been used to package thick sauces, sometimes in attractive or novelty shapes.

Mint Sauce

Mint sauce is a thin sauce but contains finely chopped mint leaves. These are steeped in vinegar, sugar, and salt and then bottled. If the mint is mold free, no trouble will occur in storage. Otherwise liquor should be pasteurized before adding the mint.

Tomato Sauces

Tomato sauces are made from tomato purée, sugar and vinegar and vary primarily in their secondary components. Because variations are almost infinite, many countries have established legal or industrial standards of definition.

Packaging.—Tomato purée is shipped in bulk to institutions or to manufacturers of other tomato products. At one time this was done in

Courtesy of Aluminium Foils Ltd.

FIG. 104.   A FOIL LABELED GLASS BOTTLE FOR A THIN
SPICY CONDIMENT SAUCE

Thin condiment sauce is shown packaged in a traditional glass bottle. The foil label allows wiping the bottle clean and gives a quality look.

Courtesy of Reynolds Metals Co.

FIG. 105.   INDIVIDUAL KETCHUP SERVING

Cellophane-polyethylene-foil-polyethylene pouches have been used
for individual servings of ketchup and similar condiment sauces.

50-gal. barrels with vinegar added as a preservative. In more recent
times No. 10 (6 lb 10 oz) and 5-gal. heavy tinplate cans have been used.
Cans are filled with hot product (about 180° F or more) and then sealed.
For retail markets 6-oz, 8-oz, 16-oz, and 20-oz cans are used with a short
sterilization at 212° F after filling and closing. Tomato paste is treated
and packaged in the same manner as the purée in cans.

Catsup or ketchup is made in the same way thick sauces are made and
is bottled hot after deaerating. A post-pasteurization at 180° F for 45
min is used if the product has been allowed to cool to 160° F.

Chili sauce is more like a chutney as it contains pieces of tomato. A
wider mouth bottle is needed. If filled above 185° F no further processing is necessary, otherwise the bottled sauce is pasteurized at 180°–200°
F as with catsup.

New Trends.—Tomato sauces are being packaged in collapsible
metal tubes in Europe and in cellophane-polyethylene-aluminum foil-
polyethylene or cellophane-polyethylene packets in the United States.
Some polyethylene squeeze bottles with dispensing caps have also been
tried. Polyethylene liners do a good job for fiberboard bulk shippers.

Rigid PVC bottles may well move into this market in the future if
problems of hot fill and oxygen barrier can be resolved. The use of aluminum foil labels has added sales appeal to the traditional glass bottle.

MUSTARD SAUCES

Prepared Mustard

Prepared mustard is made from ground brown and yellow mustard seeds, vinegar, salt, and spices (turmeric, clove, pimiento, and cayenne pepper). The ingredients are mixed to a fine homogeneous fluid paste and are packed in jars. In France and England, collapsible tubes are used as well as plastic squeeze bottles and glass jars. One of the newer packages comes with a rigid plastic pump in the lid by means of which the mustard can be dispensed.

Individual servings of mustard as well as ketchup are packaged in flexible envelopes. These are made of cellophane-polyethylene, cellophane-foil-polyethylene and polyester-polyethylene. A PVDC coating is almost always added for barrier protection. Foil based laminates are superior to the others used in product protection. A vital point in this concept is the use of a proper primer or adhesive capable of withstanding the delaminating effect of mustard oil. The inner ply is an important consideration; however, the primer must be capable of holding the laminate together. Other single service concepts include thermoformed PVC trays with aluminum foil peelable lids.

Piccalilli

Piccalilli is made by combining a prepared mustard sauce with a chopped pickled relish made from various vegetables such as cauliflower, small cucumbers, and onions. Other spices are added and some cornstarch for thickening. The product is packed hot in jars.

CONDIMENT SAUCES (OIL BASED)

Mayonnaise-type Salad Dressings

These dressings are thickened emulsions of oil in water using egg yolk as the emulsifier. Most any edible oil is suitable but vegetable oils predominate. Sugar, salt, vinegar, spices and thickeners are the remaining ingredients. Spices include mustard and tarragon. Some countries allow dried egg to be used, others artificial coloring and preservatives. Mayonnaise is usually a little thicker bodied and stronger spiced than "Salad Dressing" and the latter contains more sugar, more starch, and less oil.

Sandwich spreads are made by mixing a "Salad Dressing" with some fine chopped brined vegetables, usually onions, gherkins, cauliflower, olives and pimiento.

Since they may contain rancidity-prone edible oils, salad dressings should be kept cool and away from light and oxygen. For this reason also they are regarded as having only a 3 to 6 month shelf-life, and batches

Courtesy of E. I. du Pont de Nemours and Co.

FIG. 106.   INSTITUTIONAL MAYONNAISE PACKAGE

The"Pure-Pak" foil-lined carton is adaptable to 1-gal. capacity
institutional mayonnaise packs.

are date coded. Ingredients are premixed, pumped, emulsified in a colloid mill, and then filled at about 120 packages per minute.

Packaging.—Mayonnaise products are packed in 5-oz, 7-oz, 1/4-gal., 1/2-gal., and gallon wide mouth clean glass jars; with tightly fitted presterilized metal screw caps, with resilient coated paperboard liners. PVC and wax coatings are used. A steam injection applied just before capping permits a vacuum pack. After closing, jars are labelled then stored at 30°–40°F. Recent European trends include the use of thermoformed plastic tubs of mayonnaise with peelable aluminum foil lids.

The "Pure-Pak" foil-lined carton is being used in the institutional food market for mayonnaise. The new carton has no exposed paperboard edge along the longitudinal side seam inside the carton. This results in an airtight package and assures freshness and flavor retention.

Other Salad Dressings

There are many varieties of so-called salad dressings which are primarily oil based. These run from the simplest spiced oil and vinegar dressings to more complicated thickened French dressing, 1000 Islands dressing, and Roquefort dressing. They are variations on thick sauces, thin sauces, and salad dressings with one thing in common—high oil content. They may be thin without emulsifiers, or they may be thickened and contain emulsifiers. They are almost always packaged in glass bottles like the thick sauces, but there are already indications that blown PVC and blown polyethylene bottles may be used extensively for these products.

Horseradish Sauce

Horseradish is a pungent root. Preparation consists of trimming, peeling, and chopping to a fine particle size. This is blended with water, an edible oil, powdered skimmed milk, sugar, flour, a small amount of thickener, and mustard. The blend is boiled, homogenized and cooled. Using this as a "cream" base, a variety of sauces can be prepared. For example a relish can be made using two parts of the cream to one part of minced horseradish root with added vinegar and salt, or the cream or minced product can be blended with a vinegar diluted mayonnaise. The final product is packaged in screw-top jars similar to mayonnaise or pickled products.

A horseradish pickle relish is made by chopping the root and simmering it in vinegar, salt, and sugar for almost half an hour then filling in bottles or jars. Red horseradish is made by the addition of beets to the mixture.

BIBLIOGRAPHY

ANON. 1966. Twin-lid spice can closure and the machine that applies it. Mod. Packaging *40*, No. 3, 60.

ANON. 1967A. Polypropylene spice packs. Packaging Rev. *87*, No. 11, 36.

ANON. 1967B. Drop-at-a-time colours from drop-shaped vials. Packaging News *14*, 32.

ANON. 1968. Methodology of flavor evaluation. Special Report of the Packaging Institute, 66–1, (D 2146.2/25,847).

BINSTED, R., DEVEY, J. D., and DAKIN, J. C. 1962. Pickle and Sauce Making, 2nd Edition. Food Trade Press, Ltd., London.

GRIFFIN, R. C. 1967. Flexible packaging for pasteurized foods. Food Eng. *39*, No. 7, 76–79.

HALL, R. L. 1968. Food flavors; benefits and problems. Food Technol. *22*, No. 11, 55–58.

LOPEZ, A. 1969. A Complete Course in Canning, 9th Edition. Canning Trade, Baltimore.

MELILLO, D. 1968. Creating carbonated beverage flavors. Food Technol. *22*, No. 11, 65–68.

POTTER, N. H. 1968. Food Science. AVI Publishing Co., Westport, Conn.

CHAPTER 10

Beverages

WATER

Man has always had need for water to quench his thirst. When men roamed the earth in small numbers, pure water abounded in lakes, rivers, and streams. As he multiplied in numbers and began to live in permanent settlements he began to pollute this supply. Today, vast sums of money are spent in trying to purify used water, to find new pure sources, and even to recover it from the salty seas.

Exclusive of packaged distilled or deionized water used for scientific or industrial reasons, potable water has been canned for military or emergency use. Military specifications have been drawn up covering the procedures and requirements. In Vietnam special packages of water in double-walled plastic bags have been airdropped to front line positions.

In areas where civic water supplies are not particularly "tasty," pure mineral spring water can be bottled and sold. For the most part this is done in large bottles used for water coolers in factories, offices, or other public buildings. The main requirement is a clean, sanitary container.

Various firms bottle and sell "branch water" for mixing with whiskey. In Europe mineral water has been bottled and sold for many years. Here the connotation is a little different from that used in the United States. European mineral water ranges from a normal clear spring water as we know it to a highly flavored mineral water from medicinal springs often containing some carbonation—for example Vichy water. A recent trend has been the change in packaging of French mineral water from glass to PVC. This represents an enormous new volume for rigid PVC.

Mineral water is also being packed in 20 centiliter capacity minature bottles of PVC. These are triangular in shape and sealed with polyethylene caps. They are replacing the one piece aluminum cans that have been used previously. Air France now uses these packages and claims that they are light weight, economical and are capable of being crushed flat for disposal purposes.

Ice is sold in the United States in many stores or from vending machines. Large solid blocks are rarely if ever packaged, but chipped ice or ice cubes are bagged to prevent soilage and for easier stacking and carrying. At one time double-walled paper sacks were used with wire ties. These had the disadvantage of weakening when wet if the ice were allowed to thaw somewhat. The paper bag has been largely replaced by a low density polyethylene bag with a draw string closure. This package is not affected by melting, and gives an added advantage—the consumer

277

can see the product and be sure it does not include visible dirt.

AROMATIC INFUSIONS

An aromatic infusion is a beverage prepared by steeping a material containing coloring and flavoring matter in boiling water. Usually the solid residue is filtered or strained out, but sometimes it is allowed to settle instead as is the case with Greek or Turkish style coffee. Some infusions are used in the preparation of other products such as soft drinks, liqueurs and food flavorings. Others are consumed as beverages immediately after preparation. This latter type is sold to the consumer in essentially a dry form. It is pointless to ship water when not necessary, and in the case of infusions the proportion of water used is high. Furthermore delicate flavors and aromas are best captured when the beverage is freshly prepared. Reheating of a previously prepared infusion results in an inferior concoction. Warmed-over coffee or a reused tea bag leave much to be desired.

Coffee

The coffee bean apparently originated in Ethiopia and spread throughout the Arab world. Legend has it that a goatherd named Kaldi discovered it in 850 AD. Because the prophet Mohammed had instructed his followers not to consume wine, coffee became known as "the wine of Islam." The Arabs introduced it to the Turks in 1554 who brought it to Venice in 1615.

Today, coffee is grown in 57 different countries and imported in significant quantities by about 25 of the nongrowing countries. Over 6.2 billion pounds of coffee beans were exported in 1967–1968. Because production of coffee beans can exceed demand, in order to stabilize prices, 66 nations signed an International Coffee Agreement in 1962 which assigns export and import quotas.

Product Manufacture.—Coffee is picked as a berry looking somewhat like a cranberry. The berries may be dried and then cleaned or the seeds may be removed immediately (wet process). A fermentation solubilizes a mucilaginous layer which is then removed by washing, and the seed is then dried. Milling removes the "parchment" shell leaving a gray-green colored bean. This is the form in which it is usually exported. The green beans are usually packed in jute bags of 60 kg weight (132.3 lb). Coffees from different countries or different regions have different flavor and aroma characteristics. Green beans are imported and blended to arrive at a desired end result. Taste experts grade coffee and test blends. After blending, the coffee beans are roasted at temperatures of up to 415°F. Roasting is halted when the desired brown color is achieved. In the process the beans swell and the aromatic oils are driven to the surface. The coffee is then cooled either by water spray or forced air. Some coffee is sold in the bean and ground for the consumer in the retail market. Most coffee is ground and packaged for retail sale by the coffee blending house.

An increasing amount of coffee is now being processed by first brewing an extract from the roasted and ground coffee and then evaporating the extract to dryness. Soluble powdered coffees may be dried by drum or spray drying; however, the latter is best. Instant soluble coffees are now being marketed that have been foamed and freeze-dried.

Package Requirements.—There are four problems in the packaging of coffee. One is the prevention of losses of flavor and aroma by evaporation or migration of essential oils, one is the prevention of rancidification of the coffee acids by oxygen, another is the evolution of carbon dioxide gas from the roasted bean, and the fourth is prevention of moisture absorption by the dried roasted bean.

Packaging Materials and Package Forms.—The earliest types of coffee packages were paper sacks in which the consumer carried home the whole roasted beans. Each householder had a coffee-grinder which was used whenever a brew was desired. In-store grinding became common practice some time after the turn of the century. Later the paper sack was made double-walled; then it was lined with glassine. A major improvement came with the substitution of rubber hydrochloride film for the inner liner. The film provided an excellent grease barrier. Packaging of coffee in tins became prevalent in the 1920's and 30's with perfection of vacuum-packing equipment. Nitrogen gas flushed tins with domed lids were introduced in the late 1940's. For some reason ground coffee in glass jars has not proved as popular as the can except during wartime metal shortages. The introduction of instant soluble coffees brought the glass jar into prominence. While tinned coffee still accounts for about 85% of the total pack in the United States; in Europe, particularly Sweden, Norway, Finland, Denmark, England, Holland and Germany the trend is away from the can and strongly in favor of flexible vacuum packages.

Cans.—The standard one-pound coffee tin has evolved from a rather wide and squat key-wind strip opened can to a taller version that is opened by a standard can opener and carries a snap-on high density polyethylene reclosure lid. Vacuum pack has been the chief method.

The trend in cans is toward a larger size. The one-pound can dropped from 66% to 47% of total retail can sales between 1960 and 1967. During the same period the 2-lb can increased from 33 to 35% and the 3-lb can from 1% to 18%. Despite the preponderance of cans in the market, the likelihood is that other packages will take over in the not too distant future. These will be composite cans, plastic jars, and containers, or flexible laminates in the form of bags, pouches, or unit serving packages. Institutional, vending service and military packaging appears already well-launched in that direction. Cans will probably try to halt the trend by including easy open devices and/or interior portion packaging. In the latter instance, lighter weight cans will function.

Glass.—Regular roasted and ground coffee is rarely sold in glass, the weight and breakage factors predominate over the initial low cost and reclosure features. A recent novelty package was a glass percolator filled with instant coffee. Also large glass containers are not particularly suited to vacuum packaging, gas flushing is much safer. Instant (soluble) coffees found glass jars quite satisfactory. Here the product unit size per dollar cost was much more favorable and the reclosure feature became predominant as the container would be opened and closed many times. Glass still dominates this portion of the coffee market. Competition lies primarily in the shape of the bottle, the type of screw cap closure, and the attractiveness of the label. Most glass jars have a Saran-coated, glassine membrane closure sealed over the mouth prior to capping.

Plastic Jars or Containers.—If adequate barrier properties can be developed plastic packaging may be used as substitutes for the can or the glass jar. Straight PVC plastics may prove satisfactory for short term protection but migration of aroma may prove a problem. One major food company has reported development of a 2-lb capacity "plastic" can for roasted ground coffee with nitrogen flush and claims satisfactory shelf-life. It seems likely that the short term plastics will be used primarily as coatings, however they might be used as outer containers for inner portion packs where the latter would provide the needed barrier properties.

Composite Cans.—Several companies have advanced to test market with composite cans for roasted ground coffee.

These are foil-lined and will have either scored tops for easy opening with can openers or easy open pull tabs. Both vacuum pack and nitrogen flush packages are being tested in both 1-lb and 2-lb sizes. Consumer acceptance appears favorable. Savings of up to 25% are reported. One package maker has developed a complete system which includes multistage nitrogen flushing of the coffee right out of the grinder. So long as the residual oxygen in the coffee is brought below 2%, and preferably below 1%, the shelf-life can be expected to be satisfactory, i.e., six months or more. No problems with leakers have been reported. Conversion of can filling and closing machinery to composite cans is reported to be relatively inexpensive and involving only changes in the lid clincher chuck. Costs range from $600 to $2500 depending upon size of the installation.

Flexible Packaging.—The use of bags for ground coffee is not new, but present techniques of combining vacuum packaged coffee in bags with a printed paperboard carton are only about 15 years old in Europe and about 8 years old in the US. Vacuum-packed bagged coffee proved to have shelf-life about four times longer than atmospheric packed bags. Various bag materials have been tried including:

PVDC coated paper (duplex bag)
PVDC coated cellophane-glassine paper (duplex bags)
Polyester film-aluminum foil-polyethylene
Polyester film-PVDC-polyethylene
Polypropylene film-PVDC-polyethylene
Polyamide-PVDC-polyethylene

Since low density polyethylene tends to absorb or be absorbed by coffee oils, medium density is preferred. Ionomer resins are being tested also.

Flexible laminates with and without foil have been formed into pillow packs on form and fill machinery and into tetrahedral packages, as well as fin-sealed packets. Machinery is available from several manufactures for vacuum packing or gas flushing small packs (2–4 oz) at up to 100 a minute.

Courtesy of British Cellophane Ltd.

FIG. 107.   FLEXIBLE FILM PACKS FOR COFFEE

Instant coffee and ground coffee blends are packaged for caterers in laminates of printed "Cellophane" and polyethylene.

Courtesy of Istituto Italiano Imballagio

FIG. 108.   VALVED COFFEE PACKAGE

An Italian design for a coffee package features a valve which vents gases evolved by the coffee after packaging. Without a valve the vacuum pack (on left) would become excessively distended (package on right). Laminate is polypropylene-aluminum foil-polyethylene.

Institutional Packs.—Packages used for institutional markets range from a 20-lb capacity vacuum packed bag in a PVDC coated corrugated shipper used for the military, and a 2-lb capacity nitrogen flushed bag for restaurant brewers, to small unit portion packs of instant powder for making one cup. Nonmilitary institutional packs are used up more rapidly and therefore don't require as long a shelf-life. Two to three months is considered very adequate. A popular item for office coffee brewers is a complete kit including a coffee portion and filter, cream or noncream whitener (usually powdered), wooden stirrers, and drinking cups all in a paperboard carton. Many motels and hotels are adding single cup coffee makers which preheat water. The guest opens a fin-sealed packet of instant coffee, a packet of noncream whitener, and one of sugar and adds the boiling water.

New Trends.—Preportioned packs are now under development and beginning to reach the market place. Wilkins Coffee was the first producer to introduce a one-pound unit containing four 4-oz vacuum packs. Others have brought out a 16 oz unit for a large urn of coffee and 2-1/2 to 3-1/2 oz units for a 1/2 gal. brewer. For home percolators a ring of coffee in a filter paper container vacuum-packed ten to a one-pound coffee tin results in a 12 oz net weight. Plastic lid reclosures are furnished. Each unit is expected to make 4 cups (24 fl oz) of brewed coffee. Another approach is to pack the coffee in a single use throwaway percolator basket. Here again the preportioned units would be packed in a mother container—either a can or a composite can. Preportioned pouches are being packed in printed paperboard cartons. One popular size is six 2-2/3 oz pouches to make up a 16 oz unit. Each pouch is expected to make 8 cups (48 fl oz) of brewed coffee. The pouches have advantages over the percolator packs as they can be used in any coffee maker.

New concepts being considered also include individual one cup portions of instant coffee packed in a disposable spoon. Iced coffee concentrate might be sold frozen in metal and paperboard containers.

Tea

According to legend tea drinking began in China about 2737 BC. The first record is in 350 AD when it was described in a Chinese dictionary. Its cultivation and use spread all over China, Japan, India and Malaysia under the encouragement of Buddhist priests who urged its use rather than alcoholic beverages. Cultivation of tea was brought to Java in 1684 and later in the 19th century to Formosa. The use of tea as a beverage was first brought to Europe by the Dutch in 1610. It reached Russia in 1618, France in 1648, England and America in 1650.

It is estimated that well over 3 billion lb of tea are produced world wide. In 1967–1968, the United States imported 130 million lb of tea from Ceylon.

Product Manufacture.—The tea plant is an evergreen shrub which, if not pruned, would grow 30 ft in height. It looks somewhat like a myrtle with blossoms resembling a wild rose. It requires a moist and warm to temperate climate. Leaves are plucked at various times of the year depending upon the country, the climate, and the desired end product. Black and green teas are made from the same leaves. After picking they are dried in the sun until "withered." This takes about 24 hr. Next they are rolled about three hours to break the leaf cells. After a short pre-fermentation the product is put through final processing. Green tea is steamed without further fermentation. Black tea is fermented thoroughly before "firing" (drying). Oolong tea is a compromise between the two others. The tea is then cut, sifted, sorted, and packaged.

Product Characteristics.—Tea contains caffeine, tannin, and essential oil. The caffeine acts as a stimulant, the tannin gives "body" and the essential oils the aroma and flavor. Some teas are brewed and then the infusion is evaporated to dryness to make an instant tea.

Packaging Requirements.—The main problem in packaging tea is to protect against moisture which can permit mold growth or fermentation, and to protect against loss of aroma by evaporation.

FIG. 109. THE OLD AND THE NEW IN TEA

A metal tea caddy holds loose tea on the left, paper tea bags are
in the metal container on the right, while instant tea is in the
glass jar.

Package Materials and Package Forms.—Earliest packages
were metal-lined wooden boxes. When tea first came to Europe it was so
expensive it was weighed on apothecary scales and sold in papers (paper
wrappers) (see Chapter 1). When the price came down to reasonable
levels tea was sold in tin boxes or in glass jars. Flexible packages were
adopted as soon as metal foils were available for protection.

The packaging of tea in tea bags is one of the first examples of unit
prepackaging. The classical method for making tea involved measuring
loose tea into a tea pot of boiling water and then allowing it to brew. The
problem of keeping leaves out of the cup was at first ignored, then metal
and china strainers were introduced. The first use of a porous cloth to
hold the leaves in the pot is not known. The use of small perforated
metal "tea balls" served the same purpose and an attached chain
allowed the user to remove it when desired. Small tea balls were made
for brewing single cups of tea. The first tea bag was undoubtedly a handmade packet of cloth tied with a thread or string. This method proved so
popular it was commercialized. Eventually, porous wet strength papers
replaced the fine cheese cloth bags. Aluminum wire staples now attach
the string to the bag and to the tag. Some bags are enclosed in individual

paper packets and the tags are torn from the packet itself. One manufacturer puts the tea into a narrow bag that is folded back on itself and claims better brewing because the water can flow through. Modern transportation and distribution are more efficient, making the need for highly protective packaging less important. Simple printed paperboard boxes are used with paper liners for loose tea and paper packets for bagged tea. Many cartons containing tea bags are overwrapped with cellophane or polypropylene for added protection.

New Trends.—The latest trend in tea packaging is in instant tea powders. These are sold in glass jars and in laminated foil pouches. In 1968, mixes outsold loose tea in supermarket sales. Mixes are instant teas combined with other flavorants such as lemon, pineapple, banana, and the like. They are extremely popular for summer iced tea drinks.

Other "Tea" Drinks

"Teas" can be brewed from a wide variety of aromatic plants using leaves, roots, or blossoms. Sassafras root shavings are sold in polyethylene bags for this purpose. Other types include Verbena, Hibiscus, Peppermint, Fennel, Linden, Camomile, and Rosehip. Tea bags containing the latter varieties are often sold in gourmet stores as five bags in a cellophane overwrap for 25¢.

Beef or chicken bouillon "tea" is made by dissolving bouillon cubes in boiling water. The cubes are made from concentrated meat broths dried to a powder then compressed. Instant bouillon powders or granules are also available. The cubes are individually wrapped in foil paper laminations to protect against moisture and rancidity. The wrapped cubes are packaged in glass or plastic jars with metal or plastic screw caps. Spiral wound composite cans have also been used for this purpose. The instant product comes in a glass jar with a sealed coated paper membrane and metal screw cap for a top closure.

FRUIT AND VEGETABLE JUICES

Just as he experimented with the preparation of foods, primitive man also experimented with beverages. There is little doubt that the first beverages, other than water, were the juices of fruits or berries which were consumed together with the fruit. Just when man learned to press the juice out of the fruit to drink as a separate beverage is not known. It undoubtedly was followed by the discovery of wines made from the fermented juices.

Most fruit juices ferment so readily there has been no commercial packaging until comparatively recent times. By the mid 19th century, the knowledge gained by the canning industry was transferred to the bottling of fruit juices. Welch began bottling Concord grape juice in the

United States in 1869. The Swiss bottled unfermented pasteurized apple juices about the same time or not long after. Large scale packing of juices did not become commercial until about 1925. Domestic US consumption of pasteurized single strength juices (principally tomato, citrus, and pineapple) reached a peak of 370 million gal. in 1945, with per capita consumption the following year of 21 lb. In 1945, the introduction of frozen concentrated orange juice began to change the juice drinking habits of the nation. Canned juices continued to increase but at a lower rate. Most of this increase has been through the introduction of other juices and juice mixes. Over the years per capita consumption increased more than 36 lb. Of this amount about 4% were chilled single strength juices, about 50% were frozen citrus concentrates, and about 46% were pasteurized single strength juices. This does not include the fruit juices used as flavorings in noncarbonated fruit or milk drinks.

In recent years lower priced imitation fruit drinks have been introduced. This competion is being challenged by fruit juice processors who are bringing out formulated (modified) fruit juice and fruit juice concentrates. Vegetable juices have not had so great a popularity as fruit juices. If it weren't for tomato juice which is classified as a vegetable though botanically it is a fruit, the consumption of vegetable juices would be negligible. Some carrot juice and sauerkraut juice is packed. The remainder are either specialty diet-fad foods or used blended with tomato juice. Vegetable juices are also used as flavorants in baby foods and in soups.

Large scale production of fruit juices in Europe started after World War II. Previous efforts were small scale and sponsored mostly by teetotalers who sought to eliminate the drinking of brandy. Apple and grape juice were the first products. Citrus juices were packed mostly in Israel and Spain. Berry juices began in Germany and spread into surrounding countries.

Product Characteristics

Although many plants yield delicious juices there are also many that do not. Different varieties of a single fruit may be desirable or undesirable for juice either through flavor differences, stability, or difficulty in extracting or preparing the juice. For example, bitterness in orange juice is related to the root stock of the original tree and to the method of extraction. The flavor, aroma, and color of juices are affected by the variety, degree of ripeness, soil and climate conditions, and mechanical conditions when the fruit is harvested.

Product Manufacture

Apple and some other fruit juices are expressed from healthy ripe fruit by hydraulic pressing. Solids are removed by straining, clarification or filtration. Continuous centrifuge type processors have been developed and are being used in place of the older batch-type presses. After juices have been extracted and purified, there are several alternative further processes. If used as is, they are known as single-strength juices. These can also be diluted and processed into fruit juice beverages. Otherwise, the juices can be concentrated and frozen, or concentrated and dried. Concentrating can be done by vacuum evaporation or by freezing. When evaporation is used, volatile flavoring oils are stripped out and collected for later readdition.

Tomato Juice

For tomato juice preparation, firm and ripe tomatoes should be used. After arrival in the plant, the tomatoes are sorted and washed by soaking. This removes any residual dirt and foreign matter. In soaking, a continuous supply of fresh water is essential and steam or compressed air may be added to agitate the fruit.

After the soak tank, the tomatoes are washed with high pressure sprays and then sorted to remove any undesireable fruit. They are then trimmed to remove rot or mold.

Hot or cold extraction methods may be used; however, hot extraction produces a superior quality juice. In hot extraction, the fruit may be chopped or crushed prior to heating or the whole tomatoes may be heated. The tomatoes are inserted into large tanks equipped with an agitator. Steam coils are used to provide rapid heating. For cold extraction, the tomatoes are scalded prior to extracting, chopped and then extracted without heat.

Extraction involves either screw-type or paddle-type extractors. A shaker screen may also be used to remove any foreign material. Any dissolved or occluded air is removed by a deaerator. In some cases, the juice is homogenized at 1000–1500 psi at 150°F. This results in a thick bodied product with minimal setting of the solids.

Presterilization precedes final canning. This is necessary to kill *Bacillus thermoacidurans* which causes flat sour spoilage. The process involves heating the juice to 250°–275°F. The juice is then canned.

Standard fillers are used for filling tomato juice in cans. The cans employed should be plain hot dipped or electrolytic tinplate bodies and enameled electrolytic tinplate ends. Prior to filling, the cans are spray washed to remove any dust.

Salt may be added to the canned juice by tablets. The cans are then closed under vacuum at 190°F on a Panama or CR type closing machine. After closure, the cans are washed with a hot water spray.

Processing conditions depend on whether the juice is presterilized or heat treated below 212°F. For presterilized juice, the minimum holding times are:

Closing Temperature	Minimum Holding Time
190° F	3 min
200° F	1 min

The cans are then water cooled until the temperature of the contents reaches 100°F, labelled and packed.

Packaging of Juices, Beverages and Concentrates

Single strength juices and fruit beverages are sold refrigerated on a local basis, and pasteurized in cans or bottles for wider distribution. Refrigerated juices may be packed in glass jars, glass jugs, or PE-coated paperboard cartons. HTST pasteurized juices are filled hot into clean enameled or lacquered cans or clean glass bottles or jugs. Closures can be screw caps, vacuum caps, or crown caps. The heat of the juice pasteurizes the container. Bottle and can sizes vary from 4 oz to 54 oz and jugs up to gallon sizes.

Heat processed orange juice has also been packaged in PVDC and foil-PE flexible pouches. These pouches offer superior protection over polyester-PE or cellophane-PE pouches. Juice in pouches made of the former maintain a quality comparable to canned juice during storage for at least five months at 36°F.

Concentrated beverage syrups are packaged in a wide variety of containers from cans, plastic jugs, and "bag in box" concepts to huge drums and tanks.

Concentrated frozen juices are packaged in tinplate cans, aluminum cans, or composite cans. The latter are largely replacing the others due to lower costs. Easy open tear strip devices are used extensively.

New Trends.—The various containers used in Europe for milk packaging have also been tried for single strength juices. These include the "Zu-pak," the "Tetrapak" and various bag in box combinations.

Nonalcoholic cocktail mixes are now being sold in bottles and tinplate cans as well as paper-PE-foil-PE single service pouches. The consumer adds water and/or the alcoholic beverage. Several packages of cocktail mixes have also appeared in the thermoformed plastic single service tubs with peelable seal membranes.

A new glass beverage container combines a tear drop shaped glass container set in a preprinted polyethylene base.

Small bottles and cans of juice are being cluster packaged in PVC shrink film overwrapped bundles of six units. Paperboard stiffeners are used to divide the two rows of three and give backbone support to the

Fig. 110.   Bottled Fruit Juice

This bottle featuring elegant foil top and body labeling protects
the pure apple juice from light by using dark green glass.

package. Prune juice is being marketed in a four-4-oz bottle unit paperboard carrier. Bottles are closed with twist off caps.

A recently introduced package for frozen juice concentrates consists
of a paperboard-PE-foil carton with plastic coated sealing strips. The
carton features a polyethylene extrusion on the foil inner ply and
"safety shields" heat-sealed over the internal surface of the flaps. The
shield on the consumers end is constructed from polyester-PE and the
other shield is pouch paper-PE. The carton is opened by zipping the
strip from the end and lifting a perforated major flap which, when torn
off, peels with it the plastic "safety-shield."

INSTANT POWDERED BEVERAGES

Children have had flavored syrups to make beverages from for many
years. The use of flavored powders or granules has not attained great

commercial importance until relatively recent times. Adult instant beverages began with coffee substitutes like "Postum" in the 1920's and 30's. The popularity of instant coffee, instant tea, and instant fruit beverages has coincided with the development of improved methods of drying and dehydration perfected in the 1950's. Dried soups and sauce mixes created industrial demand for dried tomato juice and dehydrated fruit and vegetable purées. Such children's drinks as "Fizzies," "Kool-aid," and "Tang," popularized dried fruit flavored beverages. In the latter instance it was soon discovered the drink appealed to adults as well. Similar products have been introduced not just as breakfast drinks or refreshers but also as exotically named cocktail mixes. Printed paper-polyethylene-aluminum foil-PVDC heat seal coating laminates are formed into fin sealed pouches with extremely high barrier protection. Multiple packets are sold in folded paperboard cartons.

Some of the instant drinks are sold in glass jars with screw cap closures. PVDC coated glassine membrane closures are sealed across the mouth of the jar for added protection. Instant orange juice substitute is packaged in 18-oz and 27-oz jars as well as smaller sizes. It also is sold in 4-2/3-oz easy open pop top cans. Tablet type drinks are packaged in formed foil packets with paper or paperboard laminated tops.

Methods of drying and dehydration are discussed fully in Chapter 2.

MILK DRINKS

Chocolate and vanilla flavored milk drinks and egg nogs have been available for many years. The major problem in developing fruit flavored milk drinks has been the action of fruit acids on milk casein. Destabilization of casein during pasteurization renders the beverage unsaleable. Recent studies have indicated that addition of dibasic potassium phosphate will stabilize fruit flavored milk beverages during pasteurization (30 min at 145°F). Shelf-life of at least two weeks is possible when properly refrigerated.

For localized delivery milk drinks are packaged in the same type of containers used for whole fresh milk (see Chapter 5).

Some chocolate drinks are bottled and canned. If suitable stabilization techniques can be perfected, a wider line of milk drinks will become available.

CARBONATED SOFT DRINKS

The first carbonated beverages were effervescent mineral spring waters. When Joseph Priestley discovered how to make carbon dioxide gas in 1772, he began experimentation trying to duplicate natural fizzy waters. Carbonated water was bottled commercially in Switzerland and England about 1790. Similar experiments were conducted in the United

FIG. 111. MILK SHAKE MIXES

Chocolate flavored milk shake mixes are packaged in individual serving pouches of wax-laminated white opaque glassine paper. Ten pouches are packed in printed cartons (left). A special pack (center and right) includes four pouches and a plastic shaker.

States by a doctor who thought the product might have curative powers. The chemist that he hired did not like the flavor of plain soda water and added some fruit juice. This was the first carbonated "soft drink." All others since produced are fundamentally the same. Yet this basic formula is now a 2.97 billion dollar industry in the United States with over 2.7 billion cases of bottled or canned soft drinks sold annually.

Product Characteristics

A carbonated soft drink is principally water and therefore much care is expended in selecting good quality water. Carbonation consists of dissolving carbon dioxide gas in the water. The gas may be from any pure source. Modern processors utilize pressurized cylinders of purified gas. The amount used varies depending upon the type of beverage. Ginger ales are highly carbonated with up to 4-1/2 volumes of gas whereas fruit flavored drinks may only contain 1 volume of dissolved gas. (A 5 volume solution requires a pressure of about 5 atmospheres and produces about 1% solution of CO_2 in water). The lower the temperature; the more carbon dioxide can be dissolved at a given pressure. The dissolved carbon dioxide not only contributes flavor and "fizz," but also inhibits growth of microorganisms.

Sugar used to be a major ingredient of soft drinks because it was a major component of the fruit syrups used for flavoring. Fruit syrups normally contain about 50–60% sucrose. About 1-1/2 fl oz of flavored syrup is used in a 7-oz bottle of soft drink beverage. In recent years many "diet" or "low calorie" soft drinks have been introduced in which artificial sweeteners have been substituted for the high calorie sugar carbohydrate.

Natural fruit flavorings were the original additives. Today, numerous synthetic flavorants have been developed by using the technique of gas chromatography to identify components.

Soft drinks can be made clear or cloudy, colored or colorless, and with or without natural fruit juices added. Where juices are included, homogenization is usually employed to stabilize particle size and distribution of pulp. Clear beverages are filtered or otherwise clarified. Coloring can be caramel, fruit colorings or certified food colorants. Although noncarbonated beverages may be pasteurized to prevent spoilage, carbonated beverages depend upon the cleanliness of the container, the acidity of the formulation, and added sodium benzoate preservative to prevent spoilage. Long term storage can result in flavor changes due to the action of light, heat, oxygen, enzymes, or impurities introduced by the water or sugar used. Off-tastes and precipitation of solids are the usual symptoms.

Packaging Requirements and Packages

Carbonated beverages require a container that will hold pressure, not deteriorate, and not contribute off flavors. Traditionally this has meant the use of glass bottles, but in recent years special can coatings have been developed which permit the use of tinplate and aluminum cans. Bottles were more economical if used repeatedly, consequently the industry developed a system of collecting used bottles by refunding a deposit paid at point of purchase. Used bottles were washed to remove labels and sanitize, rinsed, sterilized with chlorine solutions, and then refilled, capped and relabeled. In very recent times lighter weight one trip glass bottles have been developed which are discarded after one use. It is expected that with the continuing increase in labor costs the one trip bottle will ultimately displace the multitrip bottle. Flint, amber, and green glass have all been used for different types of soft drinks. An English ginger beer is bottled in a ceramic squat bottle with a crown closure.

FIG. 112. CARBONATED SOFT DRINKS

This popular U.S. package is now being used in the United Kingdom. It features an all-aluminum easy open top.

Product differentiation has been attained through distinctive shapes, distinctive labels, distinctive product names, and intensive brand advertising.

In early days there were a variety of closures used for the bottled carbonated beverages. One consisted of a glass marble which was forced against a rubber ring in the neck by the gas pressure. A protruding rigid wire loop was used to pull the marble into position. To open, the loop was struck a sharp blow. This released the pressure with a loud "pop." The bottles become known as "pop" bottles and the contents became "soda pop." The crown cork closure has already been described in Chapter 1. Screw stoppers relied on a rubber ring for tight closure. Another popular closure was the swing lever stopper. Here a porcelain stopper was held down against a rubber seat by a strong wire lever. Such bottles are now collectors' items.

The soft drink industry was a very strong influence in the development of labels and labeling machines for both bottles and cans. Bottle labels not only had to be attractive and resist coming loose in ice water, but also they had to disintegrate readily in the hot alkaline label-removing baths. Some soft drink bottlers used fired-on glass pigments for permanent labels and others used raised letters formed in the bottle blowing operation.

New Trends

As already mentioned there is a trend toward the one trip bottles. New twist off crown bottle caps are being used. Aluminum and tinplate cans are featuring easy open "pop top" lids. Labels are dispensed with in favor of direct printing on the can metal. To date there has been no plas-

Courtesy of Reynolds Metals Co.

FIG. 113.   PALLET SHIPMENTS OF SOFT DRINK BOTTLES

The "Reynolon Pallet Overwrapping System," developed and marketed by Reynolds Metals Company, enables one man to over-wrap pallets of glass bottles and other commercial and industrial palletized loads in just 3 min.

tic container developed that will hold pressures of carbonation. It is not out of the question that an aluminum foil composite lamination may be utilized in a unique carbonated beverage package yet to be developed.

The "Tapper" aluminum keg developed by Reynolds Metals Company for beer (see later) is adaptable to dispensing soft drinks. Two approaches are possible. One involves filling the "Tapper" with a single strength carbonated beverage. The other involves filling it with carbonated water. In the second instance the user puts a measured amount of flavored syrup concentrate into a drinking glass and dispenses chilled carbonated water from the "Tapper." Since the standard "Tapper" designed for the household refrigerator holds 288 fl oz, about 48 individual 8-fl oz servings could be dispensed.

ALCOHOLIC BEVERAGES

Fermented Wines, Beers

The invention of beer is lost in antiquity. Mead was produced from fermented honey before the middle Stone Age period.

It has been estimated that 40% of all grains raised in primitive agriculture was for brewing into beers or wines. Wine from the grape is known to be older than 4000 BC and probably predates 6000 BC. Fermented date juice was popular in the Mediterranean and African areas. Fermented milk was of Asian origin, as was also rice wine. The South American Indians manufactured a wine from the cassava called "paiwari" and the Aztecs made a drink called "pulque" from the agave plant.

The excellent table wines of today date back to the labors of the Cistercian monks in the Medieval monasteries, who perfected many arts in the growing and processing of foods and beverages. They established vineyards all over Europe and particularly in the German and French river valleys where climate and soil were most suitable. Some French vineyards date back to 630 AD.

The method for making carbonated wines was known for a long time but no one could bottle it. Finally in 1670 a Benedictine monk Dom Perignon invented a cork that would hold the pressure.

Manufacture of Beers and Wines

Generally speaking beers are made from grain and have a low alcoholic content—around 5% by volume whereas wines are made from fruit and range as high as 14% alcohol. There is some overlapping of this definition for there are rice wines, barley wines and ginger beers. Fermented honey is called *mead;* fermented milk, *koumiss;* fermented apples, *cider;* and fermented pears, *perry.* Although other fruits have been used to make wine, the grape is by far the most popular and the most often used.

World production of wines is estimated at over 8 billion gallons of which better than a third is produced by France and Italy. US production is only about 1/40th of the total.

US beer production is over 110 million barrels. Per capita beer consumption is up to 17.2 gal. and increasing. In 1967, US purchases of alcoholic beverages of all description exceeded $12.6 billion.

Beer.—Barley is the chief grain used in beer. Rice and corn are used where a pale color is desired. Barley malt is made by soaking the grain in water and allowing it to sprout. This develops the enzymes which convert starch into sugars. The malt proteins contribute flavor. The green malt is dried to halt further sprouting. After the dried malt has been milled to crush the husks, it is combined with water and cooked. Corn and rice mash are also added. The cooking action converts the starches to sugars. The converted mash is called wort. The sugar laden wort is strained and boiled with hops and dextrose to coagulate hydrolyzed protein and albumin and to destroy enzymes and sterilize it. The hot wort is strained, cooled, and mixed with a pure yeast. The yeast fermentation reaction converts the sugars to alcohol and carbon dioxide. The beer is cooled to 39°F and filtered. Next it is carbonated with carbon dioxide gas. With some beers a secondary fermentation is allowed. Carbonation may consist of several purgings and take two weeks in order to insure oxygen displacement and complete absorption. After a chill-proofing and a final filtration the beer is ready for bottling, canning, or casking. Lager beer is aged after fermentation. Bock beer is darker and sweeter because the malt is roasted and caramelized. Most regular beers are made with bottom fermenting yeasts. The yeast sinks to the bottom of the tank. Ale, porter, and stout are made from top fermenting yeasts.

Wine.—The type, characteristics, and quality of wine are determined by many factors—the type of soil, the variety of grape, the location of the vineyard, the climate, the ripeness of the grapes, their sugar content, their acid content, how they are crushed, how they are fermented, and how the wine is aged. Grapes are harvested, inspected, crushed, destemmed, and pumped to fermentation vats. The crushed grapes called "must" may be treated with sulfur dioxide to kill wild yeasts and then fermented by adding a carefully prepared pure culture of yeast. For white wines the "must" is pressed to remove "pomace" (seeds and skins) before fermentation, for red wines pressing is done after fermentation is nearly complete. Red wines are fermented in open vats, white wines in closed vats and at lower temperatures. If sugar is to be added it is done before or during fermentation. In both cases a preliminary period of settling or ageing is required to remove suspended material, off-tastes, off-colors, and off-odors. Further fermentation may occur. Final ageing is usually done in wooden casks or barrels to develop flavor and bouquet.

Fortified wines owe their higher alcohol content to the addition of brandy or other spirits. Sweetened wines need to be pasteurized or specially filtered to avoid spoilage.

The characteristic flavor of Sauternes is achieved by allowing the grapes to be attacked by a special mold *Botrytis cinerea.* This loosens the skin and dries the grape, increasing its sugar content. The moldy grapes must be harvested promptly and selectively. Usually only two pickings are employed for the highest quality. The mold contributes flavor and aroma and helps keep acidity down and alcohol content up. Sauternes are aged 2–3 years in barrels and continue to age in the bottle.

Sparkling wines are first completely fermented and then drawn off and blended. Sugar and a new yeast culture are added for final fermentation in the bottle. Bottled wine is held in warm rooms during the second fermentation which may last a year. When fermentation is complete, the yeast dies. It is then necessary to remove the yeast sediment. Bottles are held upside down to allow the sediment to settle on the cork, the bottle is chilled and the neck area frozen. The cork and plug of frozen wine are removed and discarded. Losses are replaced and the bottle recorked. Sugar syrup and brandy are added just prior to the final corking to adjust flavor and alcohol content.

Some wines are carbonated artificially rather than by natural fermentation, such as sparkling Rosé and Burgundy.

Vermouths are base wines that are fortified and then flavored with spices. They derive their name from "Wermut" the German word for wormwood one of the first used flavorants. Almost any kind of herb, spice or flavorant may be used from allspice and aloe to vanilla, wormwood, yarrow, and zedoary root.

Still and sparkling wines are also made from apples, cherries, plums, currants, blackberries, raspberries and other berries to produce sweet dessert wines. Other wines have been made from Concord grapes, pineapples, pears, pomegranates, oranges, grapefruit, figs, dates and honey. In all instances the basic manufacturing processes are essentially the same.

Packaging of Wines and Beers

Beer.—The classic package for beer was a keg or barrel which was tapped and drawn off into mugs for immediate consumption. With the development of modern cooperage, metal barrels replaced the wooden kegs and draught beer became more and more common as central brew-

FIG. 114. BEER PACKAGES IN GLASS AND ALUMINUM

The two piece aluminum can is a popular package for beer. This brand of beer also features a brown glass bottle with a unique shape. A simple six pack carrier for cans is also shown.

eries distributed the product. Draught beer was used up promptly and rarely had time to deteriorate or spoil. When distribution began to cover larger distances some of the beer was pasteurized. This was accomplished by heating the beer to 140° F for a few minutes to destroy yeasts or other bacteria. When a bottle was developed which could hold the pressure of carbonation, this became a popular package. The crown cap had a very strong influence on the success of bottled beer, as the closure had to be able to withstand pasteurization also.

The tinplate can was used extensively as a package for beer. It could easily withstand pasteurizing pressures. Some metallic flavor was picked up by the beer if the tinplate or inner coatings did not adequately protect against dissolved iron oxide. Cans were opened with lever type can openers usually given away at point of sale. In the 1950's a new type of can was developed with an easy open end made of aluminum which had a scored section which could be torn loose by means of a riveted pull tab. Many varieties of the "pop top" lid have since been introduced. In addition all-aluminum cans have been perfected. The steel industry countered with tin-free steel for which improved can coatings were required. Today, the brewer can choose from many combinations:

(a) All tinplate body and ends.

(b) The tinplate body with one tinplate end, and one aluminum end with or without easy opening device.

(c) A tin-free steel body with tin-free ends.

(d) A tin-free steel body with tinplate ends or one aluminum end with or without easy-opening device.

(e) Two piece aluminum cans with conventional chime.

(f) Two piece aluminum cans with necked in top chime.

(g) Draw-and-iron two piece tinplate steel cans with steel or aluminum lids.

(h) Draw-and-iron two piece tin-free steel cans with steel or aluminum lids.

(i) Cemented seam three piece cans in aluminum, or tin-free steel.

(j) Chrome coated steel.

In recent years lighter weight one trip disposable (no-return) bottles have been introduced for beer. This has also led to customized shapes. Each bottler now wishes to identify his product not only by label but by shape of the bottle. There is talk of a square bottle rather than round. Bottles are now more squat and most are being treated with metallic oxides and polyethylene (about one millionth of an inch thick) to minimize damage from scratches.

Composite cans and plastic bottles are being carefully scrutinized by researchers as packages for beer. A PVC bottle was test marketed in West Germany in 1968. The PVC bottle offers several advantages. It is lighter weight and not easily broken. Barrier properties are adequate. However they cannot be pasteurized, and there are difficulties due to

TABLE 16

CHEMICAL SPECIFICATIONS FOR BASE STEELS[1]

Element	Percentage Permitted				Beer End Stock
	Type L	Type MS	Type MR	Type MC	
Manganese	0.25–0.60	0.25–0.60	0.25–0.60	0.25–0.60	0.25–0.70
Carbon	0.12 max.	0.12 max.	0.12 max.	0.12 max.	0.15 max.
Phosphorus	0.015 "	0.015 "	0.02 "	0.07–0.11	0.10–0.15
Sulfur	0.05 "	0.05 "	0.05 "	0.05 max.	0.05 max.
Silicon	0.01 "	0.01 "	0.01 "	0.01 "	0.01 "
Copper	0.06 "	0.10–0.20	0.20 "	0.20 "	0.20 "
Nickel	0.04 "	0.04 max.	No limitations specified	"	"
Chromium	0.06 "	0.06 "	"	"	"
Molybdenum	0.05 "	0.05 "	"	"	"
Arsenic	0.02 "	0.02 "	"	"	"

[1] *Adapted from* R. R. Hartwell (1956).

leakage and to creep of the plastic under pressure. So far composite cans have not been able to hold pressure at the seams.

Plastics packaging of beer is, however, being done using polyethylene bags. These were used in South Africa to market a "green beer" which was still fermenting in the package. The gas permeability of the plastic prevented over pressurization. Rapid distribution and consumption was also a factor.

In Sweden, the Tetra-Pak AB Company has announced the development of a new beer package suitable for pasteurized or nonpasteurized beers. This is a four component package—an internal cup with round bottom, a cone shaped top, a tear off cap and a laminated paper sleeve. The cup is made from PVDC coated PVC vacuum drawn. It is seated in the cylindrical spiral wound paper-foil sleeve, which contributes strength. The top is thicker plastic and colored to filter out light. It is heat sealed to the cup and the flange of the seal fits over the cylindrical sleeve. The cap is an easy open type which is made tighter by internal pressure acting on the concave central membrane.

Also in Sweden a plastic bottle was announced which was claimed to disintegrate within two years of being emptied and discarded. Reports on market tests of this container have not been encouraging.

In the United States, a home dispenser for draught beer was developed by Reynolds Metals Company. It comprised an aluminum keg, an interior pressurizing cartridge, a plastic siphon, and a plastic dispensing (no-drip) spigot. The "Tapper" was designed for multi-trips and consequently a production line was developed for the breweries where the "Tappers" were opened, disassembled, washed, and sterilized before refilling. A coated corrugated carrying carton was designed to make it easier to carry home the filled "Tapper." The consumer makes a returnable deposit on the container and carrier and pays only for the beer. The Tapper fits into home refrigerators with ease or especially designed electric coolers can be purchased. An aluminum seal on the spigot maintains sterility until ripped off. The "Tapper" contains 288 fl oz of beer, or the equivalent of four 6-packs of 12-oz bottles. Beer is dispensed by and carbonation is maintained at a uniform level by the unique pressure-activated valves in the pressurizing cartridge. Safety devices are provided to release pressure if it should accidentally get too high. Other auxiliary equipment has been developd. These include molded polystyrene foam carriers for picnics and plastic bases to prevent rolling and to indicate amount of beer remaining.

Various other bulk containers for home use have been tried. These usually involved a large standard container, such as a can, plus an external pressurizing device or pump for dispensing the contents. The consumer purchased the devices and was responsible for their maintenance and sanitation.

The adoption of filtered beer by many breweries has made it possible to sell a nonpasteurized "draught" beer in bottles and cans. The microporous filtration medium filters out mold and yeast spores which are the primary cause of spoilage. These filtered beers have challenged the larger home dispensed bulk containers for draught beer such as the "Tapper."

A wide variety of easy open bottle closures have been introduced. These include crown caps that can be twisted off, scored aluminum caps which can be torn off by means of a ring tab, and rolled on aluminum caps which split to indicate tampering but which can be replaced for reclosure if desired.

Opening devices on cans have also been varied. Most feature a tear out strip of aluminum. Variations are mostly in the design of the tear out, of the pull tab, or of the method of inserting it. The problem with pull tabs

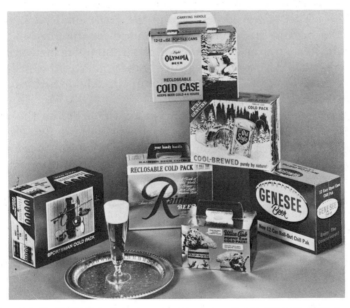

FIG. 115. FOIL LAMINATED "COLD PACKS"

Foil laminated carriers for 12 cans of beer come in different designs. Carrying handles are an added convenience. The carrier is closed on all sides to take advantage of the fact that foil keeps the contents cold longer.

is that the discarded tab can cut hands or feet. This has become a problem at public beaches.

Another area of packaging for beer lies in the many types of carriers used for six packs of bottles or cans. These have mostly been made of paperboard ranging from almost complete cartons to very open sleeves with dividers and partitions. Foil laminated carriers have been proven to keep chilled beer colder for longer periods than just plain paperboard. They have proven popular in summer. Plastics carriers have been made which snap over can tops and grasp the cans with enough force to permit them to be carried.

Another new approach developed by Reynolds Metals Company is the use of heat shrinkable PVC film to hold cans in a corrugated tray. The film is 1-1/4 mil thick which makes it strong enough to immobilize the contents, permit stacking, and to act as the hand holds. Shrink packs of this type can replace conventional corrugated shippers. Also used are molded plastic crates which serve as replacements for the traditional wooden crate used in wholesale distribution.

Wine.—As with beer, wines were once packaged in wooden kegs or casks. The corked glass bottle became the traditional package when glass bottles became cheap enough to be used for the purpose. Recent trends indicate a growing acceptance of plastic stoppers instead of corks for all wine bottle closures. In the United States nearly all wine is sold in bottles whereas in France, Italy, and Spain sizeable quantities are sold in bulk directly to the consumer.

In France, government regulations exist which define the requirements that must be met in order that a wine be permitted to use a label identifying the particular region of its origin: *(Appellation d'origine)*. These regulations known as *appellations controlees* are administered by the *Institut National des Appellations d'origine de Vins et des Eaux-de-Vie.* They specify the geographic area to which a label may refer, the varieties of grape which may be used, the minimum alcohol content, and the maximum yield per acre. Unofficially the vintners of Bordeaux have had a voluntary type of appellation controlee for centuries. The wines of the Medoc were classified in 1855 into five "growths" or vineyards. The larger vineyards were called *chateaux.* Wines bottled at the vineyard carried its name, e.g., *Chateaux d'Yquem.*

Wine is difficult to package in metal as this prevents proper ageing and adversely affects flavor, color, and aroma. Wine bottles come in a myriad of shapes, sizes, and colors. The sizes range from individual serving "splits" to five gallon capacity "demi-johns." They may be short, squat, and rounded or they may be tall, narrow necked, and slender. They may be over wrapped with straw or woven "Basketry." They may be made from clear white glass, or colored glass. Carbonated wines are made of very heavy glass to withstand the pressures.

Riesling wines and other Rhine wines are bottled in tall, narrow, long necked, light brown bottles. The finest Rhine wines came from the Rheingau, a locale on the Rhine not far from Bonn and including the village of Hochheim. The word "hock" has been used to describe German white wines and probably is derived from this village name.

Moselle wines are bottled in similar shaped bottles made of blue or greenish blue glass. Wines from the Franconia area of Germany are put in squat, broad bottomed, green glass bottles.

French red Burgundies and French Chablis wines are bottled in green or brownish green bottles similar in shape to but lighter weight than champagne bottles.

Clarets from the Bordeaux region are in dark green glass bottles with straight cylindrical sides and necks. This is also used for California red wines, Spanish sherries, and Portuguese ports although the last two may use brown glass as well. French and California Sauternes, and French Graves wines are bottled in clear white or light green glass of the same shape as the claret bottles.

Champagnes and other sparkling wines are bottled in heavy green glass with indented bottoms.

Vermouths are usually put in green bottles of the claret shape. Italian red wines such as Chianti wines are nearly always put in round bottomed green glass flasks with woven straw covers. Italian white wines are usually more conventionally bottled.

Spanish dessert wines are world reknowned. The Spanish Malaga and Sherry wines are superb. Portugal is famous for Muscatel, Port, and Madeira wines.

Wines are produced in many other countries including the United States. US wine production is centered in California, New York and Ohio. Most US wines are attempts to duplicate the finest European types, however, a few unique wines have been developed including the Californian "Angelica" and the Eastern "Scuppernong," "Concord" and "Catawba" wines.

While the glass bottle is traditional for wine packaging, the use of plastics has been tried in Europe with some success. The "Zu-pack" container has been used for this purpose. White wines employ a paper-polyethylene laminate, while red wines need the light barrier afforded by aluminum foil. In France, plastics were first tried in vending machines in factories. Here the worker could buy his *vin ordinaire* and not have to carry a bottle to and from his home. Within three years of the first trial eight major wine producers converted to blow molded PVC plastic bottles. The first plastic bottles used permitted a maximum shelf-life of six months. Certainly not a fraction of the life in glass but enough for volume consumption of ordinary quality beverages.

Small individual servings of wine in plastic containers such as the tetrahedral shaped package or the fin sealed pouch may eventually find use in aircraft where weight is of prime importance.

Paul Masson Vineyards in California have created a holiday gift package for St. Valentines Day whereby a premium line of wines can be promoted. The wines are put in heart shaped bottles decorated with antique looking labels: A folding box carries a picture of the bottle. A gift bag with carrier handle and a personalized valentine heart motif is large enough to take the bottle and box.

DISTILLED SPIRITS

Although distillation was known in Aristotle's day, the distillation of spiritous liquors is believed to have been perfected in Europe some time in the Middle Ages. (It is known that Arrack was distilled in India as far back as 800 BC.) In Northern European countries, where grapes do not flourish, fermented grains were distilled to produce whiskies, gins, vodka, schnapps, aquavit and the like. The Welsh were distilling mead in the 6th century AD. Irish and Scotch distillers were active as early as 1170 AD. In more southern climates the grape predominated and wines were distilled into brandies in Italy as early as the 9th century. Other fruits, grains, or sugar or starch bearing plants have been utilized including sugar cane (rum) potatoes (vodka), plums (slivovitz), cassava (paiwari), and cactus (pulque).

In the US and Canada over 340 million gallons of distilled spirits are consumed annually.

Brandy

Brandy, weinbrand, branntwein, aqua vini, or eaux-de-vie, is the product of distillation of wine. Several successive distillate fractions may be drawn off, blended, and redistilled. Finally, the distillate is aged in wooden casks for up to 20 years or more. Brandies may derive their color and flavor from the original grapes (and wine) or flavorants and colorants may be added. They may be consumed as a beverage directly, or used to "fortify" wines.

Whiskey and Other Grain Spirits

Whiskey.—Whiskeys are made from fermented grain mashes. The type of grain used, how it is malted, and the later treatment determines the type of whiskey. Irish whiskey is made from malted and unmalted barley and from oats, wheat, and rye. The malted barley is dried over peat fires before fermentation. The whiskey is given three successive distillations before ageing.

Scotch whiskey is made mostly from malted barley. It too is cured over peat fires. Scotch is distilled only once, then blended. Both Irish and Scotch are distilled at between 160 and 190 proof and aged in used charred oaken casks before bottling.

In the United States, whiskies are made from any grain or mixture of grains, but one grain must make up at least 51% of the mash. Rye whiskey comes from a mash containing 51% rye grain, bourbon whiskey from a mash containing 51% corn. Corn whiskey requires a mash to be 80% corn. When distillation is taken off at 190 proof or more, the product is called grain neutral spirits and contains 95% alcohol by volume. US whiskeys are by definition less than 160 proof. When exactly 100 proof and aged 4 or more years, the whiskey may be labeled "Bottled-in-Bond." It must not have been blended or diluted with water, grain neutral spirits, or other whiskies. Whiskies other than 100 proof may be labeled straight whiskey, for example straight rye whiskey. The straight whiskies must be 2 years old and may be diluted with water to desired proof but may not be blended with grain neutral spirits or other whiskies. Blends of straight whiskies are mixtures of straight whiskies with no other additives. Spirit blends or blended whiskies must be 25–40% straight whiskey with the balance grain neutral spirits. The label must include the word *blended* and cannot use the word whiskey without the qualifying adjective. Most US straight whiskies are diluted to under 100 proof. All must be aged in new charred oak barrels. A new regulation by the US government IRS Alcohol Division has permitted U.S. distillers to begin marketing in July 1972 a light whiskey made by distilling at 160 to 190 proof and storing in *used* barrels. This is intended to compete with the Scotch and Canadian whiskies that have doubled their sales in the US over the past nine years.

"Proof" is a legal definition of alcohol content. The definition differs with each country. In the United States, a proof spirit is 50% ethyl alcohol by volume with the alcohol having a specific gravity of 0.7939 at 60° F. In Holland, a proof spirit is 50% by volume absolute alcohol at 59°F (15°C). In England a proof spirit at 50°F has 12/13 the density of distilled water.

Gin.—Gin is a distilled spirit made from corn, barley, malt, and rye. Its name is foreshortened from the French word for juniper, *genievre*. In Holland and Germany it is called *schnapps*. In early days, it was called Hollands, geneva, and gin.

English gin is distilled from a mash of 75% corn, 15% malt, and 10% rye and is redistilled with added juniper berries and coriander seed. American gin is 85%, 12% and 3% mash and the flavorings are not added until the last distillation. Dutch gin contains 2 parts of rye to 1 part of barley malt. Juniper is added after the first distillation, and there are three distillations in all. Dutch gin has a more "grainy" flavor.

Gins can be sweetened with flavored syrups to produce cordials whereas liqueurs are more complex blends with flavored wines and spirits. Compound gins may contain added grain neutral spirits. Vodka is very similar to gin except it does not contain juniper flavoring and now includes potatoes as well as corn and rye malt in the mash.

Rum is made by distilling sugar cane residues and molasses. It can also be made from sweet potatoes and from sugar beets.

Packaging of Spirits

As with wines, distilled spirits first were packaged in wooden casks, later in glass bottles. The most important parts of the package were the label to identify the contents and the closure to prevent pilferage. It is prohibited by law to reuse a bottle that contained spirits and the glass is so marked. Most bottles used for spirits are colorless or pale green. The effect of light on the stability of various liqueurs was studied by Connor (1966) by storing them in sunlight for three months in flint, amber, and green glass bottles. Color was measured as a change in absorbance at 580 and 650 mμ wavelengths. Sloe Gin and Cherry Liqueur lost some of their red coloring and in general the green glass bottle offered the best protection. Triple-Sec in flint glass developed an off-flavor which may have been related to an increase in aldehyde content from 56 to 80 mg per 1.

The most recent improvements in the packaging of spirits has been in the closures used. These have evolved from cork of earlier days to molded plastic screw caps and roll on metal caps. Clear or colored plastic shrink films are used over the screw caps as a means of preventing pilferage. The metal caps have to be destroyed in part at least to remove them the first time.

Christmas gifts of spirits have been promoted through imaginative packaging. This has been through the use of decorative decanter bottles or through use of printed cartons or of printed carton overwraps. The latter have become extremely attractive through use of embossed foil laminates.

The American Can Company recently completed development of a vinyl interior coating for their cemented side seam aluminum can, which enables them now to package alcoholic beverages in cans with a shelf-life of more than a year. The cans contain mixed cocktails and each eight-oz can holds about three normal drinks. The cans have an easy open top. Heublein, Inc. is marketing the cocktail mixes.

American Distilling Company has announced it will market its Bourbon Supreme straight whiskey in a half-gallon PVC container. The plas-

FIG. 116. CHRISTMAS GIFT PACKAGING OF FINE WHISKEY

A straight Kentucky bourbon whiskey is shown in its holiday
packaging, a deluxe, cut crystal-type decanter in an eye-catching
oval shape. The gift wrap presents a flocked foil cap over a moiré
embossed foil shell.

tic container weights about 1/8th as much as the glass container it
replaces. A hollow handle makes pouring easier.

Also on the horizon are miniature PVC bottles for airline and railroad
passengers. The savings in weight as contrasted to glass bottles is substantial. Since most miniatures are bottled on order and minimal inventory is maintained, extended shelf-life is not essential. In the United
States, several states use only miniatures for bar consumption. The
overall market for these small individual bottles will probably grow.

Courtesy of National Distillers Products Co., a Div. of National Distillers and Chemical Corp.

FIG. 117.   BOTTLE FILLING LINE FOR GIN

A view of a bottling room in a distilled spirits plant showing inspection in the foreground and casing operations in the background.

BIBLIOGRAPHY

ALMARKER, C. A., and WALLENBERG, K. E. S.   1968.   New pouch method for ground coffee. Mod. Packaging *41*, No. 6, 146–148.

ANON.   1966.   Switch from glass bottle to carton for soft drinks. Package. Eng. *11*, No. 11, 27.

ANON.   1967A.   Wine cartoning gets underway—two systems now ready to go using rectangular packs. Packaging News *14*, No. 1, 2.

ANON.   1967B.   Nitrogen packaging of coffee by patented process preserves freshness for over six months. Packaging News *14*, No. 1, 11–12.

ANON.   1967C.   Bottoms up. Plastics Rubber Weekly, No. 159, 1.

ANON.   1967D.   Rapid filling system for draft beer. Mod. Packaging *40*, No. 7, 56.

ANON.   1967E.   New automatic machine for the production of tea-bags. Food Packaging Design Technol. Apr., 40.

ANON.   1967F. · Bag-in-box for sherry. Packaging News *14*, No. 4, 6–60.

ANON.   1967G.   Wine goes bag-in-tin. Packaging Rev. *87*, No. 5, 38.

ANON.   1967H.   Automation comes to tea packing. Packaging News *14*, No. 6, 1–2.

ANON. 1967I. PE bumps glass for fountain jug favor. Packaging Innovator 2, No. 3, 6.

ANON. 1967J. New beer pack to meet peak season demands. Bottling No. 173, 193–199.

ANON. 1967K. Bag-in-tin may beat bag-in-box on price. Packaging News 14, No. 7, 21–22.

ANON. 1967L. Coffee tin wins Australian export award. N. Zealand Packaging 4, No. 4, 9.

ANON. 1967M. Close up: beer. Package Eng. 12, No. 8, 75–79.

ANON. 1967N. Products requiring special packaging. (B) Beer. Verpackungs Rdsch. 18, No. 11, 1320–1322. (German)

ANON. 1967O. 'No Nonoonoo' wine is in tight bottle with reseal plug and bright label. Packaging News 14, No. 11, 3.

ANON. 1967P. Individual coffee extractor pack. Food Trade Rev. 37, No. 12, 91.

ANON. 1967Q. Beer drinkers want the one-trip bottle. Packaging Rev. 87, No. 10, 12–13.

ANON. 1968A. Unplasticised PVC jars for instant coffee. Verpackungs Actuell, No. 1, 1.

ANON. 1968B. Shake-up package in cocktail-mix market. Mod. Packaging 41, No. 3, 97.

ANON. 1968C. Portion-pack for liquor. Mod. Packaging 41, No. 3, 93.

ANON. 1968D. Sophistication in liquor half-gallons. Mod. Packaging 41, No. 5, 136–137.

ANON. 1968E. The 'Rigello' pack—a paper and plastics container for beer. Packaging 39, No. 459, 31.

ANON. 1968F. Gas flushing increases beer can shelf-life. Packaging Rev. 88, No. 6, 22.

ANON. 1968G. Problems of plastics bottles for beer. Intern. Brewers J. 104, No. 1233, 41–43.

ANON. 1968H. A review of pasteurization and sterilization of beer. Bottler Packer 42, No. 8, 53–56.

BISHOP, T. 1968. The no-deposit bottle success story. Bottle Packer 42, No. 2, 90–96.

CONNOR, H. 1966. Private communication to the authors. National Distillers Products Co., New York.

HARTWELL, R. R. 1956. Choice of containers for various products. Proc. 3rd Intern. Congr. Canned Foods, 124–148, Rome—Parma, Italy.

JOSLYN, M. A., and HEID, J. L. (Editors). 1963. Food Processing Operations, Vol. 2. Avi Publishing Co., Westport, Conn.

MÜLLER, B. 1967. PVC disposable bottle for beer and other carbonated beverages. Verpackungs Rdsch. 18, No. 10, 1186–1199. (German)

PASQUARELLI, O. 1966. Plastics in wine bottles. Mater. Plast. Elast. 32, No. 6, 659–662. (Italian)

SACHAROW, S. 1967. Market potential perks up for coffee flex packs. Paper, Film, Foil Converter 41, No. 6, 41–43.

SACHAROW, S. 1968. Films for food packaging. Packaging India, 1, No. 1, 41–46.

CHAPTER 11

Sugar, Chocolate and Confections

INTRODUCTION

Primitive man had knowledge of sweet fruits, berries and honey. Most ancient confections were mixtures of honey, fruits, and flour pastes, and often they were molded into shapes having religious significance. Ancient Egyptian temples show hieroglyphics of early candy making. Centuries of experimentation by cooks, sugar-bakers, and chocolate-millers have resulted in literally thousands of delectable combinations of ingredients.

Packaging of candies has grown from nothing, when candy was kept and displayed in open trays, to a definite and important role in protecting the product against damage, spoilage, and vermin infestation. The nature of the candy industry in modern times has limited the use of the best packaging materials. The majority of candy manufacturers have been small and regional in operation. Many have specialized in only a few types of candy. This has limited *volume* consumption of packaging materials. Also, although candy has grown to have a large overall sales volume (over \$1.5 billion worth of candy was sold in the United States in 1967) the unit price has been low. This has limited the *cost* of packaging materials. In more recent years, a major trend has developed where large diversified food processors are assimilating candy companies through mergers. This is resulting in a careful scrutiny of traditional packaging practices and the application of modern packaging know-how and materials to the well-defined product requirements.

SUGAR

Cultivation of sugar cane probably originated in Asia. The troops of Alexander the Great encountered it in their invasion of India three centuries B. C. In the seventh century A. D. Chinese and East Indians were extracting syrup from the cane, but the Arabs and Egyptians are credited with the first recrystallization processes. Arab caravans and Venetian merchants built a trade in Eastern sugar, which by the 13th century had become a virtual monopoly. Returning Crusaders, who acquired a taste for sugar in their travels, helped create and build a demand for the product. The Venetians perfected methods for refining sugar and became famous for their sugar loaves and for their artistic sugar sculptures. From the 15th century other nations began to compete in the cultivation of sugar cane. It was spread around the world by the

311

Portuguese and Spanish explorers, and brought to the Americas by Columbus.

Beet sugar was discovered by a chemist in 1747, though Californian Indian tribes were reported in 1775 to have been obtaining sugar in this way. Napoleon Bonaparte established the first sugar beet industry in Europe. Other sources of sugar include sorghum, cane, and sugar maples.

From its early beginnings sugar has grown to the status of a commodity, and world consumption is measured in billions of pounds. In the United States alone, over 1.4 billion pounds are consumed by the candy industry, and over 5.0 billion pounds are sold annually at retail for home consumption.

Product Properties and Packaging Requirements

Sugar is refined to several purities running from white to dark brown. Molasses is a by-product and is used in candy as a flavoring. The color and flavor of the brown sugars are due to the presence of dissolved "non-sugars" in the thin layer of syrup that coats each crystal. Because of this syrup constituent, brown sugars must be kept in a moist environment, preferably 65–70% rh, to prevent caking into a solid lump. White sugar on the other hand is a dry but slightly hygroscopic product. To prevent formation of a moist syrup on the crystal surfaces, which could cause caking, environmental moisture should be kept low, preferably 30%, and the environment should not fluctuate excessively in either temperature or relative humidity.

Packages and Package Materials

Granulated white sugar is sold at retail in double-walled glued end flap paper bags containing five and ten pounds, and in waxed-paper lined paperboard folded cartons containing two pounds. Cubed granulated sugar lumps are packaged in 1-1/2 lb cartons, sometimes with a transparent window. A few specialty items, such as colored or flavored sugar crystals are sold in small glass or clear plastic containers made from polystyrene or cellulose acetate. Small paper canisters and small opaque molded polyethylene canisters containing from 1 to 4 oz have been marketed for picnic or patio use. For institutional or restaurant feeding, polyethylene coated paper packets holding about one teaspoonful have been used for a number of years, and they are now available in supermarkets, 50 packets to the folded carton. Powdered white sugars and granulated brown sugars are packaged in one-pound folded cartons with waxed paper liners. The greater the need for protection; the heavier the liner.

New Trends

Among new items moving onto the shelves are a one-pound box of "de luxe" (finer grain) granulated white, a one-pound box of pourable granulated brown, and individual servings of white granulated sugar in paper tubes. The first two contain waxed glassine liners and the brown sugar has a film overwrap. Several films have been test marketed including coated cellophanes, polyethylene and polyvinyl chloride. The paper tubes are each about 3/8 in. in diameter and 3-1/2 in. long made from polyethylene coated paper. They come 50 to a carton, and are said to be superior to the flat packets in ease of opening and dispensing.

Artificial sweeteners are found on the sugar shelf, packaged in glass

Courtesy of Glassine and Greaseproof Div., American Paper Institute

FIG. 118. DeLuxe Sugar Requires Special Packaging

This standard one pound carton has been given a new golden image and a special liner of waxed wax-laminated white opaque glassine to provide an extra moisture barrier for the superfine sugar. The tinier sugar crystals have greater surface area and need the added protection to prevent caking and lumping.

and polyethylene bottles (liquids), paper packets (powders), and in glass vials (pills).

SWEET SYRUPS

Historically, sweet syrups predate crystalline sugar products. They can be pressed from a variety of plant juices or saps. Today, the most common are cane sugar syrup made from redissolved refined cane sugar, and corn (sugar) syrup. Additives include various flavorings and some artificial colorings. Pure maple syrup is so powerful in flavor that it usually is diluted with cane or corn syrup. Honey is marketed as is, a viscous pourable syrup, or as a whipped "butter." Recent years have seen a collection of waffle and pancake syrups with added butter.

Packages and Packaging Materials

Most pourable syrups sold at retail are packaged in glass bottles. Bottle shapes have been the means for brand identification in addition to labeling. Honey is sold in jars. Plastic bottles of blown polyethylene or clear polyvinyl chloride are now coming on the scene as they result in less breakage. Individual servings, so popular in restaurants and institutions, are now being made available for home use. These include clear polyethylene and polyethylene coated paperboard tetrahedral containers, and shallow cups molded from polystyrene, polypropylene, polyethylene, polyvinyl chloride, or ABS copolymer. These are closed with peelable membrane lids, some of which are vacuum metallized plastic films and others foil laminates. In either case, the plastic selected must be compatible with the cup to effect a seal. Finsealed pouches are also used. These can be made from cellophane-polyethylene, paper-polyethylene-foil-polyethylene, or cellophane-polyethylene-foil-polyethylene. In larger quantities syrups are being marketed in foil-paperboard-polyethylene laminates formed into containers called "Zupack." Here the laminate is formed into a finsealed pouch-like structure, which then is folded along prescored creases into a rectangular box shape. Gallon quantity blow-molded polyethylene bottles are available, and still larger quantities can be packaged in a "Cubitainer" made from a cube-shaped corrugated shipper with a polyethylene liner that has a molded polyethylene bottle neck sealed to the liner. Due to the flexibility of the latter the bottle neck is retractable.

CHOCOLATE AND CHOCOLATE PRODUCTS

Chocolate

The Aztecs, Mayans, and Incas of Central and South America raised cacao trees and produced *chocolatl* before the first Spanish conquistadores set foot on the soil of the New World. It was Montezuma's favorite drink. He never entered his harem without partaking from a golden cup, which led to the belief, still prevalent as late as 1712, that chocolate was an aphrodisiac. Hernando Cortes brought cacao beans to Spain in 1528 and taught the royal court how to prepare the Aztec beverage. With the addition of sugar the drink became a flavor sensation and soon spread to other royal courts. Before long lesser members of court circles were congregating in "chocolate houses" where the drink could be bought, despite its high cost. Through the 1600's and 1700's little more was accomplished. In 1828, cocoa was invented. The defatted product could be mixed with sugar, chocolate, and cocoa butter and form a moldable and more palatable food. Another important event was the discovery of milk chocolate in 1876. These two accomplishments made chocolate a much more popular confection, and its consumption has grown from a few handfuls of cacao beans to more than 500 million pounds of beans annually in the United States alone.

Product Properties.—Chocolate is prepared from fermented cacao beans. Milling the kernels of the bean produces chocolate "liquor" from which all other products are prepared. In the manufacture of cocoa the fat (cocoa butter) is pressed out. The fat is used as a candy additive, as an ingredient in the making of other types of chocolate, and in the manufacture of pharmaceuticals and cosmetics. Baking chocolate is primarily cooked and hardened chocolate liquor. Sweet and bittersweet chocolates are made by adding sugar, cocoa butter, and flavorings to chocolate liquor. Milk chocolate also contains whole milk solids. Lecithin is an optional additive to help emulsify fats. Sweet and unsweet, flavored and unflavored chocolate is sold in solid chunks and bars, in drops or "kisses," and in other molded shapes. Bits or chips of chocolate are added to cakes, cookies, and ice creams, or the chocolate may be sugar coated so it will "melt in the mouth instead of the hand." A wide variety of ingredients are dipped in chocolate from ants and grasshoppers to fruits and nuts. Many candies and complex confectioneries are dipped or enrobed in chocolate.

Although chocolate fat (cocoa butter) is not susceptible to rancidification, it does readily absorb odors and flavors. Thus, chocolate candies need protection against odor or flavor transfer. When chocolate is exposed to excess moisture or to sudden changes in temperature which cause condensation, sugar bloom results. Sugar bloom is a light gray coating of sugar crystals. Storage at humidities above 78% for milk chocolate or above 85% for plain chocolate is not recommended. Fat bloom is the formation of crystals of fat on the surface of the chocolate. Packaging is therefore concerned with protecting chocolate from the effects of heat and humidity, and from odor and flavor contamination.

Table 17

STORAGE TIME REQUIRED FOR PERCEPTIBLE TASTE TO APPEAR IN WRAPPED SWEET
MILK CHOCOLATE BARS DUE TO ABSORPTION OF ODOR FROM OUTSIDE ATMOSPHERE[1]

	Time to Taste	
Description of Wrapper	Turpentine	Pepper-mint
Aluminum foil (0.007 in.) coated with heat-sealing lacquer and laminated one side to moistureproof cellophane; coated surface heat-sealed along side seam and ends.	20 days	19 days
Aluminum foil (0.0007 in.) coated with heat-sealing lacquer; wrapper heat-sealed along side.	6 days	7 days
Plastic bleached greaseproof paper $22^1/_2$-lb.; side seam and folds not sealed.	5 hr.	$3^1/_2$ hr.
Opaque greaseproof paper 30-lb., side seam and folds sealed.	5 hr.	$3^1/_2$ hr.
Control—no wrapper.	$^1/_2$ hr.	1 hr.

[1] Alcoa aluminum foil—its properties and uses (1953). Chocolate bars stored in glass desiccators, one containing cotton moistened with turpentine and the other containing peppermint candy.

The amount of protection which is built into packages for chocolate and chocolate confections is dependent upon how much cost the product can carry, the anticipated delay between manufacture and consumption, and the anticipated severity of conditions during shipping and storage. The most effective materials for providing maximum protection are aluminum foil and foil laminates, and the best chocolates are given this protection.

Packages and Package Materials.—Solid chocolate chunks or bars or chocolate covered bar candies, especially those containing nuts, are commonly wrapped in 25-lb per ream printed glassine paper. Larger bars may use a heavier weight 30- to 35-lb glassine. Where staining may be a problem, two 25-lb glassine sheets may be used with the inner an unprinted sleeve and the outer a complete wrap. Most flat bars use a glassine wrapper and a printed 50-lb paper outer sleeve. Aluminum foil is used for form-fitting wrappers on novelty items such as molded chocolate animals and eggs, and we all are familiar with the foil wrapped chocolate "kiss." Some chocolate bars in the 10¢ and higher price category are being wrapped with foil-wax-paper-wax laminations, and some that require even longer shelf-life are being wrapped in a heat-sealed, foil-paper-polyethylene lamination.

Specialty holiday items are not only foil wrapped, but often packed in attractive cartons with various types of closures, windows, or overwraps.

New Trends.—A new concept in Europe is to cast chocolate directly into foil containers. US candy makers are taking a good look at this and are experimenting with direct casting into molded polyvinyl chloride or polystyrene trays. Barrier properties are improved by adding PVDC coatings to the trays and sealing a barrier film or foil laminate across the

FIG. 119. DIE CUT CARTON SHOWS WRAPPED BARS

A typical example of a 10¢ complex candy bar made of marshmallow, peanuts, fudge, and chocolate covering. Covering can be either milk or dark chocolate. Bar wrapper is printed opaque glassine to hide unsightly chocolate stains and help protect aroma.

top. Pressure sensitive adhesives are also being tested, because heat sealing always risks melting of chocolate components.

Candies with chocolate exteriors are not wrapped in transparent films for two reasons. The less expensive films do not give full protection against foreign odors and some contain fat soluble plasticizers. More important however, is the visibility factor. Finger pressures from handling causes chocolate to scuff or smear. A transparent film would allow this unsightly condition to be seen.

COCOA

Powdered cocoa is a relatively stable product that needs protection from moisture to prevent caking. It is sold in metal end paperboard canisters. Individual servings in polyethylene coated paper packets are now also available.

CHOCOLATE SYRUPS

Chocolate syrups contain so much water that they are susceptible to mold formation and bacterial contamination. For this reason they usually are canned. Refrigeration is recommended after the package is opened. Unit portion packages have been used to supply chocolate syrup toppings for ice cream or other desserts. Printed polyester-polyethylene film containing 1-1/2 ounces in form-and-fill pouches can be heated in boiling water to soften the contents prior to use.

CANDY CONFECTIONS

Candies or confections are combinations of ingredients with one basic common property—a primarily sweet flavor. Cane sugar, other sugars, and honey have for centuries been the primary sweeteners. Artificial, low calorie, sweeteners are relatively modern and low calorie candies have moved from the medicinal or dietetic shelf to close proximity with regular candies only in the past decade. Other primary ingredients of candies include butter, starch, milk, cocoa fat, chocolate, gelatin, and eggs. A wide variety of secondary additives are used to contribute flavor, color, aroma, texture, and stability, and finally there are components which are added for contrast such as fruits or nuts.

Candies can be grouped into a number of basic types and often are sold in a simple form of one candy type. However, there are hundreds of possible modifications of the basic types and thousands of possible combinations of two or more basic types. Chocolate, for example, is a basic candy, yet it is frequently used in combination with other candies.

Product Properties Simple Candy Types

Hard Candies (Boiled).—Hard candies are made almost entirely of sugar. Sugar is dissolved in water and boiled with or without vacuum until the desired concentration is reached—usually judged by boiling point. Additives, including colorants, flavorants and medicinals, when used, are put in during cooling to minimize evaporation. If poured immediately into molds, clear hard candies result. If pulled or kneaded, the product will be not quite so clear and slightly higher in moisture.

Medium Hard (Chewy) Candies.—Toffees and caramels are mixtures of sugar and glucose boiled together with added color and flavor and some fat to lubricate, emulsify, and prevent the candy from sticking to the teeth. Toffees are hard because they are cooked to a higher temperature (lower moisture). Caramels contain flour. Both candies can be pulled, cast or cut. Modified caramels may include fruits or nuts.

Medium Soft Candies.—*Fondants and Creams.*—Sugar creams or fondants are like hard candies, but more water is left in and they are cooled rapidly to induce formation of a fine suspension of crystals of sugar in a saturated sugar syrup. Milk can be substituted for the water. If cream or butter is added it is called a modified fondant. Colors, flavors, fruits, nuts, or vegetable products may also be added. Fondants are used as center components of coated or dipped candies such as chocolates, bonbons, or candy bars.

Fudge.—Fudge is a "cross" between a caramel and a cream. It is a basic grainy fondant cream with further additions of cream, milk, or butter. Nuts, fruits, and flavorings are added.

Nougats.—Nougats contain honey, butter or cocoa butter, sugar, corn syrup, gelatin, and egg white, and are cooked until like caramel.

Gums and Pastilles.—These candies are made from sugar, gum, water, and flavoring and put in many shapes. The pastilles are usually softer than the gums due to a higher water content, but they are not as soft as jellies and are made rather rubbery by the addition of gelatin. The familiar gum drop represents the other category.

Jellies and Pastes—Jelly candies are made from water, sugars, fats, flavorings, color, and thickening agents. Their consistency depends on the percentage of ingredients, cooking time, and holding time in starch molds. They include such products as jelly beans, Turkish paste, and various gum drops or strings.

Soft Candies.—*Marshmallow.*—Marshmallow contains sugar, corn syrup, gelatin, and egg white. Originally it also contained part of the root of the mallow plant. Now it contains a substitute vegetable colloid. It may be cooked or not, but it is always whipped to introduce air and develop a light consistency. Texture may be varied from short and crumbly to chewy or even semiliquid.

Marzipan.—Marzipan is a marshmallow formula plus almond paste. It is used extensively to mold and sculpture highly individualized decorative items.

Product Properties Compound Candy Types

Filled hard candies are prepared by forming clear, pulled, or grained hard candy into a hollow jacket, inserting any one of a variety of fillings, pinching the jacket shut and encasing the whole in another layer of hard candy. Fillings can be jam, preserves, fruits, berries, chocolate, peanut butter or any other desirable concoction. This category could also include chocolate-nut bars, peanut and other nut "brittles," candy coated popcorn, sugar glazed fruits and nuts, and enrobed centers.

Chocolate Coated Centers

Chocolate coated centers consist of various creams, fruit jellies and nuts. In coating roasted nuts, the nuts must not be overroasted since a bad taste will result. All the creams used should be fresh and of high quality. The fruits employed should be top grade glazed or preserved varieties.

In chocolate coating (enrobing), both machines and hand dipping methods are used. Machines are used for chewy, hard centers and cream

FIG. 120.   FOIL PROTECTS PEANUT BRITTLE

Aluminum foil heat sealable laminated overwrap not only provides an attractive package but also provides the vitally needed moisture barrier protection.

centers cast in starch. Fruits, nuts and hand-rolled centers are best hand coated.

The modern enrobing machine requires an experienced operator. Constant adjustment is necessary to insure a quality product. Basically, the chocolate is melted at 120° F in a steam bath or in a regular chocolate melter. It is then cooled to 86° F and taken over to the enrober. The machine operator then cools the chocolate further prior to dipping the centers. In order to prevent streaky or gray chocolates, the chocolate coating room should be maintained between 60°–66° F.

Hard chewy centers may be machine coated by a 24-in. enrober with stringing and packing attachments. Stringing is used for the centers before they go into the machine. A cooling tunnel is also part of the in-line operation. After the center is dipped, the coated centers move to the cooling tunnel. They are then packed in stock boxes as they move along the belt after leaving the cooling tunnel.

Delicate cream centers, nuts, fruits and cordials are hand coated. Dipping tables are used to accomodate a largely female work force. Hand dipping assures a firm finish as well as a sufficient coating laydown.

The packaging of chocolate centers is a delicate job. All chocolates must be handled carefully to avoid scratching. They should also be cupped. Setup boxes in a two layer box, are traditionally used for quality chocolates. The top layer is fitted and supported with soft padding to avoid crushed product. Also, the top layer is more tightly packed than the bottom one since the box tends to stretch more at the top.

Either thermoformed plastic trays are used to partition the chocolate or zig-zag paperboard inserts. The box is then checked for correct weight and a soft cushion mat is placed on top. If a wax paper liner is used, it is then folded over and the box is closed. Overwrapping consists of either a clear moisture-proof cellophane or polypropylene film. Some lower priced confections are sold in small (less than 1/2 lb) units in small set up boxes or in plastic bags.

Product Properties Complex Candy Types

This category includes complex confections, for example, a baked cookie, with a layer of nougat topped with a layer of soft caramel and the whole enrobed in chocolate; for example, an Easter egg with a marshmallow center containing chopped candied fruits and nuts covered in chocolate and decorated with sugar icing, marzipan, or jellied ornaments.

Packaging Requirements

The packaging requirements of simple, compound, and complex confections are dependent upon several factors: (1) the moisture content of the candy and hence its susceptibility to atmospheric moisture; (2) the fragility of the candy; (3) the susceptibility to rancidification (those containing nuts or butter); (4) the susceptibility of chocolate coatings to fat bloom, sugar bloom, odor pickup, or flavor pickup; (5) susceptibility to loss of aroma or color; and (6) the need for protection against soilage, pilferage, or infestation by insects or vermin.

Simple candies can be classified with respect to moisture sensitivity, other candies have to be judged individually under end use conditions. A convenient classification for simple candies is according to their relative vapor pressures, which are in turn proportional to the residual moisture content and the hygroscopicity of the ingredients.

Confectioneries with Low Relative Vapor Pressure.—Hard candies have a relative vapor pressure (RVP) of less than 30%, toffees run about 48%, and caramels a little higher. Low RVP candies absorb moisture from the air. This causes graining at the surface and can make them sticky; consequently, the package should provide a moisture barrier.

Confectioneries with High Relative Vapor Pressure.—The fondants, fudges, and marzipan have relative vapor pressures ranging from 68% to 84%. This is definitely higher than atmospheric moisture and the products tend to dry out. Packaging needs to prevent or slow down this action, however. In the event of sudden temperature changes, moisture sweating can produce stickiness or promote mold growth. Therefore the package must have some ability to breathe.

Confectioneries with Medium Relative Vapor Pressures.—Relative vapor pressures of the gum candies range from about 51% to 66% whereas the jellies range from about 59% to 75%. Since, except for the highest extreme, these are close to average atmospheric humidity, the candies do not tend to dry out, and hence require little moisture protection. In fact, they are better off in a package that can breathe. Marshmallow and nougat candies are also in the range of medium relative vapor pressure. They require a little more protection than the gums and jellies in order to prevent drying, but the package also should be able to breathe a little as insurance against "sweating" or condensation of moisture.

Packaging Materials

In the application of packaging materials as direct wraps, bags, box liners, or box carton or tray overwraps, the packaging engineer has a wide selection of materials from which to choose. Table 18 lists some typical examples along with a general classification of their performance characteristics.

To show further the complexity of the problem of selection, Table 19 shows typical recommendations to the confectionery industry by manufacturers of cellophane films.

Polyethylene and polypropylene films are challenging cellophane in various end use areas. While PE generally does not have the clarity or gas barrier characteristics of cellophane, it has excellent impact and tear

TABLE 18

PACKAGING MATERIALS FOR CANDIES

| Material | | Rating as a Barrier to | | |
	Moisture	Odor and Flavor	Oils and Fats	Heat Sealing
1.5 mil aluminum foil	Excellent	Excellent	Excellent	No
0.5 mil aluminum foil– 1.0 mil polyethylene	Excellent	Excellent	Excellent	Yes
1.0 mil aluminum foil– 1.0 mil polyethylene	Excellent	Excellent	Excellent	Yes
Coated 1.0 mil aluminum foil	Excellent	Excellent	Excellent	Yes
0.75 mil PVDC (Saran) film	Excellent	Excellent	Excellent	Yes
1.0 mil rubber hydrochloride film	Very Good	Very Good	Excellent	Yes
1.5 mil coated cellophane	Very Good	Good	Excellent	Yes
2.0 mil polyethelene film	Good	Good	Very Good	Yes
0.35 mil aluminum foil– tissue paper	Good	Good	Very Good	No
42 lb/ream bleached Kraft paper–8 lb/ream polyethylene	Good	Fair	Good	Yes
Glassine paper	Poor	Poor	Good	No

TABLE 19

CELLOPHANE TYPE VERSUS CONFECTIONERY PRODUCT[1]

(DIRECT WRAP)

Type	Product								
	Bar Goods	Caramels	Chewing Gum	Fondants	Hard Jellies	Lollipops	Lozenges	Candy Coated Popcorn	Sticks & Canes
MS (Avisco)	X								
MSD (DuPont)			X		X	X	X		
MST (Olin)									X
RS (Avisco)	X								
K (DuPont)				X		X	X	X	
V (Olin)									X
M (Avisco)	X								
MD (DuPont)				X	X				
MT (Olin)						X (twist wrap)			
P (Avisco)	X								
PD (DuPont)									

[1]Courtesy of Candy Industry.

TABLE 19A

CELLOPHANE CODE (CONFECTIONERY)

Type	Supplier		
	DuPont	Avisco	Olin
Manufactured by the specific firm (transparent)	D	—	T
Colored	FC	C	—
Saran coating (polymer coated)	K	RS	V
Plain (noncoated)	P	P	P
Moisture-proof	M	M	M
Heat Sealing	S	S	S
Vinyl coating	—	R-18	—
Anchored (water resistant)	A	B	A

Courtesy of *Candy Industry.*

strength, excellent heat-sealability, and is an excellent moisture barrier. It is softer than cellophane and therefore not quite as machineable. However, older equipment types can be modified to run polyethylenes and newer types are being designed to handle the softer films.

Another major advantage of polyethylene is its low temperature flexibility which allows storage and handling at temperatures below $-50°$ F. On both a weight and area basis, polyethylene enjoys a price advantage over cellophanes, which accounts in no small part for the degree of its acceptance despite the shortcomings mentioned.

Polypropylene films are now receiving serious attention from the candy industry as they have the optical clarity desired and can be made with more stiffness and better machineability than polyethylene. They are slightly less costly than cellophane on a weight basis and enjoy a substantial advantage on an area basis. Their major difficulties have been in their lack of good sealability. PVDC and vinyl coatings have been used to avoid this problem; however, composite films of polyethylene and polypropylene seem to be the answer. Polypropylenes have good ageing and low temperature properties, about equal to polyethylenes. They are superior in abrasion resistance, which is important where sharp sugar particles are present. Some polypropylenes have been specifically designed for twist-wrap as they have the ability to lock in position after twisting due to the stretching the film in the twisted area.

Shrink films are being used to an increasing extent for overwrapping candy boxes or for bundling multiple units. These include heat shrinkable polyethylene, polypropylene, and polyvinyl chloride films. Aluminum foil laminates are being used more and more frequently as bar wraps, particularly for the most expensive products, or those requiring a high degree of protection. Foil-wax-paper or foil-polyethylene-paper-polyethylene combinations are used.

Package Types

Individual Wrappers.—Hard sugar candies and toffees are wrapped individually to protect them against graining or stickiness due to moisture pick up. Many softer candies such as caramels or chocolate covered confections are individually wrapped to avoid their sticking together or to prevent losses or transfers of flavor and aroma. There are several kinds of wrappers and most of them today can be applied by automatic equipment. One of the most commonly used individual wraps is the twist wrap.

Both ends of the wrapper sleeve are twisted to effect a closure. It is adaptable to various shapes and sizes, and inner sleeves of contrasting or

Courtesy of Extrudo Film Corp.

FIG. 121. POLYPROPYLENE TWIST WRAPS GIVE TIGHTER TWIST

A comparison of cast polypropylene film twist wraps on the left with cellophane twist wraps on the right shows that the former holds the twist better.

protective material can be inserted easily prior to closure. Over 600 million pounds of candy are twist wrapped annually in the United States. Heat is sometimes used to help insert the twist. From 1-1/2 to 2-1/2 full turns are the usual range of twists, and candies can be so wrapped at more than 1000 pieces a minute. Original twist wraps were plain tissues. These were supplanted by waxed tissues and then by cellophane films. Polyethylene and polypropylene films are now challenging cellophane.

Round or oval shaped candies are individually wrapped in a bunch wrap where the ends are gathered underneath the piece and heat-sealed. The extra layers of wrapper insulate the candy from heat damage; however on larger pieces a board stiffener or tray may be inserted.

Rectangular pieces of candy can be wrapped with the caramel wrap. Here the ends are parcel folded and the ears tucked back and heat sealed underneath where the body overlap insulates the candy. The satchet wrap is a combination of the caramel wrap on one end and the twist wrap at the other. Larger bars of candy are often placed on a stiffener card or a U-board made from 18-point board with about seven pounds of wax coating. The wrapper can be similar to the caramel wrap or it can be made like a bag with a long body heat-seal and two closely trimmed end seals.

Round hard candy lozenges are often assembled in a cylindrical roll that is overwrapped and the ends folded or twisted. Ends are sealed shut with a hot melt or an end label. A printed paper sleeve is used for added strength and protection.

Strip pouches are multiple units of single pouches that have not been completely severed. The user tears off individual units as desired along preperforated tear lines. This is popular for medicinals and for children's lollipops; however, the latter are also very often twist wrapped.

Hard candies which have been individually wrapped usually require additional moisture protection. In earlier times this was provided by glass jars or metal tins with friction lids. Such packs are still available but usually as specialties with higher prices. The modern solution is to place the wrapped candies in an outer package which is most often a sealed bag made of moisture-proof cellophane. Though here again polyethylene and polypropylene films are now being employed. Polyethylene bags in fiberboard boxes are also being used as wholesale shippers for hard candies and toffees.

Polyethylene bags .001 to .0015 in. thick are being used as multipacks for candy bars in the 5¢ price range. They are closed with clips, heat-sealed paper headers, or plastic tape twist ties.

Cartons of various types are used for candies. They range from 15-point to 22-point paperboard. The board may be wax or polythylene coated, or the carton may have a plain or waxed glassine liner. Window

Courtesy of E. I. du Pont de Nemours and Co.

FIG. 122.   RECLOSABLE FILM BAG WITH PIGGY BACK LABEL

This new form and fill bag for candies features a pull apart top seam plus a pressure sensitive label which can be used to reseal the bag after opening.

cartons may be used to give the buyer a view of the candy. Windows are made from polystyrene film, cellulose acetate, or polypropylene. Thicknesses run from .001 to .0015 in. Some cartoned candies have a sealed cellophane overwrap with a tear tape opening device. This is very popular for cough drops.

Overwrapped cartons are used for six-pack and ten-pack units of candy bars. These are limited to higher priced (over 5¢ per bar) candies.

Piece candy assortments where chocolate is a major component are aristocrats of the candy world and can run to whatever price the consumer will pay. Consequently, packages can run a gamut from rather simple setup boxes to extremely ornate constructions. Traditionally, each piece was placed in a 25-lb glassine fluted paper cup. Candies would be arranged in layers with sectional dividers, trays, and padded liners to prevent damage. The modern approach is to use molded polyvinyl chloride trays with individual cavities for each piece. These are less

Courtesy of King Candy Co.

FIG. 123.   SET-UP BOX PIECE CANDY

Holiday motif is reflected in this heart shaped box for piece candies. Note that some
candies are foil wrapped for extra protection. This package also features molded polyvinyl chloride trays with individual cavities of varied depth to separate the candies
and to present a uniform level appearance.

expensive, permit automatic loading, and give better protection against
damage.

Very aromatic or highly flavored pieces such as mints are often individually wrapped with plain foil to protect against flavor or odor transfer. Other candies such as caramels or fudge also are individually
wrapped to protect against rancidity, or desiccation. Quality chocolates
are further protected by a plastic overwrap, which when sealed provides

additional barrier protection as well as preventing soilage, scuffing, and pilferage. Cellophane, again, has been the standard (140 to 195 gauge), however, again, .001 in. polyethylene and .009 in. polypropylene films are being used.

There are some chocolate assortments marketed in tinplate or aluminum cans and some in glass, but this is not very widespread. A few are being marketed in rigid 0.075 in. cellulose acetate trays, and a few are sold in thermoformed polyvinyl chloride or polystyrene trays. Trays are usually placed in cartons or boxes with film overwraps. Some specialty seasonal items are placed in expanded polystyrene foam trays with shrink film overwraps.

Convenience and Novelty Packages of Candy.—Because candy is a cheerful product it is associated with children, holidays, and special occasions. A wide variety of novelty packages has been developed to attract special impulse purchases, or to introduce new packages via premium offers. Cartons can be printed to represent railroad cars or village buildings. Little plastic Santa Clauses can be filled with tiny candies. Glass animals filled with candies can become penny savings banks. Polystyrene hearts are brought out on St. Valentine's Day and baskets of Easter eggs come in a wide variety of shapes and sizes. Thanksgiving sees turkeys and cornucopias. Halloween sees witches and cats and pumpkins. Sugar candy or chocolate "cigarettes" and foil-wrapped "gold" chocolate coins are great favorites at all times. One popular Dutch chocolate treat is a molded apple which breaks apart into segments like a navel orange. Plastic (PG or PVC) shrinkable film wraps can be used for multiunit bundling or for piggyback packaging, e.g., a candy cane attached to a bottle of dishwasher detergent. Colored foil containers with snap on polyethylene lids make attractive candy dishes for parties. Even spiral wound paper tubes can be overwrapped in bright foils and filled with candies. The limit is only the imagination of the package designer.

Frozen Confectionery

Because candies tend to be seasonal in demand, it would be advantageous if the manufacturers could store their product for long periods. This would enable them to stabilize production rates yet meet peak seasonal deliveries at Christmas, Valentine's Day, Easter, Mother's Day, and Thanksgiving. Long term storage would of course require that the candies would remain fresh, free of staleness, rancidity, mold, or unsightly sugar bloom or fat bloom. The only type of storage that meets these requirements is one where the product is brought to below 32°F and preferrably as cold as −10°F. The freezing of confections does however introduce some additional problems. Where many candies are totally undamaged, some crack during freezing and some show an increased incidence of sugar blooming.

A considerable amount of work was done by Dr. J. G. Woodroof and others at the Georgia Experiment Station on these problems. They reported that spun candy chips, some chocolate-coated nuts, nut brittles, toffees and some creams may crack, but only the spun candies are seriously affected (shattering). The others return almost to normal condition when thawed. Whatever cracks are still present are extremely difficult to detect. Selectivity is the only way this can be avoided. Woodroof reported over 60 candy types that were not damaged by freezing, over 30 that were improved, and about 20 that were impaired by freezing. Sugar bloom could be avoided by using moisture-proof wraps on the candies and thawing them slowly. Moisture-proof wraps were also needed to prevent drying out.

Packaging of candy that is to be frozen should therefore be capable of surviving freezing temperatures without cracking and must provide good moisture barriers. This rules out the conventional waxed papers and glassines. Although moisture barrier properties could be improved by using PVDC coated papers, or cellophanes the cold temperature conditions dictate a material that will not fracture in handling. This would have to be polyethylenes, polypropylenes, or aluminum foils, or suitable combinations. Heat sealability is very desirable for the overwrap. Some heat shrinkable PVC, polyethylene, and polypropylenes have been tried. Where economics permit and the individual candies can withstand the temperature, the individual wrappers may also be heat-sealed. The degree of protection required will be related to the types of candies, how much handling they will receive and how they are to be thawed.

There seems little doubt that freezing of candies will be done on an increasing scale as manufacturers become more and more familiar with its advantages. A few supermarkets have tried selling chocolate assortments directly from the frozen food cabinet.

Other Confections

Candied Fruits and Nuts.—Fruits and nuts are frequently used as components in composite confectioneries. Over 125 million pounds of peanuts and 35 million pounds of tree nuts are used annually in the United States for this purpose. It has already been indicated how they can be dipped in chocolate. A few other techniques are also of interest.

Nuts can be "panned," that is they are coated with a glaze made of sugar. Several layers usually have to be added. For thick coatings, as found in Jordan almonds, an occasional tumbling is inserted between coatings to knock off rough spots. Final layers may be colored, and then the product may be wax polished to a high luster.

Because of their high fat content, nuts are sensitive to oxidative ran-cidity and to extremes of moisture. Too much moisture will encourage mold, too little will embrittle the nut meat and encourage flavor loss.

Normal moisture is considered to be 3.5% for pecans and 2.0% or less for peanuts. If the nuts are allowed to absorb as much as 5-1/2 to 6% moisture, they become soggy, tough, and have a poor shelf-life.

If the aspergillus type of mold is permitted to develop due to improper moisture level and storage conditions, a poison called aflatoxin can develop. While the poison is not particularly dangerous to humans it can kill poultry. There have been cases where candies containing aflatoxin poisoned peanuts were condemned by the FDA. The Brazilian govern-ment has set up quality control laboratories to test for aflatoxins in exported Brazil nuts.

Techniques of nut processing and roasting can help in establishing

Courtesy of E. I. du Pont de Nemours and Co.

FIG. 124. PEANUTS GAIN ADDED PROTECTION FROM COEXTRUDED
FILM BAGS

These preformed bags are made from a coextruded film containing
plies of polyethylene and "Surlyn" A ionomer.

ideal moisture levels. Dipping the nuts in acetylated monoglyceride coatings help retard moisture build up. Refrigeration storage helps also to slow down the development of rancidity in nuts. For ideal shelf-life packaging materials must be selected for essentially zero oxygen permeability and the product should be vacuum packed or nitrogen flushed and held under refrigeration as much as possible. At low moisture levels, nut meats are brittle. The package must protect against physical damage during handling.

Ultraviolet light accelerates the oxidative rancidification of unsaturated fats and oils, consequently a package for nuts will give better protection if opaque to these wavelengths of the spectrum.

Packages for bulk shipment and storage of nut meats are commonly polyethylene-lined drums or shipping cases. Vacuum packed tins are used for deluxe nut-meat assortments and brittles. Nitrogen flushed glass jars are also employed. Flexible packaging normally requires aluminum foil to attain the light and oxygen barrier needed; however, for peak volume holiday "quick turnover" sales nut meats are often sold in small sealed cellophane or polymer coated cellophane bags. This is one area where polyethylene will not readily compete due to its high permeability to oxygen. However, composite films containing PVDC coatings will prove satisfactory for short term storage.

When the nut is incorporated within a compound or complex confection the degree of protection required will depend upon whether the nut is exposed. If it is encased by the other candy components, particularly chocolate, less protection will be needed, but if the nut meat is partially or totally exposed, the entire confection should be packaged to prevent deterioration of the nut meat.

Whole fruits such as cherries or kumquats and fruit parts such as citrus slices or citrus peels can be candied and/or glazed. Candying is done by removing objectionable parts, such as pits, seeds, bitter citrus oils and the like and then gradually replacing the water in the fruit with sugar by soaking the fruit in successive baths of increasing concentration. After the final step the fruit is cooled and the sugar solidifies. Invert sugar can be added to prevent graininess. Pieces can be dipped in colored granulated sugar to reduce stickiness, or they can be glazed by dipping in a highly concentrated sugar syrup, draining and cooling. Whole candied cherries are often encased in fondant and dipped in chocolate. Brandy or liqueurs can be added.

Packaging of candied fruits is similar to hard candies. Excess moisture will make them sticky. Cellophane overwraps are needed, and if there are no inner liners in the package the overwraps should be a good moisture barrier cellophane. Polypropylenes will also prove superior particularly if sharp sugar crystals are present.

JELLIES, JAMS, AND PRESERVES

Jellies, jams, and preserves are sweet concoctions by which traditionally the flavors of fruits can be captured for winter consumption. Solid fruit components are strained out of jellies and the clear mixture of fruit juice, boiled sugar, and fruit pectins congeals on cooling to a clear gel. Jams are most frequently prepared from berries. The whole fruit is included and cooked until its texture and form have been more or less broken down. Preserves contain whole small fruits or cutup pieces of larger fruit. They are cooked only enough to avoid spoilage, the original form of the fruit is preserved as much as possible. A conserve is about the same as a preserve but frequently will contain nuts or vegetables as well. Marmalades are something between a jelly and a jam in consistency and are made from orange or grapefruit with thin slices of the peel added for additional zest. English marmalades are also made from limes, lemons, and ginger.

Because fruit jellies and jams are acid, they are heated to 220.5°F

Courtesy of British Cellophane, Ltd.

FIG. 125.   JAM JARS TAKE ON MODERN LOOK

The old and the new are brought together in these multipacks of jams in glass jars. The product is low-sugar dietetic and the multipack features a shrinkable polyethylene overwrap and a thin white plastic base with recesses for the jars.

which is sufficient to kill all vegetative microorganisms and then are packaged hot. The package serves to prevent loss of moisture. Traditionally, they have been packed in glass jars and small glazed ceramic crocks. Tops must be sealed to prevent reentry of mold spores. Since the contents are rarely consumed completely, a reclosure feature is desirable. Many jars have screw caps which can be replaced, rather than the older pry-off lids.

Plastic jars are already making inroads on this market because of their lighter weight and resistance to breakage.

High density polyethylene, polyvinyl chloride, and some of the newer acrylics have been used for this purpose.

Unit servings of jellies and jams are made from molded plastic cups or formed aluminum containerettes with smooth walls and flanges. Both types are closed with peelable membrane lids. Plastic caps are molded from polystyrene, polyethylene, polypropylene or polyvinyl chloride. Closures can be foil laminates, clear plastic film, or metallized plastic film. The type of plastic used must be capable of heat-sealing or the membrane must have a suitable peelable adhesive or heat-seal coating, the selection of which will be governed by the material used to form the

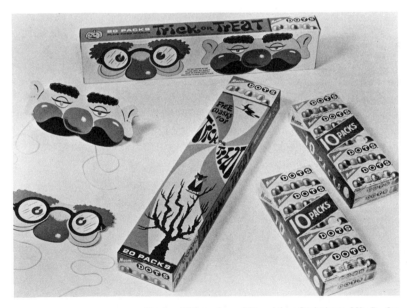

Courtesy of Mason, Au, and Magenheimer Confectionery Mfg. Co., Inc.

FIG. 126.   HALLOWEEN MOTIF FOR CANDY CARTON

This candy carton not only packages a familiar candy product but also provides cut out masks to delight the children.

Courtesy of The Metal Box Co., Ltd.

FIG. 127.   CHRISTMAS PACKAGE COMPOSITE CONTAINER

This spiral wound composite container utilizes a specially printed label to convert the simple container into an attractive children's game.

cup. In the selection of a suitable heat-seal formulation, note must be taken of the temperature of jelly fill. Jellies are filled hot (180°F or higher) and the actual degree of peelable seal does not become evident until it has "set" for 24 hr. Problems revolving around membrane formation, i.e., heat-seal coating separating from the inner foil ply, are a frequent plague to the material supplier. The foil used must be clean and free from oily residues to prevent separation of the laminate plies.

PEANUT BUTTER

The United States produces over 1.1 million tons of peanuts annually of which 63% of the edible grades go into peanut butter. Peanuts were known as early as 950 BC and probably originated in the Brazil or Peru regions of South America. From there they traveled to Africa and were imported back to North America with the slave trade.

Peanut butter was invented sometime around 1890. It was mostly made in the home at first, but later commercial processes were developed.

Peanut butter is a fine grind of roasted mature peanuts plus some salt, some hydrogenated fat, some dextrose, and a few other additives such as glycerin, lecithin, or antioxidants. Emulsifying agents are added to prevent oil separation, and the antioxidants to prevent rancidity.

Color and flavor are principally controlled by the selection of peanuts and the temperature used in roasting. Grinds vary from fine or smooth, to chunky (contains grainy particles). During grinding it is usually desirable to exclude air. Vacuum packaging helps prevent rancidity. The fatty portion of the oil solidifies when chilled. For best consistency refrigeration should be avoided except for initial processing.

Because of the high oil content peanut butter is relatively stable. If humidity does not get too high even mold is unlikely. The main problem is rancidity and staleness, which will gradually ensue during storage. Protection from oxygen and light is helpful, particularly the former. Most peanut butter is now packaged in glass jars. However tin cans, aluminum cans, composite containers, plastic tubs, and small flexible packages (for individual servings) have been marketed. Bulk shipment to candy makers and bakers may be in large tins or lined fiber drums.

A fairly new innovation is a combination of peanut butter and jelly in a "swirl" design. The product is packed in glass jars with screw tops for easy use and consumption.

CHEWING GUM

Chewing seems to satisfy an inner craving in man. He will chew on a straw, a blade of grass, or a twig. Various aromatic plants such as the sassafras yield twigs or roots that are pleasantly flavored.

The North American Indian chewed spruce gum—a highly aromatic resinous exudate. South and Central American Indians chewed the coagulated latex of the sapodilla tree, which they called chicl. The chewing of paraffin wax was common in this country in the early 1800's and traveling peddlers carried balls of the wax to sell as treats for the children on their routes. Just before the Civil War, Antonio Lopez de Santa Anna, former general and president of Mexico (better known in the United States as the commander of Mexican forces at the Battle of the Alamo) came to New York to try to raise money. Fortunes were being made in the infant rubber industry. General Santa Anna hoped to promote chicl as a commercial rubber product. In New York he met Mr. Thomas Adams, an inventor. Adams tried unsuccessfully to utilize the chicl rubber. But later he remembered seeing the general break off a piece and chew it. After some experimentation he produced a palatable mixture which he called "Adams New York Gum." Originally he sold it in chunks but later found it easier to sell if pressed into long thin strips. After considerable success with the unflavored product Adams brought out a licorice-anise flavored "Adams Black Jack Gum"—the oldest product still on the market. Other entrepreneurs soon jumped into the picture. Dr. Edward Beeman added pepsin and advertised "Beeman's Pepsin Gum" as an aid to indigestion. William White added corn syrup to gum formulations. This made a wider variety of flavors possible. He introduced the first peppermint flavored gum and founded American Chicle Company. In 1892 Wrigley was selling "Spearmint" and "Juicy Fruit" and was pioneering new techniques in merchandising and advertising. Candy coated "chiclets" were marketed by Fleer in 1897 and Fleer also invented "Dubble Bubble" gum in 1928. Penny candy coated gumballs were the specialty of Leaf Brands Company founded in 1921.

Product Characteristics

Although the first gums were chicle based, other materials were added for various reasons—such as other gums, waxes, sugars, glycerin, water, corn syrup and flavorings. Today, synthetic formulations are being used in which polyvinyl acetate largely replaces the more expensive natural ingredients.

The most popular flavors were traditionally peppermint, spearmint, and wintergreen, but many others have been tried including cinnamon, cloves, and more recently, citrus flavors.

The most popular shape has been the stick form which usually is sold in a five-stick pack. Candy coated gum, a rectangular lozenge, is second in popularity. Ball gum is also a candy coated product but it is exclusively sold as a penny item usually from automatic vending machines. Bubble gum comes in many shapes, sizes, and colors, nearly always in novelty packages with premiums to attract children.

The sales and consumption of bubble gum has more than quintupled in the last ten years. More than $100 million worth of bubble gum is sold each year. A large amount of US bubble gum is also being exported to foreign lands such as Kenya, India, and Japan. One of the major problems facing the bubble gum industry is the British reticence toward gum-chewing. The English consider the habit ungentlemanly and unladylike.

Packaging Requirements

Because the final operation in candy covered gums is to give them a wax coating, the candy is well protected against moisture. Packaging is principally used as a convenience in dispensing and to protect against soilage and physical damage. Stick gums require much more careful packaging as their large surface area permits rapid loss of moisture, flavor, and aroma. Too little moisture results in a hard and brittle texture. Too much moisture produces a soggy, tough product. Because of this critical requirement manufacture of gum is conducted in air conditioned rooms from the first batching to the final packaging.

Package Forms and Materials

Penny ball gums are usually shipped in bulk containers for vending machines. The vending machines traditionally had glass globes, but today the glass has been replaced by less breakable polycarbonate plastic.

Bubble gum packages give a minimum of protection because the product is low priced and has a fast turn-over. The clientele is less fussy about flavor and texture than about the size of the bubble that can be blown and the type of premium received with the purchase. Simple printed waxed paper wrappers are used to enclose both the slab of gum and the premium "card."

Candy coated lozenges are packaged in folding paperboard cartons with cellophane windows. Cellophane overwraps are optional and used not so much for moisture barrier as for a means to prevent pilfering.

Stick gums are individually wrapped in aluminum foil-wax-tissue paper and inserted in a printed paper sleeve. Foil is usually .0003 in. thick or less and the paper about 15 lb per ream machine glazed tissue. The five-stick pack takes five of the individual sticks in a stack and overwraps them on a foil laminate wrapper. At one time the wrapper was a reverse printed cellophane-wax-aluminum foil-wax-25 lb per ream paper. The ends were package folded and sealed with a wax stencil. A cellophane tear tape was used to help open the outer wrapper. Today, the cellophane has largely been replaced by a printed coated litho paper wax laminated to foil. The paper is usually about 30 lb per ream. The printing is overcoated with a high gloss coating to simulate the former appearance of the cellophane lamination. Tear tape opening devices are still used.

Recent trends in stick packages include novelty Christmas packs where a number of five-stick packs are put in a printed box with a cellophane sleeve overwrap to allow full visibility. There also have been larger packs introduced such as a seventeen-stick pack. These differ from the five-stick pack primarily in dimension. It is difficult to introduce drastic changes in the chewing gum package as the industry has been struggling to hold down costs and maintain profits. Nearly all operations are automated and production rates of 1200 sticks per hour are common.

The total sales of candy, confections, chocolate, chewing gum, jams, jellies, preserves, sweeteners, and flavorings in 1966 was $4.35 billion. Of this amount about $1.5 billion was candy and chocolate, another $1.5 billion was chewing gum, sugar and sweeteners accounted for over $1 billion, and the remainder was jams, jellies, and preserves. It is not likely the overall consumption will increase rapidly, but there is a definite need for improvement in packaging through the use of modern materials. The use of foamed polystyrene for candy assortments that are to be frozen is relatively unexplored, for example. With intelligent imagination packaging can play a greater part in the successful merchandising of these products.

BIBLIOGRAPHY

ANON. 1966A. Chocolate wrapper handles different shapes. Packaging Rev. *86*, No. 11, 29.

ANON. 1966B. Two-stage chocolate wrapper. Packaging Rev. *86*, No. 11, 27.

ANON. 1966C. Chocolate bar wrapper has air tight version. Packaging Rev. *86*, No. 11, 29.

ANON. 1967A. Versatile new candy packager. Mod. Packaging *40*, No. 6, 56–58.

ANON. 1967B. A combination of aluminium foil, paper, and polythene. Paper Packs, June, 17.

ANON. 1967C. Sugar turns to plastic. Mod. Packaging *41*, No. 3, 101.

ANON. 1968A. Tube of sugar. Mod. Packaging *41*, No. 3, 101.

ANON. 1968B. Confectionary wrapping. Confectionery Prod. *34*, No. 4, 238–240.

CAVALETTO, C. G., and YAMAMOTO, H. Y. 1968. Criteria for selection of packaging materials for roasted macadamia kernels. Food Technol. *22*, No. 1, 97–99.

DAVIS, E. G. 1966. Texture changes of salted peanuts in flexible film and tinplate containers. Food Technol. *20*, No. 12, 92–94.

FASSBAUER, E. 1967. The problems of vacuum and gas packaging of peanuts, nuts, and similar tropical products. Gordian *67*, No. 1605/6, 34–38. (German)

FEIKE, B. 1968. 100 years of the German Sugar Institute; 100 years of packaging sugar. Neue Verpackung *21*, No. 5, 642–646. (German)

JOSLYN, M. A., and HEID, J. L. (Editors). 1963. Food Processing Operations, Vol. 2. Avi Publishing Co., Westport, Conn.

PRESTON, L. N. 1967. Flexible packaging. VI. Sugar confectionery. Food Packaging Design Technol. April, 20–25.

SACHAROW, S. 1966A. Packaging confectionery in flexible packages. Mfg. Confectioner. *46*, No. 6, 93–100.

SACHAROW, S. 1966B. Confectionery packs demand attention. Paper, Film, Foil, Converter *40*, No. 11, 93–96.

SACHAROW, S. 1967. Packaging confectionery in flexible packages. Alimenta, *6*, No. 20, 60–64.

SACHAROW, S. 1968. Protective packaging inhibits nut meat deterioration. Candy Ind. *130*, No. 5, 17–18.

WOODROOF, J. G. 1956. Protective packaging of candies. Package Eng. *1*, No. 5, 9–13.

CHAPTER 12

Cereal Grains, Bread, and Baked Goods

INTRODUCTION

Where and when grains were first cultivated for food is unknown. Records indicate that wheat was used as food seven thousand years before Christ. To biblical people, grain meant life and vitality. To "sow tares among the wheat" was the ultimate revenge. Cereals were probably the first food crop to be cultivated. The name is derived from Roman ceremonies known as "Cerealia" which were celebrated in honor of Ceres in mid-April.

Wheat was unknown in the Western Hemisphere. Columbus found corn (Indian maize) in Haiti and introduced it to Europe. A very primitive form of corn has been found in caves in Yucatan indicating that many thousands of years of cultivation were required to bring it to its present stage of evolution. Oats are of mysterious origin, believed by some to have been a troublesome weed that grew in cultivated barley fields. A form of oats was grown by Slavic tribes of Europe during the Bronze and Iron Ages. Yet it was not regarded as an important crop by pre-Christian writers.

Barley was cultivated by New Stone Age men in Egypt. It was widely used as a porridge, to make bread, and to brew beer. Wild barley grew in Abyssinia and in Sikkim and Thibet—these may well be the ancestors of today's cultivated varieties.

Rye is a popular bread grain but its origin, like oats, is unknown. It probably also was a weed crop. Even as late as 1925 farmers in Afghanistan and Persia tried to kill the rye that grew in their wheat fields and were given tax incentives to do so.

Sorghum is a grain believed to have originated in Africa at least as far back as 700 BC. It is believed to have traveled to the United States in slave ships.

Rice had its beginnings in Southeast Asia where it was grown at least as far back as 2800 BC. Sophocles mentioned it in his writings in 495 BC and it became an important European crop in the Medieval Period when the Saracen invaders brought it with them.

Millet is a relatively minor crop probably originating in Egypt and Arabia before history began. It was popular in Medieval times but when other grains became available it dwindled in popularity.

Other cereal grains include the wild rice grown by the American Indians, and "Adlay" a grain of excellent food value grown in Brazil and the Far East. Some other plants produce a seed which is used in making a

341

flour useful in baking. These include buckwheat and soy bean.

Almost 70% of the world's harvested land is used to grow over 900 million tons of grain annually. The United States ranks first among the wheat exporting nations. The per capita consumption of grain products varies with family income, age levels and geographical region. Low income families tend to use more grain products than high income families. Cereals are used largely by children and older persons rather than middle aged adults. Consumption of corn meal and hominy grits is greater in the South than any other geographical region.

Baked goods, not including bread, have increased 65% over a ten year period (1955–1965). Bread consumption declined by 7-1/2% in this period. In 1965, each person consumed 1.32 lb per week of bread and 1 lb of baked goods, other than bread.

Dollar values for baked goods and baking needs purchased in the US groceries in 1966 showed an increase over previous years and were as follows:

White bread	$2,839,870,000
Other bread	790,430,000
Rolls	509,890,000
Crackers, biscuits, cookies	1,227,260,000
Popcorn (unpopped)	29,090,000
Pretzels	118,170,000
Cakes and pastries	1,374,520,000
Pies	263,180,000
Doughnuts	480,000,000
Cereals, flour, pasta	2,135,540,000
Baking needs	34,220,000
	$9,800,170,000

Three basic classifications of grain products exist. *Whole grain products,* where only the husk of the grain is removed, include brown rice, whole grain meals, popcorn, rolled oats, cracked wheat, shredded and puffed grains, and most hot breakfast cereals. *Milled grain products* are made by removing the bran and usually the germ of the seed and then crushing the kernel into various sized pieces. They include wheat flour, bulgar, seminola flour, hominy, corn grits and meal, rice and rye flour, parched barley, some hot and cold breakfast cereals and thickeners used for soups, gravies and other prepared dishes. Soy flour is used as a grain substitute in these products. Buckwheat groats are also included in this category.

Since breakfast cereals take many forms, are made from both

"whole" grains (partly milled) and milled grains, are served hot and cold, and may be previously cooked or uncooked; they will be discussed together as a separate classification.

Baked goods form the third category. They are made mainly from flour or meal and include yeast breads, rolls, quick breads, biscuits, doughnuts, crackers, cakes, cookies and pastries.

The fantastic diversity of products stemming from grains requires an almost limitless variety of packages.

WHOLE GRAIN PRODUCTS

Brown Rice

Brown rice consists of rice where not more than 50% of the hulls are removed. Wild rice is an entirely different plant and its seeds are narrow and long and purple in color.

Most of the rice consumed in this country consists of parboiled and converted types; however, packages of brown rice do appear on supermarket shelves in paperboard cartons with cellophane windows. There appears to be a trend in rice packaging toward the use of polyethylene bags. Protection from excessive moisture and insect infestation are important requirements of a rice package, but most packaging ignores the latter. At present, paperboard boxes represent about 50% of rice packaging; however, form-and-fill bags of polyethylene are becoming more and more popular. There is now a trend toward gourmet meals where several packets made from paper-polyethylene-foil-polyethylene or similar laminates are put in a printed paperboard carton. One packet contains rice, another a sauce mix, and the third a topping or spices. The final recipe has an exotic name like "Rice Keriyaki," "Rice Provence," or "Rice Milanese."

Whole Grain Meals

Whole grain meals such as wheat, corn and sorghum are not usually found on supermarket shelves. They are more often packed in large jute, paper or plastic bags for commercial use and export distribution. In many foreign lands, jute packages are used since they are economical and strong. The most critical packaging requirement is protection from insects and rodents. Jute bags have been treated with insecticides; however, the risk of grain contamination has retarded this development.

In export, whole grains such as wheat are delivered in bulk; rice is received in bags. At certain Indian ports, wheat is unloaded from the vessels with slings and then bagged. At most other ports, bagging is done within the holds of the ship. Storage usually consists of stacking the bags of grain in damp-proof well-ventilated warehouses. Methods of disinfestation currently used consist of bag dusting with various chemicals and

FIG. 128. WHOLE RICE GRAIN PACKAGES

Polished white rice is packaged in printed polyethylene bags.

other methods of fumigation. In India, it is not permitted to mix certain insecticides with grain. Recent trends indicate that irradiation will reduce insect attack to a minimum. Since 3–5% of food grains are destroyed by insects and rodents in India, gamma ray irradiation is being considered. It is economically equal to chemical fumigation.

Popcorn

Popcorn is a small grained variety of maize which will pop on being roasted. It is graded on the basis of popping expansion, uniformity and degree of maturity. The caramelized variety should pop into a smooth, mushroom shaped grain while the "butterfly" type is used for buttering. Yellow popcorn has become more popular than white.

Popcorn requires protection from moisture and oxygen—especially if it is caramelized. Raw kernels for home popping are sold in tinplate cans and paperboard boxes. The popped variety is sold in duplex cellophane bags and other snack food films (see Chapter 13). In one convenience concept they are packed eight luncheon packs in a printed polyethylene bag. A formed aluminum foil tray containing the raw kernels is used with an expandable foil cover. As the tray is heated, the corns pop, the foil cover expands, and no cooking vessel is needed. A printed paperboard cover is used for product identification.

Courtesy of Reynolds Metals Co.

FIG. 129. POPCORN AND FOIL WERE MADE FOR EACH OTHER

The wire handle makes this foil package a convenient skillet in which the whole grain popping corn can be heated. The printed paperboard cover is crimped in place to protect the product and is removed before cooking.

Hot Breakfast Cereals

Hot cereals are made from whole grains and must be cooked before eating. The farina cereals are made from middlings that are milled from the whole wheat grain. A middling is what is left of the endosperm after the bran and germ have been removed. Wheat should be of a hard variety and particle size should be carefully controlled to be finer than 20-mesh. Vitamins and minerals often are added. Some treatments can be added to speed its cooking time. Varieties of farina can be made by including whole wheat, cracked wheat, flaked wheat, bran, or wheat-germ.

Cracked wheat is the whole grain that has been partially milled to remove some but not all of the bran.

Rolled oats are made by heating the grain at 212°F for one hour to reduce moisture to about six per cent and to dextrinize some of the starch. This also makes the hulls brittle. The grain is cooled and graded for size. Hulls are removed by passing them between two mill stones that are spaced apart enough to remove the hulls but not crush the grain excessively. This rolling action probably accounts for the name. A mixture of hulls, broken grains and whole dehulled grains (groats) is removed and screened. The whole groats are then flaked by cutting into small pieces. The smaller the piece; the faster the cooking. A whole groat may take 15 minutes of boiling to prepare and will maintain texture in a heated steam table for as much as six hours. Quick oats are made from particles about 1/4 the size of a whole groat. They will cook in five minutes but won't stand up long in a steam table.

When corn is ground in a mill a series of products of different particle sizes results. In the order of increasing fineness these are: grits; coarse meal; medium meal; cones; sharp flour; and soft flour.

The grits can be screened into different levels of coarseness also. Pearl hominy is a grits that will pass through a No. 4 wire screen but not through anything finer. It is popular in the South as a cooked breakfast dish. The coarser corn meals are used to make cooked corn meal mush and also certain baked goods (johnny cake, etc.).

Whole milled rice can be cooked as a breakfast cereal but this is not too popular.

Cold Breakfast Cereals

Around the turn of the century, various people devised ideas for changing cereals into new forms. Henry D. Perky invented a method of forcing boiled whole wheat through shredding rolls to form filaments of dough and pressing them into pillow shaped biscuits. The product was "shredded wheat." Today, other shredded grains in bite sizes are now sold as breakfast foods.

In 1902, Alexander Pierce Anderson discovered that kernels of rice which were heated to 500°F in a test-tube would expand to ten times their original size. This process was tried on other grains and the famous "shot from guns" puffed wheat and rice breakfast cereals were born.

Flaked cereals also were first made around the turn of the century when the Kelloggs made whole wheat flakes in a barn in Battle Creek, Mich. Whole kernels are moistened and partially flattened, cooked with water, sugar, salt, malt, and caramel then further dried, flaked, and roasted. During this processing starches gelatinize, proteins and sugars turn brown, enzymes are inactivated, and the flake is made crisp due to very low moisture content.

Corn flakes start with a coarse grit which is pressure cooked together with sugar, malt, salt, and water for about two hours at about 250°-260° F. The individual particles are separated and dried to about 20% moisture then flaked under great pressure and roasted. Roasting browns them, blisters them, and makes them crisp due to low (less than 3%) moisture.

Granulated cereals are made by baking a loaf of bread made from wheat, malted barley flour, salt, yeast, and water. The loaves are then fragmented and rebaked. A final crushing and sifting finishes the job. Fines are put back into the process as part of the bread loaf formula.

Oven-puffed cereals are made today in addition to the gun puffed cereals already mentioned. Rice is parboiled and pearled, then cooked with sugar, salt, and sometimes malt. After breaking up the lumps the kernels are dried to 25-35% moisture, tempered, broken up again and dried to 18-20% moisture. After heating to 180°F they are flattened, cooled then puffed by passing through ovens at 450°-575°F.

Composite dough products some flavored with cocoa, some with fruit flavors are made from corn cones, oat flour, sugar, coloring, flavoring and water and forced through an extruder into a strand. The strand is cut into pellets. Pellets may be flattened or left spherical. They are then gun-puffed.

Many types of breakfast cereals are coated with sugar by tumbling while a hot sugar syrup is dripped on. Sometimes the syrup is flavored.

Roasted brown buckwheat groats are sold whole and as coarse, medium, and fine grits on the "kosher" shelf. A recipe on the package recommends their preparation as a side dish by heating with raw egg then adding to boiling water. Extra seasoning is optional.

Packaging Hot Breakfast Cereals.—Hot breakfast cereals require protection from moisture, insects and dust.

In the late 19th century, oatmeal was a widely used breakfast food. The product was sold from barrels and often, it was infested with vermin and mice droppings. The American Cereal Company introduced

"Quaker" brand oatmeal in round paperboard containers. The use of a consumer unit eliminated most of the problems associated with the unsanitary bulk packaging. The "Quaker" brand package today is essentially the same as it was 100 years ago.

The most common retail package for farina-type hot cereals in the past was the folded carton sometimes with a paper liner (to prevent losses due to sifting) and always with a printed waxed paper overwrap. Coarser grained hot cereals sometimes were sold in cloth sacks, paper bags (like flour) or in the round paperboard container. Modern hot cooked cereal packages are much the same. The paperboard carton is still used. The overwrap is still printed paper but the coating is no longer straight wax. Polyvinylidene chloride hot melt coatings are now being used. They not only provide a means of heat-sealing the wrap but they offer good barrier protection as well. Latest trends are instant cereals in unit servings. These are packaged in paper-polyethylene-foil-polyethyl-

Courtesy of Glassine and Greaseproof Manufacturers Assoc.

FIG. 130.   GLASSINE LINER FOR READY-TO-EAT CEREAL

This oat cereal gets adequate moisture protection from a waxed
laminated glassine liner.

ene pouches. The user rips open the package adds the contents to boiling water stirs and eats. Freeze-dried fruits may be included for added zest.

Packaging Cold Cereals.—The shredded wheat products must be packaged in cartons that can breathe else rancid odors tend to accumulate. No liners or overwraps are used. High humidities will tend to destroy crispness.

In general all cold cereals must be protected against moisture absorption. This is doubly true for sugar glazed cereals. Much experimentation has been conducted on carton liners. These have ranged from waxed glassines to foil-wax-tissue laminations. The latest of the latter is a buried foil specification. New wax-polyethylene-vinyl acetate coatings on glassines and the PVDC coated papers are giving higher barrier performances and challenging foil for those cereals where moisture protection isn't quite so critical.

Some breakfast cereals—particularly the puffed variety—have been

Courtesy of Reynolds Metals Co.

FIG. 131.   SUGARED CEREAL REQUIRES FOIL LINER

This sugared cereal incorporates a banana flavor and requires aluminum foil to protect against flavor loss and moisture pickup.

marketed in polyethylene bags. Multipaks of 8, 12, and 16 boxes of assorted unit serving cold cereals are bundled together using heat sealable cellophane, polyethylene, or heat shrink polyvinyl chloride films.

MILLED GRAIN PRODUCTS

Flours

Most flours, wheat, bulgar, seminola, corn, rye, etc. are packed in paper bags for retail sale. They are available in various sizes and should be protected from infestation and moisture. Speciality flours are sold in paperboard boxes, or in novelty concepts such as flour sacks. At home, it is advisable to store flour only in small amounts. Refrigerated storage is sometimes recommended.

PREPARED MIXES

Prepared mixes are combinations of ingredients preblended so that the user need only add a few more ingredients (usually wet), mix and cook. The first prepared mix was a self-rising flour patented in 1849. It contained flour, tartaric acid, and sodium bicarbonate. This was followed by pancake mixes which were introduced after the Civil War. Before long a whole family of mixes were available for making doughnuts, piecrust, cakes, biscuits and the like. Some mixes require addition of water only; others require milk, eggs, or both. Mixes may contain leavening agents depending on the product. Their advantages are a balance of ingredients, uniform mixing, and the savings incurred from bulk purchasing of ingredients. Even more of an advantage is the savings in time for the user. Mixes were one of the first truly convenience foods.

Packaging Requirements

Since most mixes contain fats and many contain leavening, they must be protected against oxygen and moisture. Some mixes are made from specially dried flour. Over a period of storage the mixes tend to deteriorate—that is the quality of the product baked from them gets poorer and poorer. Date coding is highly desirable.

Packages

Lined and overwrapped cartons have long been used to package consumer size prepared mixes. Composition of the liners is varied depending on product requirements. Subportioning of the mixes into sealed pouches using vacuum or gas flushed foil laminates can be expected as this will tend to minimize product deterioration.

PASTA

Seminola wheat, which flourishes in northern part of Italy, produces a hard flour which when mixed with water can be kneaded into firm dough with such a consistency that it can be extruded or molded into a wide variety of shapes. Such products were manufactured in the days of the Etruscans who predate the Roman era. By the Middle Ages various forms of pasta were being consumed all over southern Europe. It is understandable that Italy remains the greatest producer and consumer of pasta products. They produce over 3.3 billion lb of pasta annually and per capita consumption is 66 to 77 lb. The United States produces about 1.4 billion lb but per capita consumption is only about 8 lb.

Classification

There are two broad categories based on composition and many other classifications based on shape and size. Macaroni products are made of flour and water. A number of optional ingredients may be added in small quantities such as spices, gluten, egg white, or milk solids. Flour substitutions can be made using whole wheat or soy flour. Occasionally other vegetables are added (dried and powdered) such as tomatoes or spinach.

Products in the second broad category are called egg macaroni and include noodle products. These differ from ordinary macaroni in that they contain at least 5.5% by weight of whole egg or egg yolk solids. Modern macaroni or egg-macaroni products are nearly always enriched by additions of B-vitamins and iron. Since there are hundreds of possible shapes and sizes only a few typical examples will be listed.

Extruded Pasta.—Dough is forced through dies to emerge in the desired shape. In early times this was a batch process. Today continuous process machinery is available.

When hollow tubes are extruded, they are called macaroni. When solid round rods are extruded, they are called spaghetti (other names are used for finer diameters, i.e., vermicelli and cappeli d'angeli).

When flat strips or squared shapes are extruded they are called noodles. Both hollow and solid products can be extruded with smooth or corrugated outer surfaces, and both can be cut to long, medium, or short lengths. By means of special dies elbows can be formed. Some noodles are twisted into spirals. Other shapes are cut thin resulting in stars, alphabets, wheels, seed-like shapes or even spacemen or automobiles. The substitution of Teflon (a product of du Pont) in extrusion dies has reduced frictions and greatly benefited production. This also made it possible to extrude a wide flat sheet with excellent smooth surface which could then be cut into noodles.

Rolled and Cut Pasta.—Although noodles are extruded as mentioned above many manufacturers roll a sheet of dough to the desired thickness and then cut it into strips and lengths.

Other shapes have traditionally been made from rolled sheets. These include bows, shells, and ravioli.

Final Processing

After being formed into the desired shapes the pasta dough is dried. This is a very critical operation as it must be done fast enough to avoid mold or souring and slow enough to avoid warping or cracking. The drying operation determines the final color and texture of the product and influences the ultimate cooking time. In olden times macaroni was dried out of doors, then later in large rooms. Rooms became smaller, circulating air was added, then still smaller circulating air cabinets were used. Modern manufacturers use continuous driers.

Product Characteristics and Packaging Requirements

Properly manufactured macaroni products are dried to a moisture content of less than 13% and will keep without spoilage for years. Packaging is aimed at allowing the product to breathe so that no excess moisture will be absorbed or, worse, condensed on the product surface.

Packages

Short cuts and small-shaped macaroni products are packaged in cartons and in bags. Although the bags are paper in some countries, cellophanes, polyethylenes, and similar films are now being used. In the United States the usual unit is one pound, but other quantities are available. In Europe packages run from 250, 500, and 1000 gm up to 5 and 10 kg. Most bags used today are gusseted and made on form and fill machinery. Cartons are made from glued paperboard and frequently have a window to allow product visibility.

Older types were made from cheap newsboard with glassine paper liners and were overwrapped with printed paper. Modern cartons are made from a higher grade, lined newsboard which can be printed. This eliminates both the overwrap and the inner liner and permits the use of the window. Windows may be made from cellophane, cellulose acetate, polystyrene, or polypropylene films.

So long as pasta products are stored properly there is no reason why they should accumulate excessive moisture or be infested with insects or other vermin. For extremely moist climates however, additional protection can be afforded through use of foil laminates or closed tins.

COOKED PASTA PRODUCTS

A number of cooked products have been made from macaroni or noodles and then canned or frozen. These include such items as spaghetti and meatballs, spaghetti and tomato sauce, a number of Italian dishes such as lasagna, macaroni and cheese, tuna fish and noodles, and others. In such products the formula for the pasta is adjusted to make it less susceptible to softening during the cooking and canning procedures.

Another new concept is a gourmet line in which a printed paperboard carton is sold containing three packets. One packet contains macaroni or noodles, another contains a sauce (sometimes with meat) and the third contains spices, bread crumbs, or cheese toppings. Exotic names are given such as: "Noodles Romanoff;" "Noodles Italiani;" "Noodles Almondine;" "Noodles Cantony;" "Turkey Primavera;" "Macaroni Monte Bello;" and "Beef Stroganoff."

Still another product contains three envelopes, one containing spaghetti, one a spaghetti sauce mix, and the third a grated Parmesan cheese. The recipe requires the consumer to cook the spaghetti and to make the sauce by adding a can of tomato paste, water and oil to the sauce mix and simmering. This package, too is a printed paperboard carton. Selection of the components in the laminate from which the packets are made depends upon what is to be put inside them. The pasta products need very little protection. A simple plastic film pouch is sufficient. The sauce mixes are dry and can be packed in paper-polyethylene-foil-polyethylene or something similar. Cheese toppings may require more protection against loss of oils. Inner components may be polypropylene, PVDC coatings, or polyamide films.

OTHER PRODUCTS USING FLOUR

Gravies

Gravies made from flour, water, cooked-meat drippings and fat with added flavorings and colorings were invented in France about 1650 by the chef of Louis XIV. The right proportion of each ingredient is needed for the best consistency. Some prepared gravies are canned. A few "gravy mixes" consisting of the necessary dry ingredients are packaged in paper-polyethylene-foil-polyethylene packets. Gravies are also used as components of other prepared foods that are sold canned or frozen.

BREAD

Bread is manufactured by a variety of techniques depending on the specific type desired. The most common variety, "white bread" is produced by baking an expanded dough, composed mainly of flour and water, together with an aerating agent. The outside crust becomes dry and the inner portion retains a high moisture content. Many types of bread exist. Corn bread is made of corn meal. Brown bread is made of rye flour and corn meal. Black bread is a rye bread sometimes called pumpernickel. No matter what its form or origin, bread has long been regarded as the "Staff of Life."

The manufacture of bread by large bakeries is not an old custom. Only since the turn of the 20th century has commercially prepared bread become available to most consumers. Although records exist of bread making since the earliest times, most bread was produced in

FIG. 132.   BREAD WRAPPING HAS MANY FACES

DuPont MSD cellophane is used for fresh wrapping of Jewish style rye bread (center). Frozen products are packaged in polyethylene. Small unsliced loaves use opaque white premade bags with twist tops (left background). Sliced loaves include a cellophane wrapper placed inside a clear polyethylene bag (extreme left and right).

the home by laborious techniques. The first commercial enterprises were small family-type establishments. Even today, there are more than 20,000 bakeries in the United States, ranging from small family companies to huge commercial producers. Because of the short shelf-life (two to three days) most large firms are made up of a number of subsidiaries producing and selling locally. Distribution is almost always limited to a radius of 100–200 miles. Bread and rolls comprise over 75% of the total poundage produced by the entire baking industry.

Manufacture

More than 80% of all bread produced in the United States is made of white flour milled from the inner part of the wheat kernel. The remaining types are made from rye, whole wheat or corn flours. Vitamins are usually added to enrich the bread. Most states require bread enrichment and more than 80% of all commercially baked bread is vitamin enriched.

Flour is put into a blender and mixed with water, milk, shortening, salt, yeast and sugar. It is then allowed to rise in a fermentation room and in an additional operation, more of the above ingredients are added. The dough is then put into a machine which divides it into specific weights. The "rounding" machine takes the pieces of dough and dusts them with flour. From here they go to a molder where they are pressed, flattened, molded into loaf form, and placed in baking pans. They are then allowed to rise for an additional hour and placed in the oven. A slow moving conveyor carries them through controlled heat and after being cooled, they are ready for packaging.

Packaging Principles

The object of a suitable package for fresh bread is to conserve the moisture content, prevent staling, and keep the bread in a fresh condition for as long as possible.

Moisture Protection.—A moderately effective moisture barrier material is required for optimum bread wrapping. The inner portion of bread has an equilibrium humidity in the range of 90%. It tends to dry out rapidly and harden. On the other hand, the crust has a low equilibrium humidity and tends to become soggy under moist conditions. Too good a moisture barrier has the effect of promoting mold growth on the bread and allowing the crust to become soft. Yet, if a truly poor barrier film is used, the bread will tend to dry out and stale. The most preferred material should also be economical.

Staling.—Breads tend to become stale within 4 to 7 days of manufacture. This is an inherent property of the type of flour, method of baking and storage conditions. It is caused by the migration of water from the starch to the protein portion of the interior, the starch then becomes dry and loses texture. Since this activity is independent of the moisture content of the inner portion of bread, an effective wrapper must protect the bread until staling occurs.

The ideal bread packaging material must: (1) be attractive; (2) maintain adequate shelf-life; (3) run on automatic machinery; (4) be strong; (5) be inexpensive; (6) be an adequate moisture barrier; and (7) protect the shape of the product.

Packaging Materials—Flexible

In 1965, over $4 billion worth of bread and related products were sold. It is estimated that this will increase to over $5 billion by 1975. Close to $2 million worth of flexible packaging materials were sold to the bread and related products industry.

The earliest bread was unwrapped. Even today, about 33% of all bread sold in the UK is unwrapped. French bakers have only recently wrapped their loaves in paper bags. The trend toward larger distribution, preslicing and hygienic measures has catalyzed prepackaging. In the United States, most bread sold in supermarkets is wrapped in the factory. Only breads sold via bakery counters and delicatessen counters are unwrapped prior to sale.

Waxed Paper.—Up to about 1910, no bread was wrapped in the United States. During the era preceding World War I, wax paper began to appear as a bread wrapper. Machines were developed at that time to automatically wrap and seal wax paper around a loaf of bread. Present day wax paper for bread packaging has been greatly improved from the 1910 version. It is normally a glazed imitation parchment paper, printed and wax coated. Waxed sulphite paper and several other varieties have also been used. Wax paper is the most economical barrier to use and is rigid enough to run on automatic bread wrapping machines. It seals readily and retards a sufficient degree of moisture loss. Disadvantages are its opacity and its increase in permeability when creased. Although most early wax paper overwraps were identified by printed paper bands, present day wrappers are printed and incorporate end labels. Due to the modern trend toward consumer appeal and aesthetics, wax paper for bread wrappers has decreased drastically. At one time the major material, the market for printed waxed bread wraps has declined from $60 million in 1958 to $40 million in 1964 and is continuing to drop. It is still used on inner wraps for premium and speciality loaves.

Cellophane.—In the late 1920's, cellophane was introduced as a bread overwrap. The initial problems focused around machine limitations, high material cost and handling difficulties. Since it was the first transparent overwrap, many bakers considered it a fad and novelty item. Heat sealing problems necessitated the development of special lines for cellophane, but by 1956 wax paper and cellophane were being wrapped on the same line without any significant modifications. The development of end labels in the late 1930's coupled with improved grades of cellophanes promised a long and secure future for cellophane overwrapped bread.

Coated cellophane films of the semimoisture proof type were used. Moisture protection was afforded to the product and from 1930–1955, the major bread overwrap was cellophane. In the UK, cellophane never gained a foothold since its price differential from wax paper was too severe. At present, many speciality loaves are still wrapped in cellophane. These breads command a higher price and generally require premium aesthetics. Moisture-proof heat sealable cellophane is used for Jewish rye bread. These breads are often baked in three-pound loaves and sold in one-pound portions. The exposed end must be maintained at the proper texture, i.e., firm and yet not hard. For crusty type breads, i.e., French bread, Vienna bread and Italian bread, a less moisture-proof cellophane is required. Crust sogginess is prevented and a hygienic wrapper is obtained. Many of these breads are also packed in uncoated paper bags.

Polyolefins.—In 1958, polyethylene was introduced to the bread packaging industry. Bakers were concerned with high costs and polyethylene promised to reduce material costs. At the same time, rubber-based films appeared but due to odor and higher cost they never really penetrated the bread market.

Polyethylene film was more economical than cellophane and offered the baker a more than 30% savings. Its major disadvantage was machine problems in the thin gauges utilized. Static also presented early problems. At its introduction, polyethylene was made by blown extrusion; caliper control and optics suffered. Later cast films were tried and were significantly better. These latter did infiltrate the overwrap market. Before the entire wrapping industry converted to polyethylene, however, polypropylene was introduced.

Polypropylene.—Polypropylene offered better transparency than polyethylene and greater rigidity and machinability. It tended to shatter at low temperature ranges which caused distribution problems. The early films broke at zero degrees and below. In an attempt to improve weather characteristics, oriented polypropylene was introduced. This variety resisted low temperature breaking but was difficult to heat-seal on automatic machinery. The end seals appeared puckered and distorted.

The trend toward synthetic plastic films characterized by polyethylene and polypropylene led to the development of a coextruded variety of polyolefins. In 1962, a copolymer based on a PE-PP-PE sandwich was introduced. The copolymer was extruded from a single head and blended the feel and weatherability of polyethylene with the optics and rigidity of polypropylene. Although a bright future was predicted for these copolymers, the successful marketing of polyethylene film bags revolutionized the overwrapping industry. Whereas wax paper, cellophane, polyethylene and polypropylene had been used as a sealed bottom lap and end fold package, consumers in the mid 1960's were immediately attracted to the ordinary polyethylene bag.

Polyethylene Bags.—At present, more than 80% of all bread is packaged in low density polyethylene bags. The same package is also being used for the packaging of rolls and partially baked products. Even though polyolefins had distinct advantages as a bread wrapper, they never really ran satisfactorily on automatic overwrap machinery. Preformed bags eliminated production problems, were more costly, but immediately succeeded as a convenience package in the consumer's view. The upsurge in sales forced many bakers into preformed bags and the entire industry was swept into submission.

The major advantages of bags to the housewife revolve around reusability, ease of removal, better shelf-life and the elimination of belly bands. Most experts feel that in some respects these advantages are overridden by even greater disadvantages.

Wicketing of bag packages necessitates the use of holes which tend to leave ragged edges on the tied end. The bag must be oversized to permit the introduction of the loaf. Since bottom gusset, side-weld bags are generally used, a loose package results with evident slice separation. In addition, bags are hard to handle, transport and stack.

Most bags are composed of .00125 in.–.0015 in. polyethylene secured by plastic covered wire twist ties. As the bags are filled, ties are applied automatically from reels. There are two basic types of bread bagging equipment available. One utilizes preformed bags while the other utilizes rollstock. Economic considerations favor the use of rollstock machinery for long production runs. The American Machine and Foundry's Mark 50 is used extensively as a preformed bag machine. Bags are pulled over the bread and various bag sizes can be used. Rollstock machines such as FMC's work on the principle of pushing the bread into the bag.

Additional bags for bread are made of polypropylene with twist tie. For warm-in and fancy bread items, foil-paper laminates are used. Some overwraps also are composed of foil-PE laminates. Opaque wrappers provide better protection from thiamine loss than transparent wrappers.

Recent Trends

During the last two years, the bread industry has entered a period of stabilization. Presently, the market is aiming toward a combination of bags and overwraps in the same package. Equipment has been developed which overwraps a bread package in the conventional manner and gathers and ties the material on the other end. This package uses a pressure sensitive label on the end which is not tied. The bag disadvantages of ragged edges, looseness and floppy gusset ears are eliminated and a neater looking package results.

A still newer trend on the horizon for bread packages is shrink films. Soft bread tends to be squashed when the films shrink; however, shrink films do offer a neat package. Recent developments enable bread to be packaged and shrunk without serious distortion of shape by directing the hot air to the bottom of the package only.

About one-half of all bread and buns are machine-bagged. Square-end bags have recently been introduced. A new automatic bun packager for large volume and smaller bakeries incorporates devices such as the double wicket which holds 500 bags and eliminates changeover interruptions. Another machine has higher speeds and continuously forward-moving arms to feed the bags. Another makes its own bags, which saves between $2.00 and $3.00 per 1,000 bags.

The bulk bin bagger makes and seals its own bag out of .001 in. polyethylene film at rates of 15 packages per minute. Automatic coders and price markers are available.

Some specialty breads have been packaged in foil laminations. Usually a foil-wax-paper or foil-wax-paper-wax-paper is used. Heat seals are achieved by bleeding wax through registered perforations.

Blister packs and PVC films are still newer concepts for bread packages. Blister packages are expensive; however they also are attractive. PVC shrink films make good tight packages for hard crust breads.

Canned Bread

Bread is packed in tinplate cans for use by campers, hunters, sportsmen, and the military. About a one-year shelf-life is obtained. If fillings are added, the problems of mold become accentuated.

Frozen Bread

Bread can be quick frozen and sold in frozen food cabinets. This concept provides for ease of supply and convenience. After the bread is baked, it is wrapped in cellophane or polyethylene and quick frozen. The wrapper should conform as closely as possible to the bread to prevent desiccation. It is sold in printed cartons or as plain overwraps. Shelf-life of three months is obtainable. Many varieties of rolls are marketed in twist-tie PE bags as frozen varieties.

Bake-In-Wrap

If bread can be baked in its wrapper, the product is completely resistant to mold spores and the wrapper provides a barrier against further contamination. Weight loss during baking is reduced and the loaves can be baked close together insuring a more economical use of oven space.

Nylon 6 film and two specially developed heat resistent grades of cellophane have been used for this purpose.

Sandwiches

Sandwiches contain two thin slices of bread with a wide variety of fillings. They must be wrapped for display and convenience in retailing. Because of their relatively high surface area, they tend to dry out and spoil before natural staling occurs. The proper film must maintain a sufficient moisture barrier and yet not too good, or bacterial spoilage of the filling might be encouraged. Cellophane is often used in the form of bags or direct wraps. Shrink PVC, polypropylene and PE are also used

FIG. 133. SANDWICHES ARE WRAPPED WITH SHRINK FILM

Reynolon heat shrink PVC film overwraps are used to wrap bun type sandwiches in
trays for a catering service. The Reycon Case-Redi wrapping system shown here is faster
than other hand operated systems and saves up to 19% in amount of film used.

for tray overwraps. The shelf-life of wrapped sandwiches rarely exceeds
two days. Sandwiches have also appeared as frozen units wrapped in
film overwraps or packed in coated cartons. One of the latest adjuncts to
this is the marketing of a frozen sandwich filling. The product is thin
and sandwich size and comes individually wrapped in a film—similar to
the individually wrapped, process cheese slice. The consumer keeps the
product frozen, removes the wrap and puts the filling in the sandwich.
Within a few hours at room temperature it has thawed, or the sandwich
can be toasted to thaw the filling in minutes.

Refrigerated Dough

Premixed and kneaded bread, roll, and biscuit doughs are sold in
spiral wound paper canisters with metal ends and easy open devices

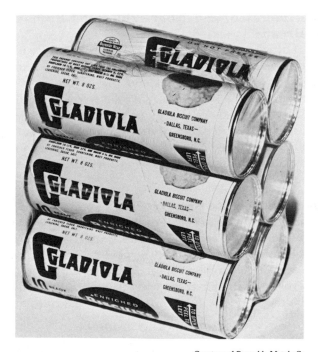

Courtesy of Reynolds Metals Co.

FIG. 134.   REFRIGERATED DOUGH IN SPIRAL WOUND CANISTERS

Biscuit dough packed in canisters is another convenience food
popular with consumers. Shrink film such as Reynolon PVC film
can be used to bundle several containers for multiple unit sales.

such as tear threads. Wet strength kraft papers are used to prevent
weakening of the package in the moist refrigerator. Foil, glassine, nylon,
or polyvinylidene chloride inner liners are used as barriers against moisture loss and fat migration. When refrigerated the product has a good
shelf-life but packages are date coded to prevent possible loss in quality
due to deterioration of the leavening agent. The consumer removes the
dough, and, if necessary, shapes and allows it to rise before baking.

Frozen Bread Dough

Frozen premixed and kneaded bread dough is sold in aluminum foil
containers with polyethylene bag overwraps. The consumer removes the
batter, allows the bread dough to thaw and rise, then bakes it in the foil
container. Frozen products have a much longer shelf-life than the refrigerated doughs.

Frozen, Refrigerated, and Nonrefrigerated Prebaked Brown and Serve Breads

A variety of breads and rolls are prepared and partially prebaked. All the consumer needs to do is give them an additional browning in an oven. Depending upon the product and the degree of prebake these items can be found in the freezer display, the refrigerator display or the open shelf. Most packages are simply polyethylene bags. Some are tied shut, some are closed with saddle labels. A new trend is to put fancy prebaked rolls into shaped foil containers and overwrap or bag. The housewife finishes the baking job right in the compact metal foil container.

Miscellaneous Speciality Bread Items

Bread crumbs and cubes of bread prepared from left over ("day old") bread are sold for use in cooking. Crumbs are used as toppings for a variety of dishes, the cubes are used in poultry stuffing. Both of these products are sold in polyethylene bags. Flavored buttered toast is packaged in paperboard stiffener trays with printed cellophane overwrap. Tins of Melba toast and Zwieback are found in gourmet departments.

OTHER BAKED GOODS

Cakes, pies, pastries, sweet rolls, cookies, and doughnuts comprise about 25% of all baked goods production. Many of these items are sold frozen but most are sold fresh and unrefrigerated.

Cakes

The hundreds of different sizes, shapes and forms of cakes sold in the prepackaged form makes this product more complex than bread packaging. Most stores aim at a shelf-life for cake between 5 and 7 days. Cakes differ as to moisture content and tendency for mold growth depending on their composition. In the vast majority of cake packages, a transparent film is required for consumer appeal. In cake baking, a mixture of flour, fat, sugar, water, eggs, spices, and leavening is used.

Sponge Cake and Swiss Rolls.—Sponge cakes and Swiss rolls are essentially plain types of cake which have a low ratio of sugar to moisture, that is they are the most moist of all cake with equilibrium humidities between 80–95%. In order to prevent mold growth, some drying of the surface of the cake is essential.

The most commonly used film is a semimoisture-proof grade of cellophane. If the cake is at the lower end of the humidity scale, dehydration can be prevented by the use of a more moisture-proof grade of cellophane. The trick is to utilize the proper film in order to extend shelf-life and still not encourage mold growth. During cold weather, a fully moisture-proof grade of cellophane can be used since mold growth does not flourish. It is absolutely essential to evaluate the specific film required with its product compatability prior to use.

There are several other factors which effect the overall shelf-life. If the cake is left in the pan too long prior to cooling, moisture may be driven out from the product. The humidity in the plant also affects the finished item. All these factors have a bearing on the moisture content of the final product.

Polyvinyl chloride films are suitable for sponge cakes and Swiss rolls because they have a moderately good moisture barrier property.

Fruit Cakes and Liquor Cakes.—Cakes containing fruit additives or liquor flavorants have lower equilibrium between the range of 75–85%. They are less susceptible to mold growth and a moisture-proof

Courtesy of E. I. du Pont de Nemours and Co., Inc.

FIG. 135.   FRUIT CAKES IN METAL TINS

Film wrapped fruit cakes are packaged in metal tins for long term protection.

cellophane film can be used satisfactorily. Other films useful include coated oriented polypropylene, polyethylene-polypropylene-polyethylene combinations, cast polyethylene, and wax-laminated glassines. The choice is dictated by machinability and economics required. Fruit cakes have the longest shelf-life of most baked goods. Laminates of polyester-polyethylene are used as well as metal cans. Generally, the package appearance is better with polyolefin films at the end of the shelf-life period than with cellophanes.

Iced Cakes and Fancy Items.—There are hundreds of varieties of fancy cakes containing different icings and flavorants. If the cake contains fruit and no icing, its equilibrium humidity will be low enough to permit the use of a moisture-proof barrier. Moisture-proof cellophane, coated oriented polypropylene, polyethylene, or foil-paper laminations can be used. If the cakes are iced, a moisture-proof barrier presents problems. The atmosphere within the package is humid and the icings absorb water and become sticky. A chocolate cake with chocolate icing has a different hygroscopic factor than white cake with white icing. Too good a barrier will cause the icing to run.

Icings and glazings complicate cake packaging. Most contain at least 80% sugar. Sugar icings are extremely hygroscopic and if a high moisture barrier film is used, the icings absorb moisture from the product and atmosphere. They tend to loose stability or trap the moisture between the icing and the product. The net result is that the icing

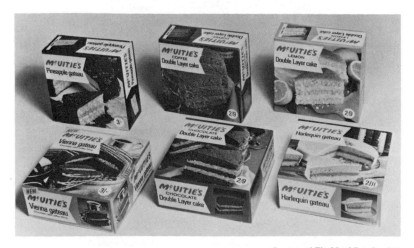

Courtesy of The Metal Box Co., Ltd.

FIG. 136. ICED CAKES PACKED IN CARTONS

The graphics on these cartons present appetizing sales messages. Cakes are put in boxes automatically.

becomes soft, it sticks to the film and separates from the cake. This problem area is under attack by researchers who are seeking stabilizers. In addition, the time before packaging and icing laydown must be properly controlled. If too long a time elapses, the icing may tend to crack and chip during packaging. Polypropylene is better than cellophane in icing release by about 50%.

Doughnuts.—Doughnuts are very similar to cakes except they are fried in deep fat. The greasiness of the product demands grease-proof liners in packaging. Some doughnuts are dusted with powdered sugar,

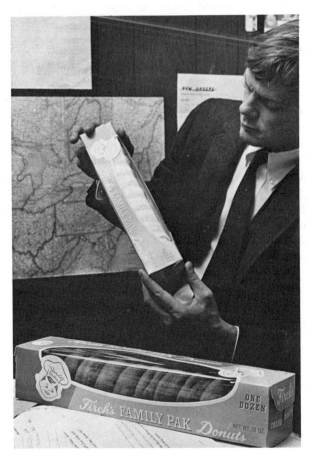

Courtesy of E. I. du Pont de Nemours and Co., Inc.

FIG. 137. DOUGHNUTS IN WINDOW CARTONS

Consumers like to see doughnuts and appreciate the transparent film window carton illustrated here.

others are glazed with sugar or honey and still others are iced. Sugary types must be protected against excess moisture getting to the sugar either from the interior of the doughnut or from the atmosphere. A moderate moisture barrier is needed.

Fancy Baked Pastries (Cake-Like)

Sweet rolls sometimes called Danish pastries with fruit fillings and iced toppings are generally baked to a low equilibrium humidity. They tend to absorb moisture and become soggy under normal conditions. When jam is added, it contributes a high humidity. Since both of these opposing effects are present, these products are very difficult to package properly in order to prevent sogginess, staleness, or mold. Shelf-life is about 2 to 3 days.

Pies, Tarts, and Puff Pastries

Unlike the fluffy Danish pastries, pies, tarts, and puff pastries contain no leavening (yeast or baking powders). Pastries are made from flour, water, and shortening with small amounts of sugar and salt for flavor. Sometimes milk solids are added. Flaky crusts are obtained by not thoroughly mixing the flour and fat. Mealy crusts are thoroughly mixed. Ingredients are kept chilled to keep the fat in discrete particles. Puff pastries are obtained by rolling the dough into thin layers and spreading shortening between the layers. Some of the best pastry doughs contain a high percentage of lard. Other fats such as butter, margarine, vegetable shortening, chicken fat, or oils may be substituted in part. Pie crusts are therefore susceptible to rancidification and will stain absorbent materials. The amount of moisture protection required depends principally upon the fillings. Some will dry out faster than others, some

Courtesy of British Cellophane, Ltd.

FIG. 138.   FROZEN PUFF PASTRY PACKAGE

This coated "Cellophane" package for frozen puff pastry is reverse printed and vacuum metallized then polyethylene coated.

are more susceptible to mold than others. Since the pie crust is low in moisture and the fillings are usually high in moisture the two tend to spoil one another.

Packaging of Cakes, Rolls, Pies, Etc.

A direct wrap is used for most sliced cakes with and without paperboard inserts for rigidity. The film is heat-sealed at the bottom by hand or by machine. Transparent bags are used for many pastries. If the cake is iced, contact with the film is to be avoided. Paperboard boxes with cellophane windows are used to prevent crushing and window contact is prevented by making sure the box is taller than the cake.

Pies used to be sold in tinplate pie pans. Now they are sold in paperboard or aluminum foil pans. Window boxes or overwrapped paperboard trays are used for added protection as with cakes. The amount of protection provided by the package is governed by the type of filling used. A variation of an actual paperboard box consists of an open-top carton with a transparent film overwrap. It is particularly useful for iced cakes and meringue pies since the overwrap does not contact the product. Doughnuts use glassine liners or overwraps to protect from grease stains.

The coming trend in baked goods packaging seems to be blister units. Thermoformed blisters made of polyvinyl chloride or polystyrene are the top of the package and the bottom consists of either a foil tray or paperboard insert. The packages provide both aesthetics and protection.

Many sweet goods are also found in polyethylene bags. This is an off-shoot from the revolution in bread bagging.

Prolonging of Shelf-life for Cake-Type Products

Most cakes have a shelf-life of about 14 days. Several methods are used for obtaining longer keeping quality. These include chemical preservatives, post bake sterilization, post bake pasteurization, and freezing.

Preservatives.—In many cakes and rolls, shelf-life extension is provided by the incorporation of a mold inhibiting agent. Sorbic acid or its salts are used. When added to the mix 0.03%–0.125% of the weight of batter is recommended depending on pH, or the sorbate may be incorporated in a wrapping material, 2.5–5.0 gm per 1000 sq in. being adequate.

Post Bake Sterilization.—Cakes may be placed wrapped in an oven at 275°F for 20 min or longer. The material must be able to provide a hermetic seal and be able to withstand the high temperatures used for sterilization. Drying out must also be prevented. Films used include polyethylene, polypropylene, and certain polyamides. Shelf-life of up to one year is possible.

Post Bake Pasteurization.—Another fairly new technique developed in France consists of infrared pasteurization in order to produce cakes with more than seven months shelf-life. Cakes and pastries are cooked in aluminum foil containers and then packed in polyamide film bags. After sealing, the cakes pass into an infrared tunnel for exposure to radiation which destroys surface mold spores. The stabilized product is then packed into printed boxes.

Post Bake Quick Freezing.—Although the practice is only about ten years old, more than 10% of the output of wholesale bakeries is being frozen. This amounted to approximate volume of $175 million in 1965. These bakery products vary as to packaging prior to or immediately after freezing. Most baked goods are frozen after packaging. The unwrapped product would lose too much moisture in the freezer and become dehydrated. This is especially true for most yeast doughs, which require more time to reach sub-zero conditions.

Although freezing will preserve a cake there are some undesirable features. Chocolate cakes lose volume during freezing and storage. Devil's food cakes become a little more crumbly. Although flat or fondant type icings are not affected by freezing, condensation during thawing can cause undesirable wetness. Few problems are encountered with frozen pies as they are harder than cakes and easily wrapped with automatic equipment. Frozen pies are principally sold unbaked. The consumer cooks the product. Tarts (small pies) are sold frozen and unbaked. A popular item is a thin rectangular tart in a pouch. The tart is cooked in a household toaster.

Muffins, waffles, pancakes, Danish and French pastries, cheese cake, and some doughnuts can be frozen successfully.

Freezing Conditions.—The actual freezing time for cakes depends primarily on the type of cake and its moisture content. The minimization of ice crystals is extremely important. With pies rapid freezing is essential to prevent breakdown of the fruit.

Packages and Package Materials.—A proper package for frozen baked goods should be: (1) attractive; (2) protective against freezer burn; (3) grease proof; (4) free from odors; and (5) nonabsorbent.

Six basic package types are used for frozen baked goods: (1) waxed paperboard; (2) foil containers; (3) waxed paper; (4) cellophane; (5) polyethylene; and (6) laminated products, (a) paperboard laminates, (b) film laminates, (c) foil laminates (rigid and flexible).

If paperboard cartons are used, it is important to select materials capable of resisting subzero temperatures. Most cartons are coated with polyvinylidene chloride, wax, polyethylene, or wax-resin blends. If there are sticky icings on the product silicones are used for release properties. The window material on a carton, dependent upon the product and storage temperature, may be either polyester, oriented polypropylene or cellophane.

Aluminum foil trays are very popular since they may be reheated in the oven directly from the package. One process used extensively commercially uses cake batters which are placed in foil-pan packages. The cakes are cooked and iced, then blast frozen. Cartoning occurs as the last step prior to palletizing.

Muffins are sold loose in polyethylene bags, though some specialty items are set in deep cavity aluminum foil trays. Waffles and pancakes are put in individual polyethylene bags. The consumer removes them and thaws and finish cooks them in the household toaster. English muffins are sold frozen (already split in half). To use, the halves may be put in a toaster or under a broiler. They are packaged in a polyethylene bag with and without a paperboard tray insert. The toaster tart is packaged in a paper-polyethylene-foil-polyethylene heat-sealed pouch. Multiple packets are over packaged with a printed paperboard carton.

A number of combination packaged products are now being marketed which in a sense could be called a cooking kit. One example is a printed paperboard carton containing about 14 oz. of a fancy "streusel." The carton has hot melt glued end flaps with a tear strip opening device. Inside are four other packages. One is a spiral wound paperboard canister with a foil liner, metal ends and a printed foil-kraft paper label. The canister contains pastry dough rolled up into a cylinder. The consumer unrolls the sheet of dough onto a greased cookie sheet. A second package contains a fruit filling, which is spread over the dough. The fruit filling is packaged in a plastic film such as oriented polypropylene which has been formed into a tube by means of a fin-seal and closed at each end by gathering and applying a metal clip as is done for a chub pack. The third package is a form and fill pillow pack made from an opaque imitation glassine with a PVDC coating, and a hot melt has been used in the seal area. This contains the streusel topping which is highly flavored. The final package is a small "tetrapak" made from a plastic film and containing a creamy icing. This is a prime example of how modern packaging technology has been used to put together a convenience package for the consumer that otherwise would be quite impossible.

COOKIES AND CRACKERS

Cookies (called biscuits in England) are prepared from doughs containing flour, fat, sugar, water and flavoring or coloring agents. The dough is rolled and cut into desired shapes. It is then baked and packaged. The composition of the dough and the type of flour used determines whether the cookie will be soft or hard. Sugar wafers are made by pouring a very thin batter onto a heated mold.

Courtesy of British Cellophane Ltd.

FIG. 139.   CREAM FILLED COOKIES IN CARTONS

These cookies are packaged in "Cellophane" overwrapped printed
cartons with a tear strip opening device.

Many varieties of cookies and biscuits exist. Some have high sugar
contents, others high fat contents. Some have chocolate or flavored
fondant fillings, others are enrobed. The primary characteristic of all
cookies and biscuits is their very low relative humidity. In order to
prevent the absorption of moisture from the air, a high moisture barrier
is needed. Since many cookies contain fat, a grease-proof material is also
demanded. Light may also be detrimental to colors or cause fat oxidation. A suitable packaging material must be able to run on automatic
machinery and protect the brittle product from physical damage. Cookies containing fruits may be susceptible to mold. Those containing nuts
are susceptible to rancidification.

Crackers are similar to cookies except they contain little or no sugar.
They are flavored with salt, cheese, or other nonsweet materials.

Packaging Materials and Package Types

Moisture-proof cellophane, coated oriented polypropylene aluminum
foil laminates, polystyrenes, waxed glassines, and corrugated glassines
are used in cookie packaging. Polyvinylidene chloride coated cellophane
offers an excellent moisture barrier, heat-sealability and gloss. It is a
fairly tough film and capable of resisting the sharp corners of many
cookies. Oriented coated polypropylene offers increased economies;
however, it is a little more difficult to run on automatic machinery. The
introduction of polyvinylidene chloride papers have led to their use in
cookie packaging. Paper is easily machined and easily coated.

The shelf-life for most cookie items is about three months. Laminates
of aluminum foil provide for a longer shelf-life under moist conditions.

Courtesy of American Viscose Div, FMC Corp.

FIG. 140. FIG BARS IN TRANSPARENT BAG

Fig bars are held in place by a corrugated divider tray and the laminated cellophane bag.

Cookie Packages.—A common cookie package consists of a printed paper bag with a film liner. The bag is secured by twist ties at the top. Many cookies are also packed in paperboard boxes with printed paper or film overwraps. A fairly new package is composed of thermoformed plastic trays of polystyrene and a cellophane overwrap. The product rests nicely in compartments and the ends of the package are crimped for easy opening.

Other packages consist of decorated tins for gift packages, window paperboard boxes, direct film overwraps, and foil-paperboard overwraps. Corrugated glassine paper is often used as separations to prevent damage.

Refrigerated cookie dough has been sold in plastic film chub packs for a number of years. A new package consists of a window carton containing 36 individual slices of chocolate chip cookie dough. This package provides added convenience in that the dough is already sliced and that the package is easier to open and reclose.

Cracker Packages.—The bakers of saltine crackers have pioneered many concepts in cracker packaging. They were one of the first to use waxed glassine liners and waxed paper overwraps on cracker packages. They introduced inner packs which subdivided the carton contents into individually wrapped units. By not opening all of the package at once crispness of the product in the unopened section was protected for a longer period. Their latest innovation is to further subdivide the package into small packets of a few crackers each.

Cracker packages are either lined printed cartons or plain cartons with polyvinylidene chloride coated paper overwraps. Flavored crackers have made great inroads in the snack field (Chapter 13).

PRETZELS

The name pretzel is derived from the Latin "pretiola" meaning a small reward. The first pretzels were made about 1800 years ago as gifts to children who learned their catechisms well. The shape was symbolic of a child's arms and hands folded in prayer. As the story goes, originally they were not so hard and crisp, but one day a batch was accidentally overcooked, and the baker dipped them in salt to mask the scorched flavor. The result was a popular new baked item which remains popular to this day. In 1968, US sales of pretzels amounted to $185 million, nearly six times the amount sold in 1950.

Before 1953, pretzels were twisted completely by hand. With the arrival of mechanical twisters in 1953, the industry was introduced to modern production methods and grew accordingly.

The first commercial plant in America was the Sturgis Bakery in Lititz, Pa., which opened its doors in 1861. Today there are 81 pretzel bakeries throughout the United States.

Product Characteristics

A pretzel is either soft or hard. If it is soft, it will generally be large, thick and doughy. This type of pretzel is baked only for local consumption because it stales quickly and has a very short shelf-life.

The hard or "brittle" pretzel is composed of combinations of hard and soft wheat, butter, shortening, salt, malt and yeast. These ingredients are uniquely formulated to give the pretzel a distinctive flavoring and supply the product with strength and endurance properties. The formula for a durable pretzel will have about 25% soft wheat and 75% hard wheat. Stronger gluten properties in the hard wheat are known to reduce product breakage.

Hard pretzels are found in at least 20 different varieties. Some of the more common are Thins, Rods, Sticks, Nuggets, Dutch, Beer, Big Boys, Butter, Teething and Cocktail Pretzels.

Since the pretzel is considered to be stronger than the potato chip and will travel extensive distances without breakage, it is distributed by some bakers on a national basis. This means that although the concentration of pretzel bakers are in the Pennsylvania area, it is still possible for them to compete in markets throughout the country.

Packaging Requirements

The two biggest factors that produce undesirable product are breakage and moisture. Although returns are generally less than one per cent, it is evident that both these areas are troublesome and costly.

It is important that the pretzel retain a moisture content in the vicinity of 3-5%. If the moisture content becomes too high, the pretzel has a tendency to stale at a more rapid rate. Some moisture is necessary to give the product strength and make it less subject to breakage.

It should be noted that unlike the potato chip, pretzels have a low shortening content and are not generally affected by oxidative rancidity problems. Still, the flavor of the pretzel is delicate and should be protected with a packaging material that provides an adequate odor barrier.

With the exception of the oxygen barrier properties needed for potato chip packaging, a film for pretzels would be basically the same.

A film for packaging pretzels should have the following characteristics: (1) grease resistance; (2) durability for a packaging shelf-life of eight weeks; (3) coating—one or two sides depending on the application; (4) machinability—should handle well on existing equipment; (5) MVTR—not more than 0.4 grams; (6) appearance—low shrinkage when sealed; (7) resistance to puncture and tear by the pretzel's salt crystals; and (8) stiffness—similar to cellophane to support package.

Winter breakage is a problem in pretzel packaging. Although industry returns are less than one per cent, there is room for improvement. Current indications are that the manufacturer is over-packaging to offset breakage.

Humidity can work against the pretzel, and for this reason it is not uncommon for the baker to use single wall bag constructions during the winter months and double wall constructions in the summer periods of high humidity.

Packages and Materials

Pretzels are packaged in cans, cartons and film bags. Most cartons are wrapped with an overprinted cellophane film. A more recent trend has been the use of printed cartons with a more protective inner wrap, i.e., waxed glassine.

Segments of this industry believe that the bagging of pretzels offers better protection than a box because the bag will give with the product where a box will not. In box packaging, pretzels are forced against rigid sides of the carton which can contribute to breakage. Other industry

Courtesy of American Viscose Div. FMC Corp.

FIG. 141. PRETZELS

Pretzels sometimes are jumble packed. This package uses a molded
polystyrene tray to position the pretzels in the cellophane bag
to reduce breakage.

leaders, however, note that the box is a stronger package. There is, therefore, less chance for the product to be crushed in transit to the retail
outlet and finally to the customer's home.

Both sides seem to have valid arguments but bagging is believed to be
winning preference because of product visibility and its relation to
impulse sales. With substantial ink coverage on a pretzel bag, the broken pretzels can be adequately hidden while in a box the consumer will
sometimes shake the contents to determine if any pretzels are broken.

The sharp salt crystals on pretzels can puncture and cause packaging
films to tear. These sharp crystals are one of the reasons the pretzel
bakers use the heavier gauge films for packaging.

Glassine.—Because of their low shortening content, a packaging
material that protects pretzels from sunlight is not necessary. Glassine
bags alone are therefore not thought to be a factor in this market. Glassine is, however, used extensively as an inner wrap or fractional package
in pretzel boxing.

Polyethylene.—Prefabricated polyethylene bags are now being used
to package pretzel rods by some bakers and at least one is using polyethylene for large size bags which are made on a Wright form-fill machine.

Cellophane.—Transparent films are used in pretzel packaging as
bags, carton overwraps, tray overwraps and fractional packages within

another package. Polymer coated cellophanes are used extensively for these applications because these films have proved to be stronger and more durable than the nitrocellulose coated films.

New Trends

A company in New Jersey started what could develop into a major trend in box packaging of pretzels. By replacing a waxed-liner box with a cellophane twin-pak box, sales were doubled six months after introduction.

Another area of pretzel packaging that is cited for growth is the use of the six-pack. Until very recently, this method of packaging had been used for potato chips and other snacks, but not pretzels. If this trend continues, it will open new opportunities for additional film overwrap usage in the industry.

Many pretzel bags are now being made from laminations of cellophane and oriented polypropylene film. The composite takes advantage of the properties of both films. It is resistant to salt abrasion, provides an excellent moisture barrier and improved flexibility at low temperatures.

BIBLIOGRAPHY

ALVI, A. S. *et al.* 1966. Effect of different types of wrappers on the loss of moisture and the growth of molds in indigenous baked products. Sci. Ind. (Pakistan) *4*, No. 4, 317–320.

ANON. 1966A. A new pastry pack. European Packaging Dig. No. 40, 6.

ANON. 1966B. Keeping the crust on wrapped bread. Biscuit Maker *17*, No. 11, 820–825.

ANON. 1967A. Breakaway for Sara Lee. Mod. Packaging *40*, No. 8, 172.

ANON. 1967B. Longer life for crusty bread. Packaging Rev. *87*, No. 7, 32.

ANON. 1967C. Heat-sealed bags choice for flour shipments. Food Process. Market *28*, No. 9, 81.

ANON. 1967D. Cartons for cake baking. Packaging Rev. *87*, No. 10, 48.

ANON. 1967E. Close up: Cookies, crackers and cakes. Packaging Eng. *12*, No. 10, 108–111.

ANON. 1967F. Flour packing line. Australian Packaging *15*, No. 10, 51.

ANON. 1967G. Swedish biscuit makers employ cartons instead of wrappers. Packaging News, *14*, No. 10, 22.

ANON. 1967H. Nitrogen filled soft pretzel container overcomes short storage life problem. Quick Frozen Foods *29*, No. 12, 76–78.

ANON. 1967I. Freezing of bakery products. Amer. Inst. Baking Bull. July, 8.

ANON. 1967J. Bread is baked in film bag. Packaging News *14*, No. 12, 1–49.

ANON. 1968A. Reusable expanded polystyrene trays for Swiss rolls. Food Trade. Rev. *38*, No. 1, 83.

ANON. 1968B. Swiss rolls stay fresh longer in polypropylene film. Packaging News *15*, No. 6, 46.

DRODGE, A. F. J. 1967. Cellulose films for bakers. Bread Baker *155*, 40–43.

EFFENBERGER, G. 1967. Sterilization of bread after packaging in rigid PVC film. Verpackungs. Rdsch. *18,* No. 4, 392–400. (German)

HUMNEL, C. 1966. Macaroni Products, 2nd Edition. Food Trade Press, London.

PRESTON, L. N. 1967. Flexible packaging. IV. Bread. Food Packaging Design Technol. Feb. 32–35. *Ibid* V. Cakes and pies. Mar. 7–13.

REDMAN, B. G. 1966. The packaging of cakes. Federal Trade Rev. *36,* No. 10, 124–127.

SACHAROW, S. 1966. Flexible packaging for bakery products. Svensk Emballagetidskrift *32,*, No. 10, 24–27. (Swedish)

SCHÖNEWALD, G. 1966. Packaging materials for baked goods. Brot u Gebäck *20,* 227–231. (German)

SEILER, D. A. L. 1968. Prolonging the shelf life of cake. Bread Baker *156,* No. 7, 25–26.

TERSER, R., and HARRCUN, C. 1966. The flexible bread box revolution. Baking Ind. *126,* No. 27, 47–55.

WALLENBERG, E. 1967. Packaging of Bread. Packing No. 9, 25–28. (Swedish)

WILLIAMS, K. S. 1967. Advances in bread packaging. Food Mfg. *42,* No. 9, 42–44.

CHAPTER 13

Snack Foods

INTRODUCTION

The meteoric rise in development of snack products has been one of the most impressive trends in recent food history. Increased leisure time coupled with consumer affluency has sponsored a total snack food market estimated at over two billion dollars annually. The term "snack food" embraces an enormous range of products. From potato chips and pretzels to chocolate covered ants—consumers are being exposed to new snack products at a rate of almost 115 in 1966 alone, each requiring specialized packaging.

A snack food is a product not normally used as an essential part of a meal. It is something eaten between meals and usually consumed under conditions of recreation, sports or relaxation. By definition, the term includes potato chips, pretzels, nuts, snack crackers, cheese spreads, many cookies, biscuits, puffed snacks, cracker sandwiches and additional products. The packaging of these many different snack foods involves differing principles but considerable overlap exists. This chapter will be devoted to those snack foods cooked in oil as well as the new range of boxed snack foods currently being sold in folding cartons. Pretzels, crackers and nuts will be found in those chapters involving baked goods and confectionery products.

POTATO CHIPS, CORN CHIPS ETC.

The retail sales of potato chips in 1965 amounted to over $600 million. It is expected to reach 1 billion dollars by 1970 and the per capita consumption rate has grown by ten per cent annually since 1942. Between 1956 and 1964, the sales of potato chips rose nearly 70% compared to 14% for all goods.

In 1965, over $200 million of sales were recorded for other chip-like products, i.e., corn chips, potato sticks etc. Since all chip-type items represent similiar packaging problems, they will be treated in one category. More development has been reported on potato chips than any other chip item; thus, the following data is representative of all chip-like foods.

Product Characteristics

Potato chips are prepared from thinly sliced potatoes which are cooked in hot oil. Nutritionally they consist of 2.0% water, 6.7% protein, 37.1% fat, 51.1% carbohydrates and 3.0% minerals and other substances. The best potato for chip processing is round with a two- to three-inch diameter. It should have a high solids content since this reduces tendency to absorb cooking oil and yields more chips per pound. A low content of reducing sugars gives a lighter and more flavorful product.

Potato Chips

In the processing of potato chips, the best results are obtained from second crop potatoes grown in a dry, sandy soil. Useful varieties include Idaho Russet and Colorado.

Preparing the potato for frying involves peeling by machine, cutting out the eyes and slicing by machine into a tank of cool water. The amount of soaking time is quite important and varies depending on the particular type of potato. Times used are between 1/2–2 hours. After removal from the water, the potato slices are drained for several minutes and fried.

The chips are lowered into the frying oil by means of wire baskets. Oil temperature is between 200°–230° F. For every pound of sliced chips, there should be about 25 lb of fat. Frying time is between 3–5 minutes depending on the degree of browning desired.

After frying, the chips are thoroughly drained either by hanging the wire basket above the oil or by using a metal draining shelf. The chips must be kept warm during draining. After the chips are drained of excess fat, salt may be applied. The amount of salt depends on taste. Chips are then cooled prior to packaging. This prevents sweating and sogginess.

Packaging is generally conducted via automatic "form-fill-seal" machinery of the vertical type. The cooled chips are loaded into a hopper and filled into the bag via an inner tube. As one bag is filled with chips, another is being automatically formed around the mandrel. The filled bag is sealed, severed, and conveyed to the packing room.

Package Requirements

Important functions of a suitable potato chip package are to protect against rancidity, moisture, loss of odor or entry of foreign odors, and product crushing. Additional factors imposed by marketing requirements include attractiveness, no oil wicking or staining, easy opening, and machinability. Potato chip packages exist in many sizes ranging from the small 5¢ bag to the one pound family size unit. Each package size influences the selection of materials and package structure and the price at which the product is offerred for sale. Due to the ratio of surface area to product volume smaller bags require the highest barrier protection but are least able to justify the cost. Many chippers compromise on the properties of small bags depending upon quick turnover of the product for assurance of freshness.

Rancidity.—Rancidity may occur because of the large quantity of oils used in processing potato chips. The oil content of potato chips ranges between 35-40% and in some cases, even higher. Since the overall shape of a chip involves a large surface area, rancidification becomes a critical problem. BHA and BHT antioxidants may be used up to 0.02% by weight of the residual oil to reduce tendency for rancidification.

Oxygen and light produce oxidative rancidity in the oils. This action is accelerated further by heat, free circulation of air, presence of trace quantities of copper and blue or ultraviolet light. Red and yellow light in the visible spectrum have minor effect. Ozone is particularly a bad actor as are any other peroxidic compounds.

The most suitable transparent chip package should have an oxygen barrier less than 1 cc per 100 sq in. per 24 hr at 1 atm and 75° F. Since opaque materials block off light, the oxygen barrier properties of these materials would be suitable in the 20-25 cc range. The relatively poor light barrier properties of transparent film packages for chips is a major reason for the increasing use of aluminum foil laminates. In addition, glassine provides a fairly good light barrier and requires less oxygen barrier to provide adequate protection.

Moisture.—Moisture protection is required to prevent the potato chip from becoming soggy and tough. Potato chips containing moisture in excess of 3.5% are tough and unpalatable. The preferred MVTR of a packaging material for chips should be below 0.4 gm per 100 sq in. per 24 hr at 100°F and 95% rh. The development of moisture pickup by potato chips is a greater problem to the chip industry than oxidative rancidity.

Fragility.—In successful merchandising, the extremely fragile nature of potato chips makes package stiffness essential. The use of vertical racks and display cabinets also requires a stiff package. If the material is too flexible, a poor package results.

Odor Permeation.—A good barrier to odor permeation is required since the product is oily and odor permeation causes aroma or flavor alteration.

Wicking and Staining.—Oils and grease tend to stain a material having an inadequate grease barrier. Delamination and/or ink deterioration may result from such staining, in addition to its unattractive appearance.

Other Properties.—Since the package may be shipped to various temperature zones, cold weather durability is essential. High gloss is needed for sales appeal and easy opening features are mandatory. In all cases, the material must be capable of being handled economically on the packaging machinery at high production speeds without material deterioration.

Package Making and Filling

There are two basic methods used to fabricate packages of potato chips. Either pre-made bags or automatic form-and-fill machinery may be used.

The success and rapid growth of the snack food industry during the late 1940's led to an increased use of automatic form-and-fill machinery. Since waxed glassine and other materials for pre-made bags did not function on these newly developed machines due to technical and economic reasons, the need for better packaging materials became apparent.

Development of materials suited to automatic machinery started with cellophanes which graduated from the first small 5¢ bags to the present day one-pound sizes. Now a wide range of laminated transparent materials are available for use on automatic machinery. Opaque materials have developed at a slower rate. At present, several glassine and many aluminum foil laminates are used.

Pre-Made Bags Versus Form-and-Fill Machinery.—Economics offered in the use of rollstock often is the deciding factor in the selection of automatic form-and-fill machinery. For large volume processors, automatic machinery is an economic necessity. Small volume chip producers or large volume producers for small volume products will still order pre-made bags but over 85% of the chip industry will use automatic machinery in future years.

Transparent Versus Opaque Packaging.—Chip manufacturers have been divided into two different philosophies with respect to transparency. Some favor transparent packages and others favor opaque packages. Based on the preferences of certain geographical locations, chippers are often reluctant to change a successful package. The history of the potato chip industry is based on regional manufacturers serving selected geographical areas. For various reasons, the first processor in a region often sets marketing policy. Consumers are slow to change and processors do not argue with success. In America, transparent packages are widely used in the East, Southeast and Northeast. Opaques are widely used in the Southwest, West, Midwest and Canada. Total sales are about evenly divided between opaque and transparent packages.

Technically, opaque packages are superior to transparent packages.

When exposed to light, potato chips are unsaleable in six days. The same chips shielded from light remain fresh much longer. Since oil randicidification is a major source of off-flavor in potato chips, the ultraviolet light barrier present in opaque materials provides superior flavor protection. Other advantages in the use of opaque materials are the nonvisibility of brown and broken chips. If the potato used in chip making contains too high a content of reducing sugars, brown chips result. Poor handling, cooking and consumer misuse cause broken chips. Opaque packages mask these problems and add to the overall saleability of the product.

Specific advantages in using opaque aluminum foil laminates are their positive barrier properties and aesthetic appeal. In addition, foil packages may be heated for a different taste treat in snack foods.

Transparent packages offer product visibility which may enhance merchandising appeal. Transparent packages also are "sparkly" or "glossy." Newer opaque laminates now are available which match the gloss of transparent materials. Reverse printed films or high gloss heat resistant laquers provide satisfactory aesthetics. High quality printing techniques are able to print vignettes on opaque packages which depict the product inside.

The growing trend toward material standardization among chippers seems to favor opaque materials. Regional processors will still market packages aimed at their specific markets. Larger processors will probably standardize and opaque materials do offer greater product shelf-life.

Packaging Materials

Pre-Made Bags—Opaque.—The earliest pre-made opaque bags used by the potato chip industry consisted of waxed glassine and aluminum foil-paper laminates. Duplex waxed glassine bags, wax-laminated glassine bags and aluminum foil-paper laminates were used in the 1940's and 1950's. Improvements in material machinability and economics are the major reasons for the continued sale of opaque pre-made bags.

The largest selling pre-made bags were made from waxed glassine. At first they were filled manually and stapled shut. Later the "thermotop" bag was developed which provided a heat-sealed closure and eliminated the staples. Protection was related to the quantity of wax used; ultimate protection was obtained by the use of double walled waxed glassine bags.

Other structures used were coated glassines and duplex combinations of waxed sulfite paper. Pre-made aluminum foil laminate bags have been tried by several chippers, and this trend is growing.

For packages containing "twin packs," the inner pouches are two glassine bags. The outer bag may be sulfite paper or glassine. New developments in pre-made opaque bags include the introduction of PVDC-coated glassine, PVDC-coated glassine laminated to oriented polypropylene and sulfite paper laminated to oriented polypropylene.

Although the overall use of pre-made opaque bags is expected to decline, the economics of a waxed material is an important factor in their continued use. As further development continues in the area of opaque form-and-fill materials, newer laminates are expected to appear.

Pre-Made Bags—Transparent.—The most popular type of material used in pre-made transparent bags are duplex cellophane combinations. When less rigidity and protection will suffice, single wall cellophane bags are used. Barrier protection is improved by means of polyvinylidene chloride coated varieties; however, earlier duplex cellophane bags were nitrocellulose coated.

Duplex coated cellophane bags offer excellent protection to the potato chips but suffer from low flex endurance, poor cold weather flexibility and marginal stiffness. In addition, many manufacturers prefer a single film.

Courtesy of Glassine and Greaseproof Manufacturers Assoc.

FIG. 142. OPAQUE GLASSINE POTATO CHIP BAGS ARE POPULAR IN CANADA AND PARTS OF UNITED STATES

This bag is a plasticized grade of opaque white glassine. Special coatings are applied to the glassine to improve moisture barrier and facilitate heat sealing.

Courtesy of Glassine and Greaseproof Manufacturers Assoc.

FIG. 143. "SNACK PACK" FEATURES EIGHT SMALL POUCHES
OF CHIPS IN A LARGE OUTER BAG

Coated glassine form and fill bags are used for individual packs
of potato chips. Eight small bags are then packaged in a printed
coated paper bag which can be reclosed.

Recent trends indicate that duplex cellophane bags are being displaced by cellophane laminates on form-and-fill machinery.

Form-and-Fill—Opaque.—The development of opaque materials for form-and-fill machines has proceeded at a slower rate than transparent materials. The early success of aluminum foil laminates in the chip industry was hindered by the short supply of aluminum in World War II. Only within the last five years has there been a concerted effort by both foil and glassine manufacturers to capture this market.

Opaque bags for form-and-fill machinery may be classified according to size. Small size packages (below $0.30) usually utilize a single web. Larger size (above $0.30) employ a composite lamination as the packaging material.

Two of the most used materials for small size chip packages are: (1) heat seal coating-glassine-PVDC coating; and (2) wax-EVA coating-glassine. Most small size chip packages consist of material No. 1. The use of a PVDC coating provides good barrier properties and heat sealable coating provides good gloss and back seam seals. Material No. 2, used in significantly less volume, provides fair barrier protection. The material works on small five cent bags if oxygen barrier and flex endurance are sacrificed. The low hot-tack of the coating often causes sealing problems. The major problem in running glassine materials on automatic machinery is that they tend to fracture on formers and during shipping.

Larger size chip packages normally are made from laminates of glassine or sulfite paper with transparent films. Materials such as barrier coated glassine-adhesive-oriented polypropylene, PVDC-coated cellophane-polyethylene-PVDC-coated glassine, oriented polypropylene-adhesive-hot melt coated glassine are used. The stiffness required is provided by the glassine or sulfite papers while the use of either cellophane or polypropylene as the outer ply provides package gloss.

A new market has recently become available in aluminum foil laminates for form-and-fill machines. In previous years, foil specifications which were economical enough to use would not run on automatic machinery. The major problem had been the successful running of a foil material around the severe turns in the machine forming head. The introduction of foil-paper-Elvax materials was the first economical lightweight specification that could be run successfully on these machines. Various other foil materials may be used for chip packaging ranging from cellophane-polyethylene-foil-polyethylene to foil-polyethylene-PVDC-coated glassine. While the major area for aluminum foil laminates is boxed snack packaging, an emerging market exists in corn and potato chip units. Several major processors are currently using paper-foil laminates for potato and corn chip packages.

Form-and-Fill—Transparent.—The earliest form-and-fill pouches for potato chips were made from duplex cellophane bags. The printed web was obtained from a converter; while the other web was obtained from the raw material manufacturer.

In 1961, a relatively simple process was introduced which combined two rolls of polymer coated cellophane by the application of heat and pressure. Although thermal laminates may also be prepared by adhesive bonding, the former are cheaper and easily made by small converters. Laminated cellophane-cellophane and polypropylene-cellophane materials offer greater stiffness and superior aesthetics to duplex bags.

Polymer coated cellophane-polymer coated cellophane laminations may be made by several different methods. In dry adhesive lamination, the two webs are joined after the adhesive is applied to one web and all residual solvents removed. Wet lamination is rarely used since the two webs would be combined prior to solvent removal making the chance of solvent retention very high. Thermal bonding combines both webs by heat and pressure with steam injection as an optional technique. The use of steam allows for the restoration of moisture lost in printing and laminating.

A polymer coated cellophane-polymer coated cellophane laminate offers excellent moisture and oxygen protection. It is usuable on automatic machinery and yields a stiff package. Problem areas have been the lack of cold weather durability and resistance to hot and dry climatic conditions.

Oriented polypropylene-polymer coated cellophane laminations are available in many varieties. By the addition of oriented polypropylene, cold weather durability is increased and an economical, low cost material results. A major disadvantage in the use of oriented polypropylene is its fairly narrow sealing range. In recent years, several coated varieties have been introduced to remedy this effect. Since polymer coated cellophane will not seal to uncoated polypropylene, several different methods have been used to secure a back seam seal.

The "Arbond" lamination consists of uncoated bioriented polypropylene-adhesive-polymer coated cellophane. The principal feature of this material is the use of the "cut back" edge. The oriented polypropylene sheet does not extend the full width of the lamination. It terminates between 1/2 in. to 5/8 in. from one edge of the lamination while being flush on the other side. The cellophane web is coated with the adhesive used across the entire width of the lamination. By the use of the "cut back" edge, the end user is able to produce a bag with a seemingly conventional top seam. In essence, the seal is really a cellophane-to-cellophane created by folding the extended edge back on itself. The cellophane seals to itself and the adhesive coating on the cellophane activates under heat to tack down the uncoated oriented polypropylene. Advantages include the elimination of the cost of using coated polypropylene. Disadvantages are the fin back seam, soft roll edge and poor tracking on the machine.

A thin strip of heat-sealable coating (thermal strip) can be applied to the polypropylene and it forms a back seam upon contact with cellophane. The major problem in this approach is that a strong back seam is not always obtained with a thermal strip. In recent months, the thermal strip approach has been preferred by most chippers.

The introduction of coated oriented polypropylenes has opened up new areas of development. The overlap back seam, desired by potato chip manufacturers, can readily be made using coated polypropylene. A

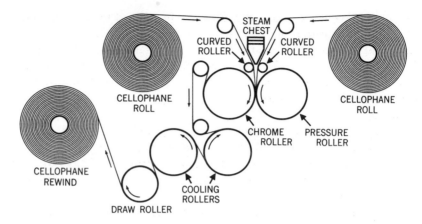

Courtesy of American Viscose Div., FMC Corp.

FIG. 144.   FMC THERMAL LAMINATOR COMBINES TWO FILMS
BY HEAT AND PRESSURE

Two webs of polymer coated cellophane can be combined on this machine by
heat and pressure to form a high barrier duplex laminate for snack foods.

heat-sealable coating on polypropylene offers many machine advantages to the chip manufacturer. Improvements are still needed on seal
strength and jaw release.

Packaging Machinery

A large variety of automatic machines can be used for filling pre-
made bags with potato chips. The use of form-and-fill machinery is a
newer and more dynamic trend.

Wright Machinery Company offers a full range of form-and-fill
machinery for opaque and transparent materials. Other firms supplying
machinery include the Woodman Company and Mira-Pak, Inc. Three
factors of great significance in proper machine production are "former"
design, electric eye positioning and design of the sealing jaws. Each of
these must be carefully evaluated prior to selecting a flexible material.

New Developments

The use of inert gas packaging represents a tool capable of extending
the current shelf-life of chip-like products. Current distribution methods
used for most chip products revolve around direct store delivery from
the regional processor. The shelf-life required for these packages is 5–7
weeks. New snack foods which require boxed cartons also require warehouse distribution and utilize entirely different package concepts. The
shelf-life requirements of these products is a six month minimum.

Many chippers are examining the use of packaging under a nitrogen

Courtesy of The Woodman Co., Inc.

FIG. 145. WOODMAN COMPANY FORM AND FILL MACHINE

This model DF-8 "Fleet-Weigh Dual-Pac" is a large bag automatic, net weigh, form, fill, and seal machine which can handle potato chips or other snack foods.

atmosphere. The significant advantage to the regional chipper is his ability to extend his market to a greater area. If inert gas packaging is to become a successful process, a shelf-life of three months would be mandatory.

Any packaging system to be used in inert gas packaging must provide an acceptable moisture barrier coupled with an oxygen barrier below one cubic centimeter. If the amount of moisture present in potato chips rises above 3-1/2% they become soggy. Oxygen contents greater than one cubic centimeter tend to cause oil rancidification.

Several chippers have experimented with inert-gas packaging concepts. The lack of specific equipment together with an economical material has retarded progress. In addition, expensive and specialized machinery would have to be purchased by the chipper.

Possible material candidates include foil laminates, polyvinyl alcohol laminates; PVDC-coated cellophane-polypropylene-polyethylene; and amylose film laminates. In all these cases, a sufficiently high oxygen barrier is required. If a material is to be PVDC-coated, the amount of laydown must be sufficient to retard oxygen permeation. At present, the use of inert gas packaging in chips is still highly developmental.

BOXED SNACKS

Due to the growing trend in snack food consumption, many major food firms are rapidly entering the snack food business. The distribution cycle of the major companies is radically different as contrasted to the smaller regional chippers. Products are distributed through warehouse channels and thus eliminate the need for a route salesman. Their shelf-life requirements are between 6–9 months.

Boxed snacks comprise a wide range of products. Most have been introduced during the last four years. They range from General Mills' "Bugles," "Daisys," "Whistles" and "Buttons and Bows" to Quaker Oats' "Salty Surfers." Rice, wheat, corn and other grains are used for these snack foods and they are sold in puffed and regular varieties. They are also made by the same procedure as breakfast cereal. Other varieties such as Keebler's "Caraway Crazys" and "Potato Piffles" are made by the same procedures used in making cracker production. In 1966, retail sales of all new snack foods have been estimated at over 25 million.

Package Form

Since the vast majority of these new snack foods have been introduced by firms involved in breakfast cereal production, the packaging closely resembles breakfast cereal, i.e., a folding carton and inner aluminum foil laminate liner. Not all breakfast cereals are packaged in inner foil liners; however, the functional needs of the new snack food demand superior protection.

The outer folding carton is composed of a printed clay coated cylinder board. A rigid package is used because of its resistance to crushing and handling damage. Additional advantages include ease in stacking on supermarket shelves and easy opening and closing features.

Desired shelf-life is obtained by the use of a higher barrier inner liner. Due to the superior protective properties of aluminum foil, laminates of foil are commonly used. Most utilize glassine-wax-foil-wax-glassine and paper-polyethylene-foil-polyethylene. An added bonus is the excellent dead-fold properties of aluminum foil laminates, which make reclo-

Courtesy of Reynolds Metals Co.

FIG. 146. BOXED AND BAGGED SNACKS

A wide variety of snack foods are packaged in bag in box concepts
which provide needed moisture, grease, and light barrier protection. Aluminum foil laminates are necessary for inner bags. Small
portion pouches of the same products also feature the superior
protection of aluminum foil packaging.

sure easy. In addition to aluminum foil laminates, glassine-wax-glassine
and duplex PVDC-coated cellophane inner pouches are used for less
sensitive products.

An even more recent development has been the introduction of boxed
snack items in ten-cent size glassine-polyethylene-foil-polyethylene
pouches. These are made by one of the large food companies and distributed on a regional basis by local processors. The aim is to offer consumers the product by means of in-store rack merchandising. Since the larger sized boxed snacks are sold on regular grocery shelves, this offers an
additional method for market saturation.

Future Outlook

New snack products have been introduced at a fantastic rate. In 1966, 117 new items were marketed. Although presently considered as specialty snacks, boxed snack foods are expected to grow at a rate equal to the population growth. Since these items are still being produced by breakfast cereal firms, the package will probably retain the cereal like form, i.e., folding carton and inner liner.

New snack food products are aimed at marketing the item as a sporty, fun-time product. The use of highly imaginative names are designed to yield a "fun image" in marketing. Normally considered the pacemaker in these new snacks, General Mills introduced their line of "Bugles," "Whistles," and "Daisys" in May, 1964. After success with initial product introduction, repeat sales did not reach expectations. The original test market results indicated that the product would be highly successful; however, repeat sales were significantly lower. Consumers had tried these new snacks once and then returned to the traditional habit of potato chip consumption. Current results indicate that while potato chip consumption is still high, the entire snack food market has continued to expand.

BIBLIOGRAPHY

ANON. 1966A. Leisure stirs push to snacks. Business Week No. 1900, 102, Feb. 26.

ANON. 1966B. Pretty chipper. Arthur D. Little, Inc., Bull. *442.*

ANON. 1968A. Notes toward a definition of snack food. Snack Food *57,* No. 1, 24–27.

ANON. 1968B. Close-up: snack foods. Package Eng. *13,* No. 3, 84–86.

BROWN, M. L. 1969. Gas packaging for chips, snacks. Snack Food, *58,* No. 3, 44–45.

DALY, C., and JENSEN, L. C. 1969. Increases in-store exposure, doubles shelf-life. Food Process. Marketing *29,* No. 3, 14–15, Chicago.

GIBLIN, J. P. 1968. USA Snack food packaging. Food Process. Marketing, England *37,* 429–432.

PERINO, D. A. 1965. Snack food sweepstakes. Paper, Film, Foil Converter *39,* No. 10, 104–106.

PERINO, D. A. 1968. Snack foods—frustration and opportunity. Paper, Film, Foil Converter *42,* No. 3, 70–72.

SACHAROW, S. 1966. Snack packs—where the action is in new laminations. Paper, Film, Foil Converter *40,* No. 8, 62–65.

SACHAROW, S. 1969A. Update on packaging snacks. Food Eng. *41,* No. 3, 82–84.

SACHAROW, S. 1969B. Foil and snack food—a happy marriage. Snack Food *58,* No. 3, 39–41.

TALBERT, W. F., and SMITH, O. 1959. Potato Processing. Avi Publishing Co., Westport, Conn.

TURNER, C. J. 1965. New developments in laminations for the snack food industry. Speech delivered at 1965 Potato Chip Inst. Intern. Conv., New York. Feb., 1965.

CHAPTER 14

Statutory and Religious Regulations
Affecting Food Packaging

INTRODUCTION

Adulteration of food and drink without regard for the health of the consumer has been a commonplace undertaking for centuries by those who sought to exploit the unwary. Milk and wine were watered, chalk or clay was added to sugar, salt or flour. Copper salts were added to green vegetables to make them greener. Perhaps the worst extreme took place in England in the 1750's when gin mills operated to extract the last penny from the poor by serving them "gin" made from almond oil, turpentine oil, alcohol, sugar, lime water, rose water, alum, tartrate of soda, and sulfuric acid.

There is little doubt that this made the consumer blind drunk. The political satirist painter James Hogarth railed against the practice with little effect. "Let the buyer beware" was the law until very recent times.

US FEDERAL REGULATIONS

Pure Food and Drug Laws

When the first Pure Food and Drug Law was passed in 1906, the federal government of the United States was given the authority to control the food supply of the American consumer. This law constituted a cooperative effort between the canning industry and federal officials to protect the consumer through intelligent enforcement of food law. Initially, the law was administered by the US Bureau of Chemistry which was the predecessor of today's Food and Drug Administration. The law of 1906 was a beginning, but it was lacking in many respects.

In 1938 the federal Food, Drug, and Cosmetic Act was passed. This law was a tremendous improvement over the law of 1906, but it did not provide for the control of food additives except where such additives were known to be poisonous or deleterious substances. In 1958, the Food Additives Amendment to the Food, Drug, and Cosmetic Act was passed. This is the amendment that has caused great concern among people in the food packaging business. The rules and regulations promulgated by the F&DA under this amendment are of significant concern to the food packaging industry. Other important amendments to the Food, Drug, and Cosmetic Act which have been passed since 1958 include the Color Additives Amendment of 1960, which establishes the rules under which colors can be qualified for use in or on foods, drugs, and cosmetics, and the New Drug Act Amendment of 1962. This drug

Courtesy of USFDA

FIG. 147.   F&DA INSPECTION

An F&DA Import Inspector of the New York District is sampling a shipment of olives on the docks in New York City. The sample will then be taken back to the District Lab. for analysis.

act is responsible for checking and approving the efficacy and safety of all drugs and includes provisions for approval of packaging materials and labels for drugs.

Food Additives Amendment.—The most important amendment ever made to the original law, as far as the food packaging industry is concerned, is the Food Additives Amendment of 1958. This law deals with food supply in the United States and takes cognizance of all chemical components of this supply, whether these components enter the food by direct addition or by indirect means, for example by virtue of migration from a food packaging material. The packaging industry is concerned with the part of this 1958 amendment that deals with indirect food additives; i.e., those substances which can become a component of the food by migration from a packaging material.

When the 1958 Food Additives Amendment was passed, the F&DA was faced with a great deal of work. It had to find a way to administer the new law as well as to bring under control by scientific evaluation the many thousands of substances involved in foods and in packaging materials for foods.

Rules and Regulations.—In order to accomplish this, the F&DA first set up a set of Rules and Regulations for administering the new law. These Rules and Regulations included Subparts A through G under Part 121—(Food Additives) of Title 21 (Food and Drugs). The collective evaluation of each of the Subparts totally defines the full scope of the law. The Subparts had the following definitive titles:

Subpart A—Definitions and Procedural and Interpretative Regulations.

Subpart B—Exemption of Certain Food Additives from the Requirement of Tolerances.

Subpart C—Food Additives Permitted in Feed and Drinking Water of Animals or for the Treatment of Food Producing Animals.

Subpart D—Food Additives Permitted in Food for Human Consumption.

Subpart E—Substances for Which Prior Sanctions Have Been Granted.

Subpart F—Food Additives Resulting from Contact with Containers or Equipment and Food Additives Otherwise Affecting Food.

Subpart G—Radiation and Radiation Sources Intended for Use in the Production, Processing and Handling of Food.

Subpart A.—*Subpart A* contains several sections and relates primarily to matters pertinent to the administration of the new law. This includes the rules and regulations for the filing of petitions, objecting to proposed regulations, public hearings, judicial review, matters of evidence, etc. This Subpart makes up the substantive law upon which the Food Additives Amendment is administered.

Subpart B.—*Subpart B* is quite short and includes only two sections, but nevertheless, affects the food packaging industry to a great extent. Under this Subpart Section 121.101 appears and lists all those "Substances that are generally recognized as safe." These substances have now come to be known as having a "GRAS" status. The forerunner of this section was the so-called "White List." The F&DA does not add new materials to this list, but often removes materials from the list when evidence comes to their attention to show that the GRAS status should not be continued.

Subpart C.—*Subpart C* includes approximately 75 sections and relates to "Food Additives Permitted in Food and Drinking Water of Animals or for the Treatment of Food-Producing Animals." It is not of great relevance to packaging suppliers except for the packaging of pet foods or feed for farm animals.

Subpart D.—*Subpart D* relates to "Food Additives Permitted in Food for Human Consumption." This Subpart includes approximately 186 sections and lists those additives which are made directly into food. It is of most value to those concerned directly with food processing; however, packagers must know which substances are permitted in a packaging material.

Subpart E.—*Subpart E* relates to substances for which prior sanctions have been granted. These substances are listed in the only section under *Subpart E,* i.e., 121.2001, which is entitled "Substances Employed in the Manufacture of Food Packaging Materials." Prior sanctioned materials listed in this section mean materials that were sanctioned by F&DA for use in food packaging prior 1958. Doctor Lehman of F&DA is well known for his issuance of prior sanctions prior to 1958. No further additions have been made to the prior sanctions list since 1958.

Subpart F.—*Subpart F* relates to "Food Additives Resulting from Contact with Containers or Equipment and Food Additives Otherwise Affecting Food." This is the most relevant subpart for the packaging industry. It contains approximately 97 sections. Within these sections most components of food packages are regulated. Section 121.2501 relates to olefin polymers which include polyethylene, polypropylene, etc. Section 121.2502 relates to nylon resins. Even though there is a regulation for a food packaging material such as polyethylene under *Subpart F* it is not enough to refer to the regulation, it is also necessary to obtain assurance from each supplier that the particular material in question meets the provisions of the regulation because not all polyethylenes are food grade. Such assurance must be obtained in writing and state that the grade of polyethylene to be used in a food packaging application is food grade by virtue of its compliance with Section 121.2501 of regulations under *Subpart F.* Similar assurances must be obtained for all other packaging components, such as lubricants, plasticizers, coatings, adhesives, etc., citing the pertinent regulation complied with in each instance.

Furthermore, several F&DA regulations under *Subpart F* require that extraction tests be run. This is required to make sure that the chemicals involved do not migrate into the food beyond the level established by F&DA to be safe.

Subpart G.—*Subpart G* relates to radiation and radiation sources in the handling of food.

Clearance.—A substance can be cleared for use as a food packaging material in any one of four ways: (1) It is GRAS, that is generally recog-

Courtesy of USFDA

FIG. 148.   F&DA INSPECTION

An inspection team composed of an F&DA Inspector (center); and F&DA Microbiologist (right) and the USDA Plant Inspector (left) are in the process of performing a joint plant inspection of an egg breaking and drying plant. They are looking at egg breaking machinery.

nized as safe *(Subpart B)*. (2) It is permitted to be used as a direct additive in food *(Subpart D)*. (3) It is prior sanctioned *(Subpart E)*. (4) It is approved under a regulation of *Subpart F. Subpart F* is the only Subpart which contains regulations which require compliance since if a material can be added directly to food, is GRAS or has been prior sanctioned, it is not regulated as an indirect food additive. Accordingly, the food packaging industry is concerned with those violations of F&DA regulations which reside primarily within these *Subpart F* regulations.

Food Additive Definition.—The definition of a food additive is of great value in fully understanding the law. The 1958 Amendment provides that a food shall be deemed to be adulterated if it contains a food additive. Once a food contains a food additive or a food packaging material is capable of giving rise to a food additive, penalty provisions of the act apply. The definition of a food additive is most important. A food additive is not any substance added to food directly or indirectly.

The Food Additives Amendment as passed by Congress included a special definition for a food additive. The legal definition of a food additive means "any substance the intended use of which results or may reasonably be expected to result, directly or indirectly, in it becoming a component or otherwise affecting the characteristics of any food (including any substance intended for use in producing, manufacturing, packing, processing, preparing, treating, packaging, transporting, or holding food; and including any source of radiation intended for any such use), if such substance is not generally recognized, among experts qualified by scientific training and experience to evaluate its safety, as having been adequately shown through scientific procedures (or in the case of a substance used in food prior to January 1, 1958, through either scientific procedures or experience based on common use in food) to be safe under the conditions of its intended use." This term does not include—pesticide chemicals, or any substance used in accordance with a sanction or approval granted before the enactment of this act.

The complexity of the definition has caused a great deal of confusion to industry.

A food additive is actually any substance that can become a component of food by direct or indirect means unless that it is known to be safe or is prior sanctioned. If the substance is known to be safe or is prior sanctioned, it is not a food additive. In other words, all substances that can become a component of food are presumed to be additives unless declared otherwise by F&DA. The F&DA has declared some substances nonfood additives by listing them on the GRAS list and has declared other substances nonfood additives by prior sanctioning. These substances appear in *Subparts B and E* respectively.

The simplest check for a questionable material is to ascertain its acceptability with the F&DA published rules. If the product is a film variation, adhesive or coating formulated by a supplier, a letter noting F&DA compliance should be requested. Any formula containing all acceptable ingredients is considered suitable for clearance.

All other materials which are used in food directly or in packaging materials for foods are treated by the F&DA in a different manner. If assurance that the material complies cannot be obtained from the raw material suppliers, a petition must be filed with the F&D administration. Extractability studies must be conducted and the data forwarded to Washington. The F&DA requires the petition to be filed by the chemical company wanting a permit to use a product in food or in a food packaging material. A petition to the F&DA must include the following information: (1) details of usage; (2) composition of the finished packaging material; (3) analytical method to determine additive; (4) identity and composition of the additive; and (5) extractability data. All extractability information must be obtained by using the final package. The suitable analytical method must be sensitive to within the safe limits of the ingredient. The sum total of the ingredients is additive as relaton to F&DA rules. After F&DA reviews the toxicological information in the petition and finds it satisfactory, the F&DA will issue a regulation for the safe use of the chemical in question. This regulation then regulates the use of the material. Regulations which cover direct additives for food appear in *Subpart D* and regulations which cover indirect additives appear in *Subpart F*. The regulations appearing in *Subpart F* are of most relevance.

By the issuance of regulation for a particular substance, F&DA has declared that substance safe and, therefore, not a food additive so long as the use of the substance complies with provisions of the issued regulation. Suppose a regulation states that migration of certain substances from a coating shall not exceed 0.5 mg per sq in. of contact surface in the presence of a certain solvent under particular extraction conditons. If, in fact, the extraction level from a sample of a packaging material containing the coating is higher than the allowable amount, then the packaging material in question would be ruled capable of introducing a food additive into the food supply. At this point, violation of the 1958 Food Additives Amendment exists and all the provisions of the Food, Drug, and Cosmetic Act apply.

It is known that if the F&DA should become aware of a food packaging material which contains an unsafe food additive, they may seize the material immediately. They may not wait until the food was packaged in the material and then prove analytically the presence of the food additive in the food itself. Under provisions of the Act, F&DA may be able to seize the packaging material strictly on the basis of violations of packaging regulations as exist under *Subpart F*. With respect to the seizing of packaging materials by F&DA which are alleged to contain an unsafe food additive, industry spokemen have questioned any basis for this claim in the 1938 Food, Drug, and Cosmetic Act.

Extraction Data May be Complex.—All extraction data must contain the amount of a single substance which migrates under standard conditions and the total contamination of the food product. In addition, toxicity data must be available for F&DA evaluation.

Migration of additive is dependent on time, temperature, area, and type of food. While extraction studies may be conducted on the specific food, most tests utilize special extracting solutions. This is due to the wide range of variables inherent in food products which cause problems in obtaining accurate results. For particularly sensitive applications radioactive tracer analysis is used. When tagged carbons are employed, excellent results are obtained. However, radioactive tracer tests are expensive and require extremely intricate equipment. Due to these factors, the process is not in widespread use.

In order to formulate a range of solvent solutions capable of simulating food extraction, several factors are significant. The various solutions types that should be included are acid, alkaline, and neutral solutions as well as alcoholic and oil materials. A representative list suggested by the United States Food and Drug Administration is: (1) distilled water; (2) 3% aqueous NaCl; (3) 3% aqueous $NaHCO_3$; (4) 3% aqueous acetic acid; (5) 3% aqueous lactic acid; (6) 20% aqueous sucrose; and (7) lard or vegetable oil.

The list of solutions recommended by the French authorities is more complex: (1) distilled water; (2) 10% ethyl alcohol; (3) 50% ethyl alcohol; (4) 95% ethyl alcohol; (5) 10% acetic acid; (6) 50% acetic acid; (7) 1% NaCl solution; (8) 5% NaCl solution; (9) 10% sucrose solution; (10) 2% citric acid solution (pH = 4.5); (11) lard; (12) arachide oil.

It is essential that all extractants be used correctly to arrive at accurate determination of the migrant. Since various migrants often require different extracting solutions, the proper test method for the specific chemical must be used. For example, highly poisonous substances require extremely sensitive test solutions.

Migration studies should actually be conducted for the entire shelf-life of a packaged food. In principle, the duration involved makes long-term tests impractical. Most tests are conducted at accelerated temperatures and the results correlated with actual conditions. US authorities state that the test conditions used must correlate to actual packaging conditions. If a film is autoclaved during a packaging operation, migration tests must be conducted at the autoclaving temperature. Other nations feel that the test conditions must be more severe than actual practice. Since it is often impossible to know the total packaging cycle, extremely stringent test conditions are usually employed.

The diffusion of migratory substances into food products increases with increasing temperatures. It has been reported that extraction is fairly uniform for the first five days at a temperature of 140°F (60°C). After the fifth day, the rate decreases and extraction ceases by the tenth day.

The thickness of the sample used should be uniform. A 1.0 mil sample is recommended since it encompasses most commonly used plastic films and sheets. Total migration should be expressed in parts/million, and a ratio of 0.3 ml solvent/cm^2 of surface is preferable for the test method.

After allowing the sample to be exposed to the solvents chosen at the specified time and temperature cycles, quantitative identification of migratory compounds is made by gas chromatography or conventional analytical methods.

Manufacturing Regulations.—Good manufacturing practice is carefully outlined in Federal regulations. *Regulation 121.2500* "General Provisions Applicable to *Subpart F*" states that the use on any and all of the additive substances regulated in the 96 other regulations of *Subpart F* is predicated upon the use of good manufacturing practice.

Section 121.2500 of Subpart F "General Provisions Applicable to *Subpart F,*" contains the definition of good manufacturing practice; Paragraph (a) (1) "The quality of any *food additive* substances that may be added to food as a result of use in articles that contact food shall not exceed, where no limits are specified, that which results from use of the substance in an amount not more than reasonably required to accomplish the intended physical or technical effect in the food contact article, etc." This means that a food additive can be contained within a food packaging material, but that such additive must be contained at the absolute minimum required to accomplish the intended physical or technical effect.

There are special places where this provision would apply in packaging operations, for example in the production of packaging materials where a lacquer coating is applied to a construction, where such coating is not properly dried, and where the construction is used as a food wrap with the coating in direct contact with food. Since the solvent retained in the coating would normally migrate into the food, the definition of a food additive applies. Any time that a packaging material gives rise to a food additive in excess of the minimal amount allowed by this provision under good manufacturing practices, industry must face application of the Food, Drug, and Cosmetic Act penalty provisions.

In other *Subpart F* regulations, reference also is made to good manufacturing practice. In the *Adhesives Regulation 121.2520* the quantity of adhesive that contacts packaged dry food must not exceed the limits of good manufacturing practice. The quantity of adhesive that contacts packaged fatty or aqueous foods must not exceed the trace amount at the seams and at the exposed edges that may occur within the limits of good manufacturing practice. In the case of adhesive lamination this means using only the amount of adhesive required to obtain firm bonding of the seams and laminate components.

At present, F&DA, is upgrading its regulations relating to good manufacturing practice. On December 15, 1967 a new proposed good manufacturing practice regulation for the sanitary processing and packaging of human foods was published in the Federal Register. This regulation has not yet been issued in its final form. The proposed new rules state that the packaging processes and materials shall not transmit contaminants or objectionable substances to the food product, shall conform to any applicable food additive regulation and shall provide adequate protection from adulteration.

It is apparent from provisions of this new good manufacturing practice regulation that food packaging materials are more and more going to be considered separately in connection with food handling and food adulteration. If this emphasis continues, sooner or later there will be a court action for alleged violations of packaging regulations. It should be pointed out however that there have been no government prosecutions as yet. For the moment, a grace period exists.

Since there will be an ever increasing number of F&DA regulations under *Subpart F*, the F&DA is entertaining suggestions as to how the review of petitions can be streamlined to bring them to issuance at an earlier date. New guidelines for the filing of petitions are expected to issue sometime in 1969.

The result of issuance of these regulations will be as in the past, the means by which the industry will be guided along a relatively safe path from a public health viewpoint. The incentive to conform to these regulations will be the fear of prosecution lingering in the background. There seems little doubt at this point, that the Commissioner of Food and Drugs will make every effort to protect the public health in accordance with provisions of the Food, Drug, and Cosmetic Act and by the authority delegated to him under this Act.

Color Additive Amendment.—In 1906, the original US Food and Drug Act provided for a voluntary system of certification of the purity and safety of dyes used in or on foods. Under the 1938 revision, limits were established on the minute amounts of lead, arsenic and other heavy metals that could be formulated in the colors. It also added a provision setting down the exact specifications of each individual color. Manufacturers submitted samples to the Food and Drug Administration for certification. Covered, under this act, were the coal tar colors but not the nonaniline colors such as iron oxide, titanium dioxide and naturally occurring colors.

The color additive Amendment of 1960 embraces all colors, both natural and synthetic used in foods or in packaging materials contacting foods. It was designed to replace the outmoded provisions which governed the use of color in foods under the 1938 act. Tolerance limitations were established when required and other conditions for safe use. The system of official certification was continued.

In essence, three groups of colors are permitted for use in coloring of packaging materials for food. Certified FD & C dyes and alumina lakes are acceptable. F&DA approved pigments and F&DA sanctioned colors are also permitted. Any colorant that has an impermeable barrier between it and the food is not subject to the Color Additive Amendment. The colorant may be nonapproved; however, the burden of proof relative to migration, offset, bleeding, etc., rests with the user.

An ink should be formulated by using certified or F&DA approved colors. It should also use ingredients that are generally recognized as safe or approved materials under the Food Additives Amendment. Lists of acceptable ingredients are available in the various periodical editions of the Federal Register.

"Fair Packaging And Labeling" Bill.—In 1966, the Federal Packaging and Labeling Act was enacted. The FPL act became effective July 1, 1967. Specific extensions have been authorized by the F&DA, but otherwise all food products must comply with the laws.

The regulations state that the following information is required to appear on the label of food packages:

(1) The food must be identified. This must appear on the principal display panel and be in bold type. It should be in lines generally parallel to the package base when the unit is displayed. The common name of the food should be used whenever possible. If the food is offered as whole, sliced or chopped, this must be clearly visible to the consumer either by a statement or vignette.

(2) The name and address of the manufacturer, packer or distributor must be identifiable. It should be conspicuous and denote the type of firm supplying the product.

(3) The net quantity of contents must be stated. The use of "Jumbo Pound" etc. is forbidden. This must be placed within the bottom 30% area of the display panel and appear as a distinct item. If the food is liquid, fluid measure is used. If solid, weight is used. For fresh fruits and vegetables, the bushel, peck system is necessary. The net contents statement must be accurate and may not include the weight of the packaging material. For packages containing between one and four pounds or between one pint and one gallon, a dual declaration of net weight or net volume is required. The first statement must show total ounces or total fluid ounces. The second statement in parentheses must express the total in pounds and ounces or pounds and fractions or decimal fractions of a pound for solids or in quarts, pints, fluid ounces and fractions thereof for liquids. A separate declaration of net contents in terms of metric units may also be added. Net contents must show the total amount of food in the package. Where part of the contents is not used as food but thrown away, such as with the brine in a bottle of olives, only the amount of solid food should be stated.

(4) A statement of the net contents of a serving is required. This may appear in units of weight, liquid or dry measure or any other commonly used term, i.e., tablespoons. If the contents are noted in terms of servings, the description of a serving size is needed.

(5) All ingredients must be mentioned. This must be legible and appear on a single panel of the label. It must appear in decreasing order of ingredients. If a special ingredient is advertised as exceptionally valuable to the food, a declaration of per cent content is required.

OTHER GOVERNMENTAL REGULATIONS

In addition to the F&DA several other governmental organizations regulate packages. The Meat Inspection Act of 1906 established all Meat Inspection Division laws under the Department of Agriculture. Labels and the overall visibility of all meat packages must be approved by the MID (now Technical Services Div. of Consumer and Marketing Services). Poultry and poultry products are also subject to Department of Agriculture approval.

The Treasury Department supervises the packaging of alcoholic beverages and tobacco. Narcotics are supervised by the Bureau of Narcotics. Imports are subject to the Bureau of Customs and postal shipments are enforced by the Post Office Department.

Often, overlapping jurisdiction occurs and a product must meet F&DA and MID regulations. In such cases, approval is obtained by each agency. It is important to note that rules and regulations differ considerably. The most stringent regulations are F&DA laws, and if a product meets these regulations, additional governmental approval is fairly easy.

A package considered perfectly legal in one state may be illegal in a neighboring state. All state laws must conform to the basic provisions of the F&DA rules; however, many states have adopted their own interpretations of some of the ambiguous parts of F&DA regulations.

Extraction and migration data is standard in all states. Variances occur with package design and labeling. Colors are controlled and the size of lettering may become a part of state law. It is important to note that the limited financial ability of most individual states serves to retard intrastate court action. Prosecution is cumbersome and expensive for state authorities.

Even cities have their own packaging rules. Large metropolitan areas often have regulations relative to deceptive packaging and product visibility. A recently passed bill in New York City ("Peek-A-Boo") states that all prepackaged meat be displayed in transparent trays. The danger now exists for suppliers of molded pulp trays as well as polystyrene foam that such legislation may spread to other cities.

EUROPEAN REGULATIONS

In Germany, substances migrating from packaging materials into food are not considered to be food additives. While there are laws governing the basic quality of foods (Food Act of 1927), only a recommended list is available for plastics in food packaging. These recommendations have no legal force and full responsibility lies with the plastics processor. If plastic materials do not conform relative to safety, prosecution rests under the Food Act of 1927.

In the United Kingdom, no specific legislation on food packaging materials exist. However, the British Plastics Federation has formulated a plan for testing materials on a voluntary basis. Migration and toxicity data are collected and published by the British Industrial Biological Research Association. It is expected that legislation governing packaging materials will soon be introduced.

Most European countries are on the verge of drafting positive lists of food additives. At present, many seek assurance by conforming as closely as possible to US F&DA regulations.

RELIGIOUS REGULATIONS

Kosher Rules

Many Jewish consumers subscribing to Orthodox Judaism examine products for Kosher approval. A recent figure stated that more than 2000 products from over 400 companies are manufactured according to the laws of "kashruth."

In the United States, about 45 large companies manufacture kosher-labeled processed meat products such as salami, wieners, and pastrami. The gross wholesale business of these firms averaged $55 million. The total retail volume of all kosher-labeled processed meats, including small plants operated by kosher butcher shops, is about $100 million. It is significant to note that a substantial portion of all kosher-labeled products are bought by non-Jews who have developed a taste for "kosher" meat products.

A kosher product must be packaged in an approved material which is subject to the same regulations as the product. Package approval is not easy to obtain but it is an absolute necessity for certain markets.

Definitions.—*Kosher*—A Hebrew adjective meaning "proper," used to describe a food or related product prepared and packaged in accordance with Jewish Dietary Law. Kosher is a religious term which has nothing to do with taste. The term kosher flavor is often confused with a garlic flavor. The two are not synonymous in any way, though some kosher foods may have a strongly spiced or garlicky flavor.

Kashruth.—The noun of Kosher. The five requirements of kashruth are the following:

(1) All pork products are inherently unable to be Kosher. The law also forbids consumption of birds of prey, e.g., owls, vultures, and eagles. Worms, snails, oysters, crabs, clams, and shrimp are also forbidden. The use of various byproducts of nonkosher animals are not permitted.

(2) Acceptable animals include the flesh of all quadrupeds with cloven hoofs, i.e., cattle, sheep, and goats. The barnyard fowl and scaly fish with fins are permissible.

(3) Fresh vegetables and fruits in their natural state are inherently kosher.

(4) The acceptable animals must be slaughtered and prepared according to certain regulations. If this is not followed, the entire food is nonkosher.

(5) No dairy and meat product may be mixed or consumed at the same time. Separate cooking vessels and dishes must be used for these products.

The above list is only a guide to the general requirements of the laws of kashruth. Other regulations exist.

The term "Glatt Kosher" is used to describe fresh meat from cattle or sheep whose lungs were checked by "bodkim" (trained religious Jews) and were found free of any adhesions. If the adhesions can be removed from the lobes of the lungs without leaving any trace of perforations, the meat is kosher. The term "Kosher-Style" on meat products is a euphemistic term for nonkosher and is very misleading. Its use is permitted in most states.

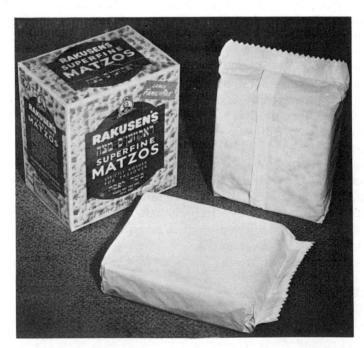

Courtesy of British Cellophane, Ltd.

FIG. 149.   THIS MATZOS PACKAGE INDICATES PASSOVER
ACCEPTABILITY OF CONTENTS

To attain this labeling extremely strict care is taken in manufacture and packaging of the food. The white opaque "Cellophane" film conveys the message of purity and hygienic preparation. Two inner packs insure freshness.

An additonal term in kosher terminology is "Parvae or Parveh." This is applied to foods which can be classified either as dairy or meat dishes. They may be eaten with either type and do not infringe upon the dietary laws. Care is needed to keep parvae utensils separate and not to cleanse them together with either dairy or meat utensils.

The term "Kosher for Passover" means that there is no leaven in the food product. Due to religious regulations, it is strictly forbidden to eat leaven during the Passover season. Labels are attached to the package denoting Passover acceptibility. These are often pressure sensitive labels which can be removed after the Passover season.

The Interpretations of Kosher Laws.—Since there is no authoritative Jewish hierarchy in the United States, an orthodox Rabbi's interpretation of the kosher laws is binding. It is important to note that not all Rabbis are qualified to judge the acceptability of a product. Judaism is divided into three basic sects, —Orthodox, Conservative, and Reform. All Orthodox Rabbis susbscribe to kosher laws and may interpret their tenets. In seeking approval, only Orthodox Rabbis should be consulted.

In the manufacture of food packaging materials and the packaging and processing of foods, a large number of chemicals are used. Each one must comply with kashruth. Cleaning materials must be free from any ingredients derived from animals forbidden by Jewish Law. Emulsifiers, stabilizers, defoamers, and lubricants should be obtained from vegetable and/or synthetic sources. Glycerol widely used in cellophane manufacture, must be obtained from vegetables and not animals.

Although all natural fruits and vegetables are inherently kosher, processed fruits and vegetables are subject to investigation. Fresh scaly fish is inherently acceptable but not frozen fish. The preservative solution used for freezing must not have been used for freezing any non-kosher product. The freezing of poultry involves an analogous situation. Kosher poultry must not be processed with any chemicals or baths used for the nonkosher variety.

Aluminum foil must be rolled using synthetic lubricants. In some cases, high rolling temperatures may destroy animal lubricants and therefore the foil would be acceptable. Certain waxes derived from tallow are forbidden. No animal glues are acceptable. All plasticizers and flexible film additives are subject to investigation.

Acceptability is based on definition and interpretation of Rabbinical law. Many issues are decided after careful study and each product is a specific case. A layman cannot decide whether a product adheres to the Jewish law. Only a fully trained Orthodox Rabbi is capable of arriving at an answer. The steps involved are simple and the results may be significant. The difficulties in relating a modern product to ancient laws may be illustrated by an interesting example.

Most Orthodox Rabbis considered nonkosher any cheese made with rennet. Since rennet is derived from the stomach of a calf and non-kosher cattle are used, they reasoned, the product is not kosher. In order to process rennet, the stomach of a young calf is cleaned and stored for a considerable amount of time. Chemical additives are then added to yield a fine powder. Rennet is also used in pudding-type desserts. In biblical days, this process was unknown. A study made by a Rabbi determined that rennet is kosher—even if derived from nonkosher cattle. His reference was a quotation from ancient law stating, "The skin of the stomach is sometimes salted and dried so that it is rendered like wood, and if then it is filled with milk it is kosher, because after the stomach has been dried, it is just like wood, and does not contain any moisture of meat."

How to get Approval.—There are three basic organizations offering converters a method of attaining kosher approval.

The Union of Orthodox Jewish Congregations of America created a national service in 1925. At the request of the interested company, a completed application detailing specific ingredients is submitted to the UOJCA central office. After an initial investigation of the product, the UOJCA sends rabbinical investigators into the plant. Supervision consists of observation of day-to-day practice and plant conditions. Packaging and processing equipment are carefully checked and investigated in order to determine compatibility with the kosher laws. All data submitted and observed is fully confidential. If the laws of kashruth have been fulfilled, the company is permitted to use a (U) symbol on the specific products. The entire study involves a fee based on the total cost to the UOJCA.

The Joseph Jacobs Organization in New York City offers another supervisory service. An investigation and visit is made by an Orthodox Rabbi in order to determine adherence to kashruth. Approval is then given to display a "K" on the product. All information obtained is confidential and the fee is based on the total cost.

In Massachusetts, the Vaad Harabonium (Rabbinical Council) authorizes the use of a "VH" symbol. Certification is granted to products only if their manufacture remains under the strict supervision of the Vaad Harabonium. Companies are required to submit a detailed list of their ingredients and an investigation is made. Once a product has been cleared, the manufacturing plant and equipment are cleaned and sterilized. Manufacture then starts under the supervision of an Orthodox Rabbi assigned to the plant's production and quality control departments. As in the prior cases, all information is confidential and a rated fee is paid.

In addition to the above three organizations, an agreement may be made with an individual Orthodox Rabbi for approval. The degree of supervision is wholly dependent on the Rabbi. Approval will allow for his signature on the package vouching for its kosher purity.

When a product has been granted approval for adherence to the kosher laws, the three organizations provide a variety of marketing and counseling services for the product. The potential increase in sales from obtaining approval may be substantial.

Other Rules

In contrast to the rules of Orthodox Judaism, no other religion has formulated a set of rules for food packaging in the United States. This may be due to the lack of immigrants from other than Christian lands. Until recently members of the Roman Catholic faith could not consume meat on Fridays, but this did not prohibit the use of packaging materials containing glycerine from animals. Kosher meat products are bought by Seventh-day Adventists and Seventh-day Baptists who observe the Sabbath day on Saturday. They do not eat any products which may contain an ingredient of pork or lard. Thus religious precepts differ not only among faiths but also in the degree of strictness of adherence to the precepts.

Although not formalized in the United States, members of the Hindu and Islamic faith follow strict rules as to food and indirectly to packaging.

Regulations Affecting the Packaging of Margarine

The manufacture and sale of margarine in its early days was undertaken by so many that inevitably some very poor quality products were marketed. It was also easy for the unscrupulous to sell the cheaper substitute at the higher price of butter. To insure product identification and reasonably good quality nearly every manufacturing country passed legislation regulating the composition and the packaging of margarine. In some instances powerful dairy industries influenced repressive legislation which prohibited the sale of margarine. Most of the prohibitive legislation has since been repealed.

It is not within the scope of this book to consider all legislation relating to the composition of margarines, but we are interested in those laws affecting the packaging of the product. Austria, parts of Australia, Argentina, Belgium, France, West Germany, Norway, and Switzerland all require retail packages of margarine to be cubical in shape. Portugal says the retail package cannot exceed one kilogram; Spain says it must be more than 100 gm, South Africa limits it to 1/2 lb. Most countries require special markings. Some require red bands on bulk packaging. Nearly all require prominent labeling with large letters showing the name margarine. In France, this must be on four sides. It will require special legislation to change these rules before newer margarine packages can be adopted.

BIBLIOGRAPHY

ANON. 1955. Customs and Traditions of Israel. Joseph Jacobs Organization, New York.

ANON. 1965. New federal regulations covering the marketing of fruits and vegetables. Quebec Lait. Aliment *24*, No. 10, 24.

ANON. 1967A. Hygienic judgment of plastics in the framework of the food law. Wbl. Papfabr. *95*, No. 7, 262–266. (German)

ANON. 1967B. Safety review of plastics in the food stuffs regulations field. Bundegesundheitsblatt *10*, No. 2, 24–26. (German)

ESTEVEZ, J. M. J. 1967. Toxicity in plastics: the problem as typified by plastic packages. Plastics Inst. Trans. J. *35*, No. 116, 448.

HIGGINS, J. C. 1968. Food packaging regulations. Paper No. 5. Flexible Packaging Conference. British Business Publications, Lt. London.

JONES, A. 1967. Toxicity, plastics, and packaging. Rubber Plastic Age *48*, No. 4, 340.

KATZ, M. C. 1968. Deception and Fraud with a Kosher Front. Private publication by the author. Endicott, New York.

KORFF, S. L. 1966. The Jewish dietary code. Food Technol. *20*, No. 7, 76–78.

MORTON, R. A. 1966. Food additives and contaminants. Chemistry, Medicine and Nutrition Symposium, April 14–15. Bristol, England.

SACHAROW, S. 1968. Plastic processors and the U.S. Food and Drug Administration. Australian Plastics Rubber J. *23*, 31–35.

SPINA, A. M. *et al.* 1966. Review of the analysis, toxicology and regulations in Italy concerning the use of plastic materials in contact with foodstuffs. Russ. Chim. *18*, No. 3, 125–135. (Italian)

TURVI, L. 1966. International legislation on packaging of food products with plastics. Materie. Plast. *33*, 289–294. (Italian)

VAN DER HEIDE, R. F. 1964. The safety for health of plastics food packaging materials. Ph.D. Thesis. Ubrecht Univ., Holland.

DEWILDE, J. M. 1967. Development of the European legislation on the acceptability of plastics materials for food packaging: problems of interaction between packaging material and product. Farbe und Lack *73*, No. 2, 153. (German)

Index